UNDERGROUND INFRASTRUCTURE OF URBAN AREAS 4

Underground Infrastructure of Urban Areas 4

Editors

Cezary Madryas, Andrzej Kolonko, Beata Nienartowicz & Arkadiusz Szot

Department of Civil Engineering, Wrocław University of Science and Technology, Wrocław, Poland

CRC Press
Taylor & Francis Group
Boca Raton London New York Leiden

CRC Press is an imprint of the
Taylor & Francis Group, an **informa** business
A BALKEMA BOOK

CRC Press/Balkema is an imprint of the Taylor & Francis Group, an informa business

Typeset by V Publishing Solutions Pvt Ltd., Chennai, India
Printed and bound in Great Britain by CPI Group (UK) Ltd, Croydon, CR0 4YY

Published by: CRC Press/Balkema
Schipholweg 107C, 2316 XC Leiden, The Netherlands
e-mail: Pub.NL@taylorandfrancis.com
www.crcpress.com – www.taylorandfrancis.com

ISBN: 978-1-138-55953-0 (Hbk + CD-ROM)
ISBN: 978-0-203-71257-3 (eBook)

Underground Infrastructure of Urban Areas 4 – Madryas et al. (Eds)
© 2018 Taylor & Francis Group, London, ISBN 978-1-138-55953-0

Table of contents

Underground Infrastructure of Urban Areas 4 – Madryas et al. (Eds)
© 2018 Taylor & Francis Group, London, ISBN 978-1-138-55953-0

Foreword

The lifestyle and work of modern city dwellers and also their expectations and requirements have led to the increased demand for high quality services, fast and convenient communication, parking spaces and the wider use of underground network infrastructure, i.e. providing communication, supplying water and energy and discharging sewage. There is an increasing demand for 24-hour cities, ones in which some areas function 24 hours a day. It is therefore obvious that the technical infrastructure, which is adapted for such purposes, must be able to not only meet the basic requirements of human existence as it currently does, but to also provide a high level of comfort and safety. This is only possible if it is managed by developed information-control systems. The development of technical infrastructure is also necessary with regards to the perspective of a society with a high proportion of older people. Moreover, it is also required due to the need to organize an increasing amount of leisure time for city dwellers, which is mainly a result of the technical amenities of civilization (computerization, wireless communication, etc.). Therefore, current social expectations are that the modernization of parts of cities, as well as their expansions, should be conducted with a consideration of greater living comfort, while also adapting the newly emerging urban infrastructure to the social, spiritual and cultural needs of contemporary lifestyles and current values. Creating an urbanized area with the above-mentioned features has already been partially taken up in the World, especially in developed countries. However, this has been done to an insufficient level. The urban renewal and urban development projects of cities must be characterized by a better use of urban space than ever before. This can be achieved by a more intensive development of underground construction and a higher degree of integration of the infrastructure system, which can be divided into three subsystems. The first subsystem can include all devices that are connected to the communication related services of the city. The second subsystem may concern all appliances related to energy, water, sewage, waste disposal and utilization. In turn, the third subsystem will include communications and information devices, which with the assumption of the need for control and also with regards to other infrastructure devices would form the basis of an urban management system.

If the proposed objectives are to be achieved, a package of administration law regulations, which is preferential for the development of urban construction development, is required. It should reflect the principles of crediting, subsidizing, commissioning, etc. for the best possible solutions. However, the most important is to develop a concept of technical solutions that would provide the basis for the creation of coherent detailed studies that meet the requirements described above. The basic premise of such studies must be the creation of a much more spacious urban space, which is not possible without intensifying the investment of the underground space of cities. Multifunctional and heavily developed underground space allows for the release of some of the functions of terrestrial space, which can then be used for other purposes (mainly for residential areas) and also be partially ecologically renewed.

The attention of planners must therefore be more focused on the wider use of underground space as a direction for improving public transport, increasing the capacity of city centres by transferring many commercial and service functions to underground, and also modernizing and expanding network structures while increasing their efficiency and operational reliability.

Underground Infrastructure of Urban Areas 4 presents a set of some of the above-mentioned problems. These studies are a continuation of the studies published in the three previous books from 2008, 2011 and 2014. It is obvious that they do not exhaust the topic. They are, however, a voice in an international scientific and technical discussion about these very important urban planning problems. Therefore, on behalf of the co-editors and myself, I express my deep hope that they will meet the interest of readers.

Main editor
Cezary Madryas

Underground Infrastructure of Urban Areas 4 – Madryas et al. (Eds)
© 2018 Taylor & Francis Group, London, ISBN 978-1-138-55953-0

Sponsors

PLATINIUM SPONSORS

HERRENKNECHT AG

GOLD SPONSORS

BLEJKAN
Sp. z o.o.

SILVER SPONSORS

BETONSTAL
Sp. z o.o.

HOBAS System
Polska Sp. z o.o

MC-Bauchemie
Sp. z o.o.

TERRA THALER
Sebastian
Grygorcewicz

Uponor Infra Sp. z o.o.

OTHER SPONSORS

Centrum Technologiczne
BETOTECH Sp. z o.o.

HABA-Beton
Johann Bartlechner
Sp. z o.o.

Dolnośląska
Okręgowa Izba
Inżynierów
Budownictwa

METRO WARSZAWSKIE
Sp. z o.o.

MOLEWSKI
Sp. z o.o.

ViaCon Polska
Sp. z o. o.

Underground Infrastructure of Urban Areas 4 – Madryas et al. (Eds)
© 2018 Taylor & Francis Group, London, ISBN 978-1-138-55953-0

Reviewers UIUA 2017

- Han ADMIRAAL, *President of ITACUS (NL)*
- Rolf BIELECKI, *President of EFUC (D)*
- Tarcisio CELESTINO, *President of ITA-AITES (BRA)*
- Józef DZIOPAK, *Politechnika Rzeszowska (PL)*
- Soren Degn ESKESEN, *Past President of ITA-AITES (DK)*
- Bernhard FALTER, *University of Applied Science-Münster (D)*
- Zbigniew GIERGICZNY, *Politechnika Śląska (PL)*
- Andrzej KULICZKOWSKI, *Prezes PFTB (PL)*
- Marian KWIETNIEWSKI, *Politechnika Warszawska (PL)*
- Eric LECA, *Vice President of ITA-AITES (F)*
- Dariusz ŁYDŻBA, *Politechnika Wrocławska (PL)*
- Cezary MADRYAS, *Politechnika Wrocławska, PSTB (PL)*
- Monika MITEW-CZAJEWSKA, *President of Polish Underground Construction Subcommittee ITA-AITES (PL)*
- Dietmar MÖLLER, *Universität Hamburg (D)*
- Daniele PEILA, *Politechnico di Torino (I)*
- Maria Anna POLAK, *University of Waterloo (CAN)*
- Chris ROGERS, *University of Birmingham (UK)*
- Anna SIEMIŃSKA-LEWANDOWSKA, *Member of Executive Council ITA-AITES (PL)*
- Ray STERLING, *Louisiana Tech. University (USA)*
- Olivier VION, *Executive Director of ITA-AITES (F)*
- Roland W. WANIEK, *President of IKT (D)*
- Adam WYSOKOWSKI, *Uniwersytet Zielonogórski (PL)*
- Andrzej ŻELAŹNIEWICZ, *Polish Academy of Science (PL)*

Underground Infrastructure of Urban Areas 4 – Madryas et al. (Eds)
© 2018 Taylor & Francis Group, London, ISBN 978-1-138-55953-0

Basic diagnostics and analysis of the safe use and repair of large diameter steel technical pipelines located in industrial areas

T. Abel
Faculty of Civil Engineering, Wroclaw University of Science and Technology, Wrocław, Poland

ABSTRACT: Issues related to the diagnostics and analysis of the safe use of large-diameter steel technical pipelines located in industrial areas is the subject of the paper. The paper presents the problems of keeping the above-mentioned objects in good condition, their diagnostics and also the analysis of the safe use of such structures located in areas that are particularly exposed to the effects of variable external loads and aggressive chemical factors. The main source of knowledge about the technical condition of large-diameter underground pipelines are direct inspections and inspections carried out in these objects during their operational breaks. In the case of steel pipelines, basic tests that can be performed by users of these networks are measurements of the thickness of the pipeline wall and measurements of their deformations. These are simple diagnostic operations that do not require high qualifications and costly equipment, but allow an overall assessment of the condition of a pipeline. Based on the knowledge obtained at the stage of diagnostic tests, it is possible to perform analysis that allows the wear level of sections of the structural objects exposed to the highest mechanical and non-mechanical external loads to be determined. On the basis of this knowledge, it is possible to plan repair works that will improve the static-strength parameters and protect the structure from further degradation that can lead to failure.

1 SPECIFICITY OF INDUSTRIAL AREAS

Areas where industrial activity is conducted, mainly chemical production, are laden with many degradation factors. These factors distinguish industrial areas from areas that lack production activity on a large scale and they include the following:

- the effects of chemical compounds in the atmosphere, surface water, ground water and soil medium,
- dynamic influences from various types of external loads,
- stray currents that have a particularly dangerous effect on underground steel structures.

Keeping technological pipelines in good technical condition is very important because of the functions they perform to ensure the continuity of production and technological processes. Any breakdown of the underground network can be a source of disturbance in the operation of the whole technological system. This phenomenon is very dangerous due to the strategic importance of most industrial plants, especially power plants and also heat and power plants that supply the energy that enables urban agglomerations to function.

The failure of hard-to-reach underground networks can cause very serious events such as power cuts, the possibility of damaging devices that require e.g. cooling with technological waters and also the possibility of other damage occurring. Therefore, it is necessary to monitor underground objects and keep them in an appropriate technical condition [Abel, 2012].

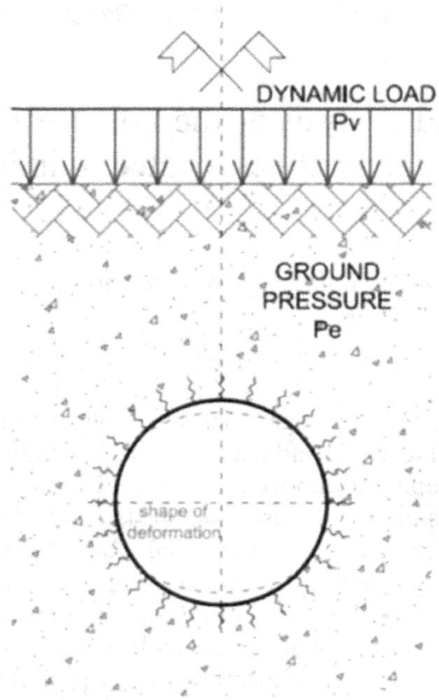

Figure 1. Scheme of acting loads.

2 CHARACTERISTICS OF THE OPERATION OF STEEL PIPELINES

Steel pipeline networks are most commonly found in industrial areas as expanded pipeline systems. They can also work as collective large-diameter conduits, which are the main elements of the infrastructure that is used to transport media such as e.g. cooling water, products made during technological processes and also by-products made during production (gases and liquids).

Networks that are currently in operation, due to the very aggressive environments in which they operate, remain in poor technical condition. This situation is a result of the synergies of several factors. These factors include the natural processes of aging of the pipelines, execution errors committed during the construction of these objects, external loads acting on the pipelines (Fig. 1) that include the previously mentioned dynamic effects, corrosion processes resulting from the aggressiveness of the media inside the pipeline and also the aggressiveness of the external environment. In the case of steel pipelines, the main factor that affects the reduction of strength parameters is the corrosion of steel. The corrosion of steel pipes mainly causes a decrease in their peripheral stiffness and therefore an increase in their susceptibility to external loads, which causes deformations that can lead to structural damage.

3 CHARACTERISTICS OF THE TESTED PIPELINE SECTIONS

The paper presents the assessment of the technical condition of steel sections of pipelines that transport technological waters coming from cooling towers, and also the assessment of the wall thickness of these pipes. The pipelines are made of pipes with a diameter of DN 2000 mm with an original wall thickness equal to 20 mm. Conduits are laid directly in the soil medium and have an outer bituminous isolative coating. The soil medium includes fine, dusty and clay sands that are interbedded with medium-grained sands and sandy and dusty clay lenses. Cohesionless soils are moderately compacted to a level of 10.5 m under ground level and below this depth they are compacted.

There is no uneven settlement in the area where the pipelines are laid. The free groundwater table was found at a level of 11.5 m below the ground surface. The cohesionless soils that occur above it are permanently or temporarily moistened by precipitation, capillary uptake and fluctuations of the water table. Below a depth of 12.0 m, the cohesionless soils are watered. This area and the whole surroundings are not affected by seismic activity and are free from the impact of mining.

In addition to the technological pipelines, the area is equipped with the following networks [Abel, 2016]:

- water and sewage networks, central heating networks;
- drinking, raw and fire water networks;
- power cables, light and earthing cables;
- telephone networks.

Within the conducted research that was carried out on site, the following characteristic factors that influence the durability of the pipeline were identified:

- significant defects of structural material caused by the impact of the aggressive internal environment—a reduction in the wall thickness of a pipeline,
- external cyclic dynamic loads caused by the movement of trains in the assessed area.

The synergy of the both above-mentioned phenomena contributes to a faster deterioration of the technical condition of the object.

In order to determine the actual state of degradation, it was necessary to carry out diagnostic tests, which gave knowledge in the area of changes in the geometric and strength parameters.

4 DIAGNOSIS OF THE TECHNICAL CONDITION OF STEEL PIPELINES

There is a wide range of available methods for testing the strength parameters of pipelines. However, in the case of active technological pipelines, consideration should first be given to the possibility of rapid and inexpensive diagnostics. This would enable the users of these objects to make decisions about forecasting repair works that would increase the static-strength parameters of a specific network.

In the case of steel large-diameter pipelines, basic tests that are possible to be carried out on site and enable technical parameters and the extent of pipe degradation to be determined should be conducted. The following operations should be performed in order to obtain the knowledge that is necessary to decide on further repair works:

- visual inspections of the interior of a pipeline—direct control conducted by qualified personnel,
- measurements of the wall thickness of a pipeline,
- measurements of the vertical and horizontal deformations of a pipeline,
- visual inspections of the connections of individual pipes,
- verification of soil and water conditions and also foundations.

5 EXAMINATION OF THE OBJECT DIAGNOSIS OF THE TECHNICAL CONDITION OF STEEL PIPELINES

Basic knowledge about the technical condition of a man-entry underground facility is provided by either a visual inspection of the pipeline or a Closed Circuit Television (CCTV) inspection. In such a case, an examination is performed by staff staying inside a conduit (Fig. 2). The simplest form of research is an inventory of damage that includes its amount, description and photographic documentation.

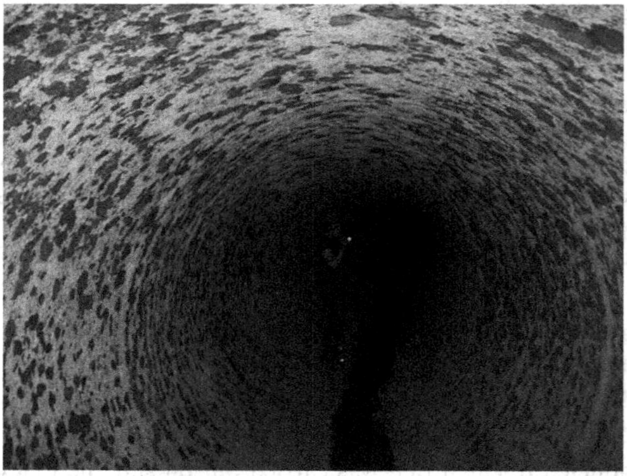

Figure 2. Examinations of an underground pipeline [Abel, 2016].

If necessary, it is complemented by measurements that locate the damage and give its geometric features, such as the size and depth of pits. Studies of such an extent can be considered as current checks. A description of the observed damage should enable the pipeline condition to be identified and classified with regards to the required repair works.

The advantage of such testing is the ability to take samples for laboratory testing and also the carrying out of specific examinations on site. It is therefore possible to achieve a level of expertise examination, which is conducted according to a specific program and enables the technical rehabilitation of an assessed conduit to be planned [Madryas]. In the case of a small immersion of an object, it is also advisable to perform local excavations in order to determine the external condition of the insulation coating.

The most commonly used test is the measurement of the wall thickness of a pipeline with the use of the ultrasonic method. Its popularity is due to the availability of equipment and an easy measurement procedure. The apparent ease of conducting measurements and obtaining results may be illusory for a person taking such measurements. This is due to the many factors that disrupt the correctness of measurements in pipelines. Ultrasonic thickness measurements are not problematic in the case of flat surfaces that are not corroded.

Well-prepared flat surfaces for the application of a device's head, a properly calibrated measuring set and a suitable coupling medium allow the results to be obtained to within ±0.1 mm under normal conditions [Deputat; EN 14127]. The use of basic structural analysis of the obtained results may provide additional information on the predicted minimum or maximum thicknesses of the measured element and may also determine the corrosion or erosion rates, etc. [Volk; PN-ISO 2602].

There are a number of measurement procedures that use the capabilities of various ultrasonic instruments. It should be noted that during ultrasonic thickness measurements, the time of transition of the ultrasonic wave through the tested material is being measured and the thickness is calculated using a coefficient i.e. the velocity of a wave in a specified material introduced as a known value or a value that results from a suitable calibration. An important assumption here is the constant velocity of the wave in the studied area. The material thickness is equal to:

$$g = 0{,}5 * (t * v) \tag{1}$$

where: g – the measured thickness of a material, t – the time of ultrasonic pulse transition, v – the velocity of the ultrasonic wave in the tested material.

The type of head must be adapted to the type of measuring apparatus, the type of tested material (attenuation) and to the range of measured thicknesses.

The basic external factors that affect the occurrence of measurement errors when using ultrasonic thickness gauges are: the quality and repeatability of the acoustic coupling of a head with the measured element, the invariance of wave propagation conditions in the examined element, the reflecting surface shape and the measurement temperature [Sozanski].

The quality of the head coupling with the measured element is determined by preparing the measuring point surfaces and the type of coupling medium. The surfaces of the measuring points must be flat, preferably machined to Ra = 6,3 μm (as is the case with ultrasonic defectoscopic examinations), and twice as large as the transducer contact surface (Fig. 4). The coupling medium (oil, water, special pastes) must provide good conditions so the ultrasonic wave can be transmitted from the transducer to the tested material. However, it cannot be too thick (Fig. 3), as this can lead to results being distorted (overestimation).

The velocity of the ultrasonic wave in the tested element should be constant. Forged or rolled metals most commonly exhibit a low attenuation and constant velocity in the analyzed direction. The acoustic attenuation of the tested material, which usually results in a reduction of amplitude or signal distortion, must be taken into account when selecting the measuring apparatus. The increase of the reflecting surface (limiting) may also significantly affect the measurement results.

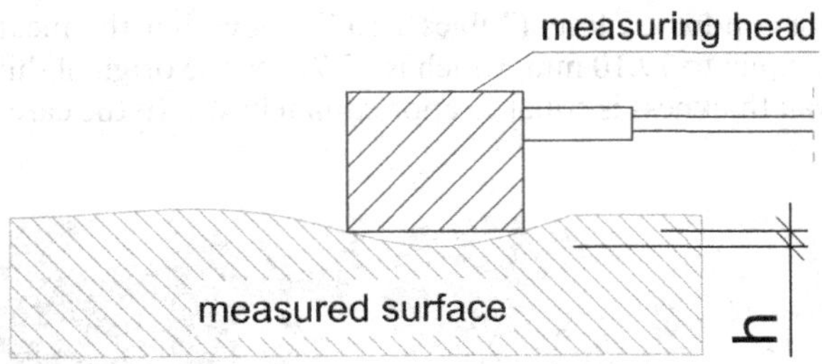

Figure 3. Badly prepared surface for the application of a measuring head: h – thickness of the coupling medium that affects the measurement results.

Figure 4. Preparation of a surface for measurements [Abel, 2016].

6 INSPECTION OF PIPELINES

The tested pipelines have been in operation for the past 45 years. An appropriate evaluation of their technical condition and the evaluation of their usefulness for further operation was possible after emptying the pipelines and conducting qualification tests and a visual inspection.

The inspection showed that the pipelines have deposits of sludge, mill scale and numerous corrosive pits (Figure 5). The pipeline qualification tests showed that the measured wall thickness losses varied (see Table No. 1, 3, 5). Crucial deformations of the pipelines, especially in places subjected to an increased live load (railroad, road), were also identified. The results are summarized in Table No. 2, 4, 6. Tests of the joints also indicated their unsatisfactory technical condition in many places [KAN-REM].

The basic tests were conducted using non-destructive methods of testing the wall thickness and deformations at selected measuring points. Due to the fact that the pipelines are not linearly evenly loaded, as was previously mentioned, sections that are exposed to the highest values of variable external loads were selected for tests. These sections (Figs. 6, 7), which are located under a railway line, were characterized by the largest deformations, while the degree of corrosion remained at a level comparable with other sections.

The tests were conducted at three characteristic measuring points. The measuring points were designated with the symbols MP1, MP2, and MP3 and were located on the section of the pipeline that is laid under the railway area where the values of external variable loads are the greatest.

The results of the conducted tests (Tables 1 to 7) show that the mean wall thickness on the tested section is equal to 19.10 mm, which is 95.7% of the original thickness. The average reduction in pipe wall thickness is equal to approximately 4%. In the case of the deformation

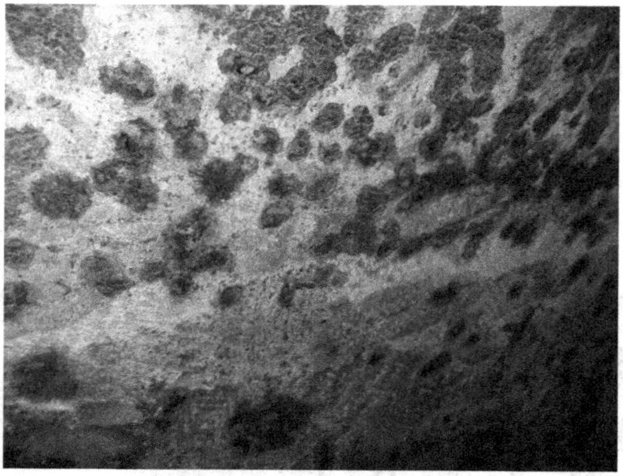

Figure 5. Pipeline wall—corrosion [Abel, 2016].

Table 1. Measuring point No. 1 – selected measuring points of MP1.

No.	Selected cross-section	Row No. 1 [mm]	Row No. 2 [mm]	Row No. 3 [mm]
1	1	19,3	19,5	19,2
2	4	18,9	18,9	19,0
3	7	18,6	19,0	19,6
4	10	19,3	19,4	19,2
5	Average	19,0	19,2	19,3

Table 2. Values of deformation—MP1.

	Vertical deformation A [mm]	Horizontal deformation B [mm]
Row No. 1	1943	2005
Row No. 2	1946	2006
Row No. 3	1931	2014
Average	1940	2008

Table 3. Measuring point No. 2 – selected measuring points of MP2.

No.	Selected cross-section	Row No. 1 [mm]	Row No. 2 [mm]	Row No. 3 [mm]
1	1	19,5	19,2	19,0
2	4	18,8	18,7	19,1
3	7	18,8	19,1	19,2
4	10	19,0	19,1	19,5
5	Average	19,0	19,0	19,2

Table 4. Values of deformations—MP2.

	Vertical deformation A [mm]	Horizontal deformation B [mm]
Row No. 1	1941	2006
Row No. 2	1942	2010
Row No. 3	1932	2018
Average	1938	2011

Table 5. Measuring point No. 3 – selected measuring points of MP3.

No.	Selected cross-section	Row No. 1 [mm]	Row No. 2 [mm]	Row No. 3 [mm]
1	1	19,1	19,4	19,0
2	4	18,7	18,6	19,1
3	7	18,1	19,2	19,4
4	10	19,0	19,2	19,6
5	Average	18,7	19,1	19,3

Table 6. Values of deformations—MP3.

	Vertical deformation A [mm]	Horizontal deformation B [mm]
Row No. 1	1940	2005
Row No. 2	1941	2010
Row No. 3	1929	2016
Average	1937	2010

Table 7. Obtained results.

No.	Place of measurement	Wall thickness [mm]	Vertical diameter [mm]	Horizontal diameter [mm]
1	MP1	19,10	1940	2008
2	MP2	19,10	1938	2011
3	MP3	19,00	1937	2010

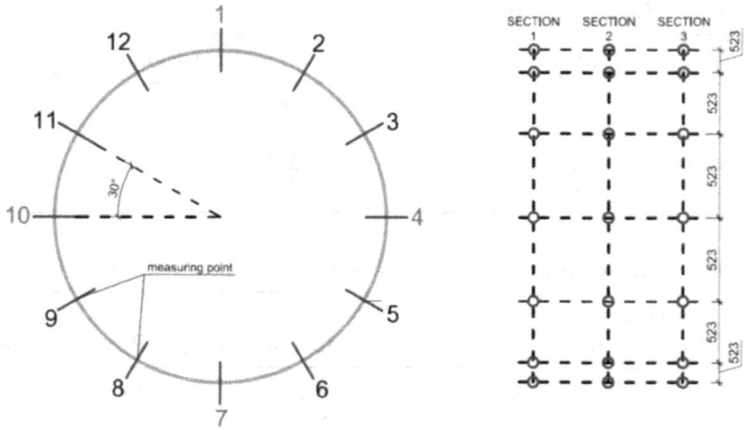

Figure 6. Measurements of the wall thickness of the pipeline [Abel, 2016].

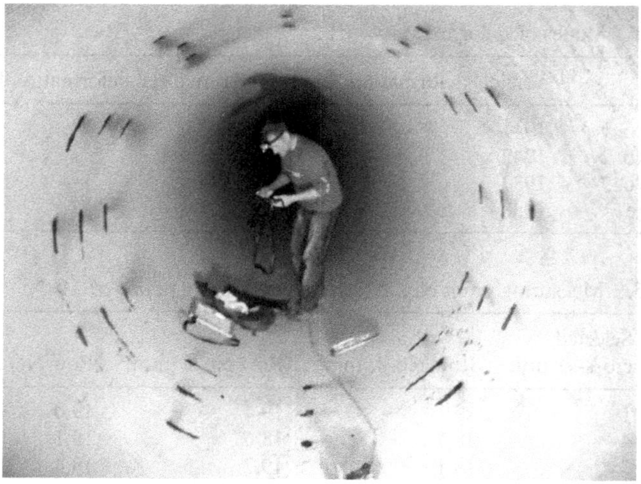

Figure 7. Location of measuring points [Abel, 2016].

of the pipeline on the section that is exposed to significant variable external loads, the mean value of the vertical deformation is equal to 60 mm, which represents 3% of the conduit diameter value and does not exceed the permissible values.

The conducted tests and inspections indicated that the object was in good technical condition. However, due to progressive corrosion processes and the necessity to stop them, it was required to make an internal protective coating. Such a coating should provide the two following functions:

- protection of the steel pipe wall against further corrosion,
- the formation of an inner shell to reinforce the pipeline and allow its further safe operation.

In addition, the used revitalization method should enable the minimum reduction of an internal diameter while providing all the strength parameters with respect to water and ground pressure. In this case, the permissible reduction of the diameters of the revitalized pipelines was equal to a maximum of 100 mm. The absolute roughness coefficient of the inner surface was not meant to exceed $k \leq 0.1$. At the same time, it was necessary to guarantee that the installation was 100% tight.

After analyzing the different technology available on the market, the decision to use the "close-fit" technology was taken. This technology combines the characteristics of the classical "sleeve" with commonly used relining techniques, in which self-supporting modules are inserted into the interior of an existing pipeline.

The used technology involved the building of a reinforced concrete construction, the interior surface of which was a high-density polyethylene (PEHD) liner (Fig. 8). This method is usually used for the renovation and reconstruction of large-diameter pipelines, including those exposed to significant loads. In such a method, based on static-strength calculations and depending on the conditions and needs, the quantity and quality of the reinforcement installed inside the pipeline is determined.

The use of polyethylene liner as an inner layer provides a coating with a very low roughness coefficient, high chemical resistance and mechanical abrasion resistance. As a result, the pipeline obtained a smooth inner surface and was strengthened in order to meet the appropriate requirements (Fig. 9).

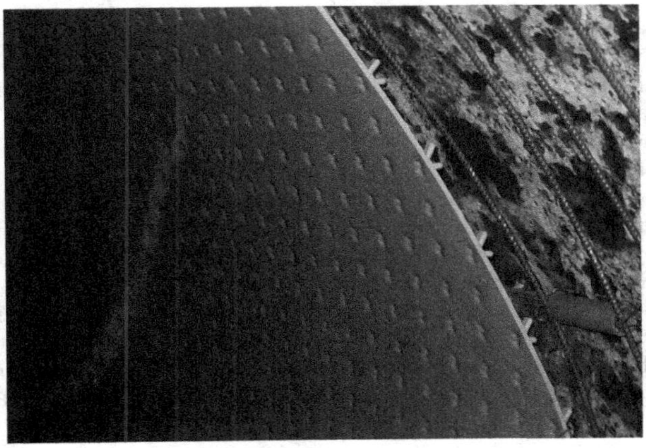

Figure 8. View of the layers of the reinforcing system before injection [Abel, 2016].

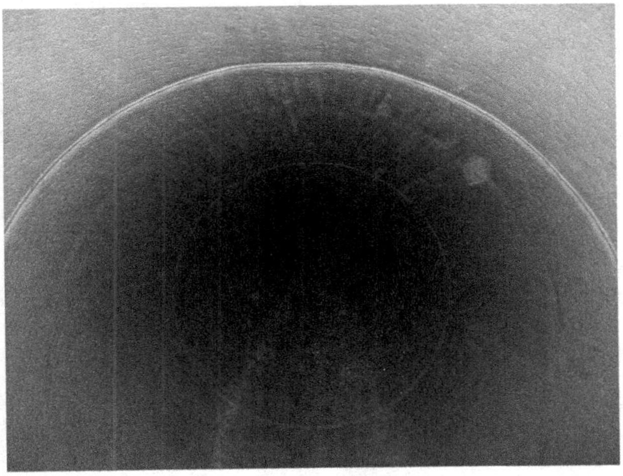

Figure 9. Pipeline after strengthening [Abel, 2016].

7 SUMMARY

Large-diameter steel pipelines located in dense industrial areas, as linear underground objects located in the vicinity of other industrial networks that are essential to the functioning of the industrial areas, are extremely important and strategic elements of the entire system. Therefore, it is necessary to carry out periodical inspections of the technical condition of these pipelines and conduct basic tests that enable the preliminary diagnosis of these facilities to be performed.

The slowing down or stopping of the destructive processes acting on a pipeline will prolong its service life, and therefore prolong its failure-free operation. This effect can be achieved through the application of the available underground trenchless repair technology. The economic aspects that may have a negative impact on the decision of whether to repair a pipeline should be compared with the possible costs expected in the case of a technological breakdown of a network. In addition, maintenance of underground networks in good technical condition eliminates the occurrence of many road infrastructure failures, which in most cases result from damage to underground infrastructure.

REFERENCES

Abel T. 2012. Application of MAXI-LINING technology with regards to the repairing of large-diameter pipelines located in industrial areas. *Infraeko*. Cracow.

Abel T. 2016. Author's own materials.

Deputat J. 1979. Ultrasonic defectoscopy. *IMŻ*. Gliwice.

EN 14127:2001 Non-destructive testing. *Ultrasonic thickness measurements.*

KAN-REM. Technical materials. 2013. Wroclaw.

Madryas C., Przybyła B., Wysocki L. 2010. Testing and evaluation of the technical condition of sewage conduits. *Dolnośląskie Wydawnictwo Edukacyjne*. Wroclaw.

PN-ISO 2602:1994 Statistical interpretation of test results. Estimation of the mean value. Confidence interval.

Sozanski L. 1993. Errors in ultrasonic thickness measurements. *Conference materials, IV Conference of Metrology in Machine Manufacturing Techniques. ITMiA WUOSaT*. Szklarska Poreba.

Volk W. 1973. Statistics used for engineers. *WNT*. Warsaw.

Underground Infrastructure of Urban Areas 4 – Madryas et al. (Eds)
© 2018 Taylor & Francis Group, London, ISBN 978-1-138-55953-0

Slag cement CEM III/B 42,5L-LH/SR/NA as a component of durable concrete

Z. Giergiczny
Górażdże Cement S.A., Heidelberg Cement Group, Poland
Faculty of Civil Engineering, Silesian University of Technology, Gliwice, Poland

M. Batog & K. Synowiec
Górażdże Cement S.A., Heidelberg Cement Group, Poland

M. Ostrowski
Institute of Ceramics and Building Materials, Glass and Building Materials Division in Cracow, Poland

ABSTRACT: In numerous application, particularly in aggressive environmental conditions, cement selection is crucial for durability of concrete. Evaluation of the durability of cementitious composites is the main objective of the corrosion resistance test because aggressive liquid and gaseous media can damage the cement matrix structure. In order to ensure the corrosion resistance of concrete, slag cements with significant amount of ground granulated blast furnace slag are being used more and more frequently. This is due to the activity of granulated blast furnace slag and its influence on shaping the microstructure of the cement matrix. Paper presents the results of durability tests of mortars and concrete made of industrial slag cement CEM III/B 42,5L-LH/SR/NA. Resistance to sulphate corrosion, chloride ions permeability, carbonation depth were evaluated. Moreover, fineness and w/c ratio influence on ground granulated blast furnace slag activity were assessed.

It was found, that CEM III/B 42,5L-LH/SR/NA cement composites are characterized with high resistance to corrosive agents, especially with low w/c ratio.

1 INTRODUCTION

The European Union's environmental policy (reduction of greenhouse gas emissions, including mainly CO_2, industrial by-products management) has inspired the cement industry to take multidirectional measures, including the use of alternative fuels mainly biomass-based, the use of alternative raw materials for the production of Portland clinker, the substitution of Portland clinker in the cement composition with low-carbon main constituents and the development of alternative technologies of binders production without clinker (Schneider et al. 2011, Hauer et al. 2010, Harder 2011, Giergiczny et al. 2014). The current cement standard PN-EN 197-1 offers great opportunities for shaping the composition of cement with the main constituents other than Portland clinker. The most common are granulated blast furnace slag, siliceous fly ash and limestone.

Practical experience and numerous research works (Geiseler et al. 1995, Torii et al. 1998, Deja 1998, Locher 2006, Małolepszy 1998, Jasiczak 1998, Giergiczny et al. 2002) confirm that Portland-slag cements CEM II/A,B-S and slag cements CEM III/A have better properties than Portland cement CEM I. This is due to the reactivity of granulated blast furnace slag and its effect on the properties of cement composites (Sprung et al. 2001). Cements containing granulated blast furnace slag are characterized by longer setting times, lower heat of hydration, better workability (maintain consistency over time), significant increase in strength over longer hardening periods and higher resistance to aggressive environments compared to Portland cements CEM I. In paper the properties of slag cement CEM III/B

42,5L-LH/SR/NA containing about 70% of granulated blast furnace slag and concrete with it with particular emphasis on resistance against chemical corrosion are discussed.

2 GRANULATED BLAST FURNACE SLAG AS COMPONENT OF COMMON CEMENT

Blast furnace slag is a by-product obtained by the process of melting pig iron in a blast furnace. The process is conducted at 1400–1600°C. As a result of the melting of the feed mixture a melted blast furnace slag is obtained, which after separation from the pig iron is subjected to a granulation process by rapid cooling by water. Granulated blast furnace slag is a component with latent hydraulic properties, i.e. as a result of comminution and activation, binds and hardens similarly to Portland cement (Giergiczny et al. 2015). High specific surface area and proper granulometry (Figure 1) is important for increasing the hydraulic activity of the slag and thus, obtaining slag cement with its high content in the appropriate strength class. This is illustrated by compressive strength results of slag cements containing 70% of ground granulated blast furnace with different specific surface area (Figure 2) (Ostrowski 2017). Comparatively, the compressive strength of Portland cement CEM I 42,5R is shown in Figure 2. The slag cements were obtained by mixing Portland cement CEM I 42,5R and separately ground granulated blast furnace slag with chemical composition shown in Table 1. The content of glass phase in the slag was 98.6%. Table 2 shows selected properties of granulated blast furnace slag with comparison to the requirements of the cement standard PN-EN 197-1.

Analyses of compressive strength results of slag cement CEM III/B show that the level of compressive strength, at each tested date, increases with increasing surface area of granulated blast furnace slag. With a specific surface area of 6,000 cm^2/g, the compressive strength of CEM III/B cement after 14 days is higher than strength of Portland cement CEM I.

Figure 3 shows impact of w/c-ratio reduction from 0,5 to 0,3 on the strength properties of cement mortars with CEM III/B. Effective reduction of the w/c-ratio through the use of the latest generation of chemical admixtures leads to early strength improvement, which greatly expands the scope of application of this type of cement in the construction industry.

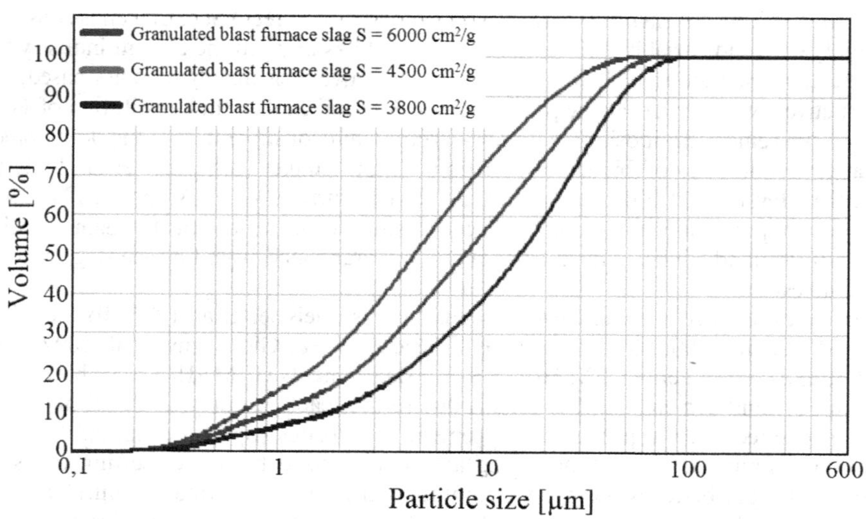

Figure 1. Particle size distribution of granulated blast furnace slag (Ostrowski 2017).

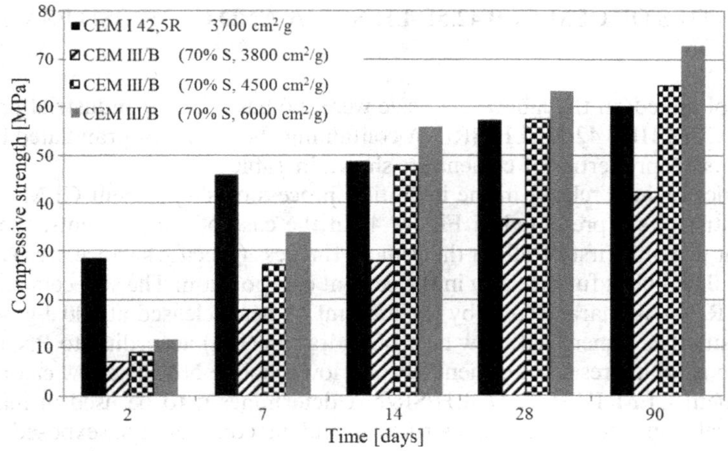

Figure 2. Compressive strength of slag cement CEM III/B depending on the slag fineness (slag content S = 70%).

Table 1. Chemical composition of granulated blast furnace slag.

Content [%]										
LOI	SiO_2	Al_2O_3	Fe_2O_3	CaO	MgO	SO_3*	S^{2-}	Na_2O	K_2O	Na_2O_{eq}
0,17	38,40	7,77	0,99	43,69	5,77	1,12	0,37	0,53	0,54	0,88

* – total content of sulphur expressed as SO_3, including S^{2-} content.

Table 2. Properties of ground granulated slag as main cement component.

Property	Content	Requirements acc. to PN-EN 197-1
Glass phase [%]	98,6	≥2/3
SiO_2 + CaO + MgO [%]	87,9	≥2/3
CaO + MgO/SiO_2	1,3	>1,0

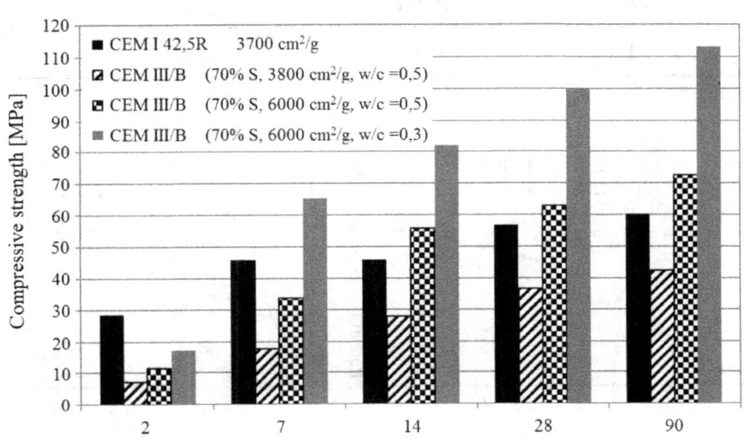

Figure 3. Effect of w/c ratio on compressive strength of cement mortars.

3 PROPERTIES OF CEM III/B 42,5L-LH/SR/NA AND COMPOSITES WITH ITS ADDITION

The results obtained on the laboratory scale were the basis for the industrial production of slag cement CEM III/B 42,5L-LH/SR/NA containing about 70% of granulated blast furnace slag. The physical properties of cement are shown in Table 3.

The kinetics of heat release in the hydration process of slag cement CEM III/A, B produced industrially are presented in Figure 4. In the case of slag cements, the amount of released heat depends first of all on the cement fineness (specific surface area) and the content of granulated blast furnace slag in the cement composition. The slag cement CEM III/B 42,5L-LH/SR/NA is characterized by the amount of heat released at 200 J/g, which allows it to be classified as cement with low heat of hydration (LH) according to PN-EN 197-1. It also refers to sulphate resistant cement SR and low alkaline NA. The low calorific value of the slag cement CEM III/B 42,5L-LH/SR/NA determines it to be used in mass concrete, hydrotechnical concrete, marine construction and in constructions exposed to chemical aggression (construction of sewage treatment plants, swimming pools, in underground mining, construction of massive foundations).

The characteristic property of slag cements CEM III, in particular with the high content of granulated blast furnace slag, is low early strength and significant increase of strength in long hardening period (Figure 5).

The slag cement CEM III/B 42,5L-LH/SR/NA is characterized by a standard strength (after 28 days) comparable to Portland cement of the same strength class with strength significantly developing with time (Figure 5). The strength development is related to the slower hydration process of CEM III/B cement compared to the hydration of Portland cement CEM I. The increase in strength is associated with the formation of dense microstructure of the cement matrix, which results in high durability of concrete structures. Therefore, the durability characteristics of cement composites (strength, frost resistance, resistance to chemical corrosion, etc.) with slags cements should be determined after 56 or 90 days.

Figure 6 shows the expansion of cement mortar in Na_2SO_4 solution, which is composed of 50 and 70% of granulated blast furnace slag. These mortars did not show significant linear changes during test period compared to Portland cement mortar with CEM I 42,5R. This is due to the reduced amount of C_3A phase as a result of replacing Portland clinker by granulated blast furnace slag and smaller amount of calcium hydroxide $Ca(OH)_2$ generated from

Table 3. Physical properties of CEM III/B 42,5L-LH/SR/NA cement.

Specific surface [cm^2/g]	Water demand [%]	Initial setting time [min]	Final setting time [min]	Soundness [mm]
5200	31,8	190	260	0,6

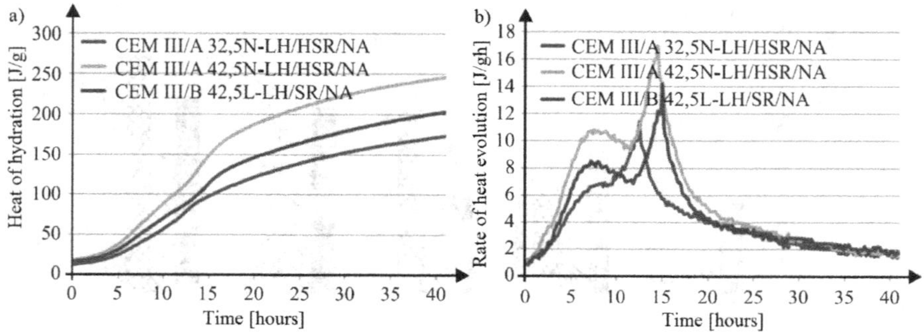

Figure 4. Heat of hydration of slag cements acc. to PN-EN 196-9 a) total generated heat b) rate of heat evolution.

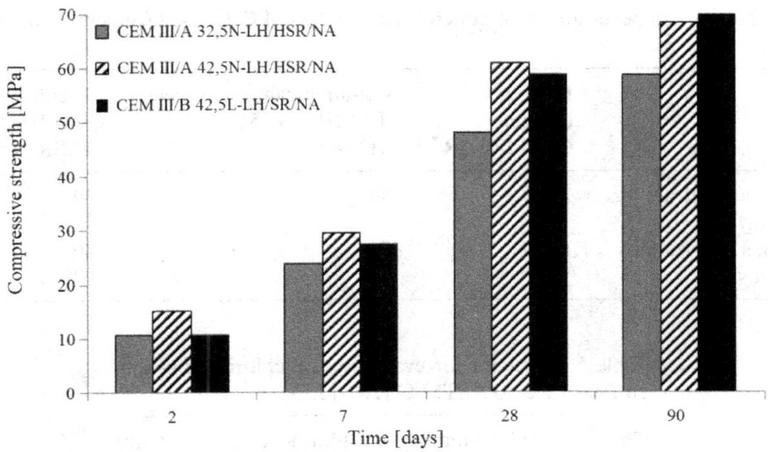

Figure 5. Compressive strength development of slag cements CEM III/A, B.

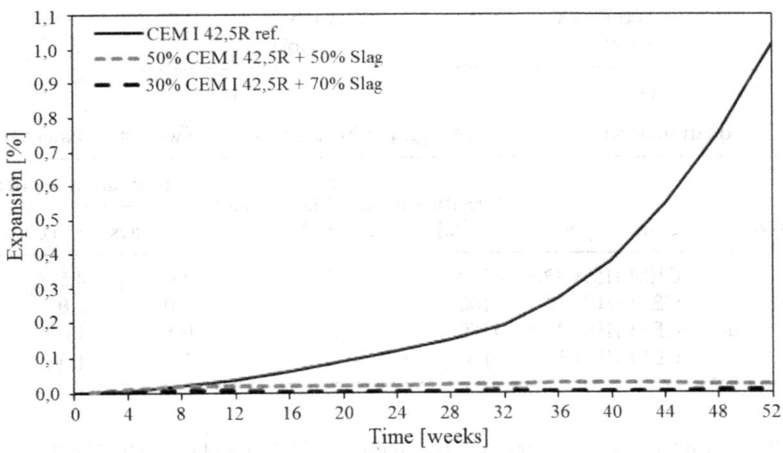

Figure 6. Expansion of cements with granulated blast furnace slag in Na_2SO_4-solution.

the calcium silicate hydrolysis from Portland clinker comparing to Portland cement CEM I. In addition, during the hydration process of CEM III/B cement, a stable CSH phase with a low C/S ratio is produced. As a result, a tight, very compact microstructure, with limited capillary porosity is created (Chłądzyński et al. 2008). Therefore, the use of cement with high granulated blast furnace slag content, e.g. CEM III/B 42,5L-LH/SR/NA is particularly recommended for the construction of structures exposed to chemical aggression in XA2, XA3 exposure classes caused by sulphate ions. The examples of such structures may be the elements of sewage systems—pipes, manholes, tanks, sewage treatment plants, foundations in aggressive grounds, elements of agriculture infrastructure, etc.

Chloride penetrability in concrete with CEM III/B 42,5L-LH/SR/NA cement is shown in Table 4. Concrete with CEM III/B 42,5L-LH/SR/NA cement is a composite with very low penetrability of chloride ions according to ASTM C 1202-12 (Table 5).

The dense microstructure of hardened cement paste with slag cement CEM III/B, especially at low w/c ratio, does not significantly carbonate even after two years (Table 6) (Lang 2005).

Slag cement CEM III/B 42,5L-LH/SR/NA is a low alkaline cement (NA), which is used to reduce the risk of alkaline corrosion in concrete with potentially reactive aggregates (Blight et al. 2011). This is due to the fact that alkalis in the granulated blast furnace slag are hardly

Table 4. Chloride ion penetrability of concrete acc. to ASTM C 1202-12 (cement content 350 kg/m^3, w/c = 0,45).

Property	Reference concrete CEM I 42,5R	Non-airentrained concrete CEM III/B 42,5L-LH/SR/NA	Air entrained concrete CEM III/B 42,5L-LH/SR/NA
Chloride ions penetrability after 28 days [C]	2730	580	500
Chloride ions penetrability after 90 days [C]	1800	370	340

Table 5. Criteria for evaluation of chloride ion penetrability acc. to ASTM C 1202-12.

Charge passed [Culombs]	Chloride ion penetrability
>4000	High
2000 ÷ 4000	Moderate
1000 ÷ 2000	Low
100 ÷ 1000	Very low
<100	Negligible

Table 6. Carbonation depth of normal and high-performance concrete with slag cement (Lang 2005).

Concrete type	Cement type	Specific surface [cm^2/g]	Slag content [%]	Carbonation depth [cm], after		
				180 days	360 days	720 days
Normal (w/c = 0,5)	CEM III/A 42,5	4550	52	4,0	4,5	5,5
	CEM III/B 32,5	5100	70	5,0	7,0	8,5
High-performance (w/c = 0,3)	CEM III/A 42,5	4550	52	0,5	1,0	1,0
	CEM III/B 32,5	5100	70	1,0	1,5	2,5

Table 7. Water soluble alkalis content in Portland cement CEM I and slag (Drożdż 2014).

Sample	Alkali content						Soluble alkali content to total alkali content [%]		
	Total [%]			In water solution [%]					
	Na_2O	K_2O	Na_2O_{eq}	Na_2O	K_2O	Na_2O_{eq}	Na_2O	K_2O	Na_2O_{eq}
Portland cement CEM I	0,28	0,58	0,66	0,1752	0,4382	0,464	62,57	75,55	70,30
Ground granulated blast furnace slag (S)	0,53	0,39	0,79	0,0068	0,0046	0,010	1,28	1,18	1,27

soluble (Table 7), which leads to a decrease in OH^- ions concentration of concrete pore solution (Figure 7) (Böhm 2006). In granulated blast furnace slag the majority of alkali is contained in the glassy phase (insoluble alkalis). Only a very small part is found in the form of readily soluble sulphates. This property makes cement with a high content of granulated blast furnace slag e.g. CEM III/B 42,5L-LH/SR/NA, very suitable for making alkali-resistant concrete (Figure 8).

Hydraulic activity of the granulated blast furnace slag leads to the production of a C-S-H phase (or C-A-S-H phase) with a lower Ca/Si ratio and higher alkali binding capacity when compared to CEM I hydration products.

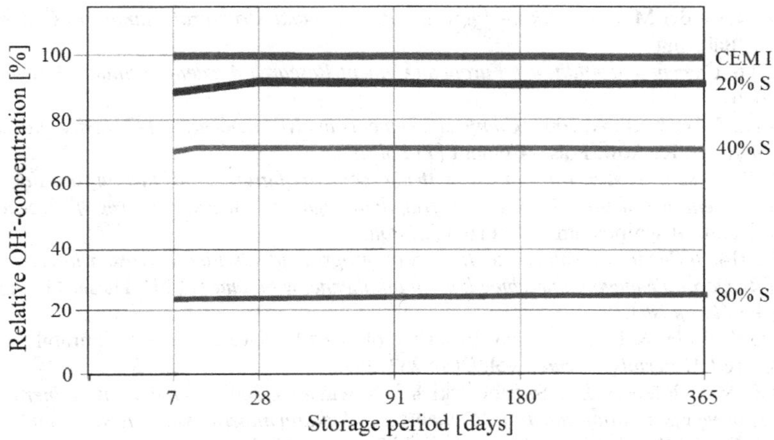

Figure 7. Influence of slag content on concentration level of OH^- ions (Böhm 2006).

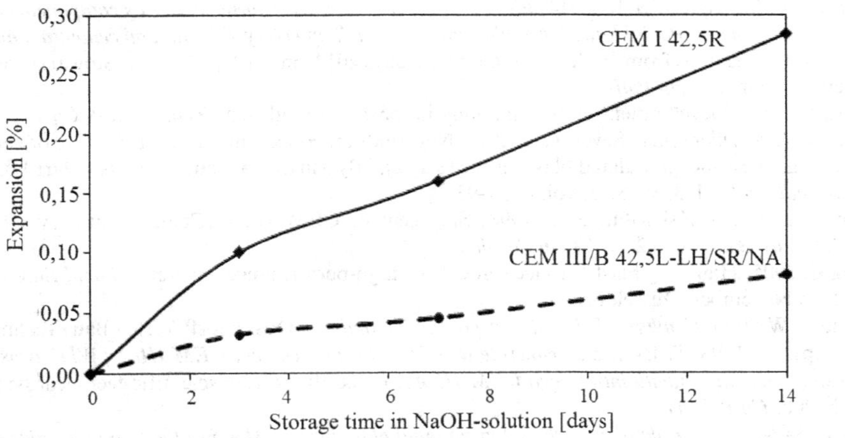

Figure 8. Expansion of mortars with quartzite aggregate and different cements (Ostrowski 2017).

4 SUMMARY

The slag cement CEM III/B 42,5L-LH/SR/NA is a high-performance binder that offers a wide range of applications in construction industry, especially using the latest generation of chemical admixtures allowing to produce concrete with low w/c-ratio.

The use of slag cement CEM III/B 42,5L-LH/SR/NA is not only ecologically and economically feasible, but it is also the construction of durable building structures resistant to various chemical aggressive media. It is difficult to imagine a production of mass concrete without the use of CEM III/B with low heat of hydration LH or a concrete for wastewater treatment plant without the use of sulphate resistant SR (HSR) cement containing mineral additives.

When making structures and concrete elements from slag cement CEM III/B special attention should be paid to proper thermal-moisture treatment.

REFERENCES

ASTM C114-04. *Standard Test Methods for Chemical Analysis of Hydraulic Cement.*

ASTM C1202 – *12 Standard Test Method for Electrical Indication of Concrete's Ability to Resist Chloride Ion Penetration.*

Blight G. & Alexander M. 2011. *Alkali-Aggregate Reaction and Structural Damage to Concrete*. Leiden: CRC Press/Balkema.

Böhm M. 2006. *Cement and Additions. European Cement Research Academy Seminar*, Dusseldorf, October 24, 2006.

Chłądzyński S. & Garbacik A. 2008. *Composite cements in civil engineering (Cementy wieloskładnikowe w budownictwie)*, Kraków: Polski Cement *(in Polish)*.

Deja J. 1998. *Corrosion resistance of cements with high content of granulated blast furnace slag (Odporność korozyjna cementów o wysokiej zawartości granulowanego żużla wielkopiecowego)*. Proceedings from scientific-technical symposium in Chorula *(in Polish)*.

Drożdż W. 2014. *Influence of calcareous fly ash on progress of alkaline corrosion in concrete (Wpływ popiołu lotnego wapiennego na przebieg korozji alkalicznej w betonie)*, PhD Thesis, Gliwice: Politechnika Śląska *(in Polish)*.

Geiseler J. & Kollo H. & Lang E. 1995. Influence of blast furnace cements on durability of concrete structure; *ACI Materials Journal.*, vol.92(3): 252–257.

Giergiczny Z. & Małolepszy J. & Szwabowski J. & Śliwiński J. 2002. *Cement with mineral additives in technology of new generation concrete (Cementy z dodatkami mineralnymi w technologii betonów nowej generacji)*. Opole: Wydawnictwo Instytut Śląski Sp. z o.o. *(in Polish)*.

Giergiczny Z. & Synowiec K. & Batog M. 2015. Slag cement CEM III/B 42,5L-LH/SR/NA—properties and possibilities of use in construction (Cement hutniczy CEM III/B 42,5L-LH/SR/NA—właściwości i możliwości zastosowania w budownictwie), *Materiały Budowlane* 10: 96–99. *(in Polish)*.

Giergiczny Z. & Synowiec K. 2014, *Fly ash and granulated blast furnace slag as components of low CO_2 emission cement (Popiół lotny i granulowany żużel wielkopiecowy składnikami cementu o niskiej emisji CO_2)*, Proceedings from conference "Concrete days 2014" in Wisła, Kraków: Stowarzyszenie Producentów Cementu *(in Polish)*.

Harder J. 2011. Development of sustainability in the cement industry, *Zement Kalk Gips* 12:26–35.

Hauer B. & Mueller Ch. & Severins K. 2010. New findings concerning the performance of cements containing limestone, granulated blast furnace slag and fly ash as main constituents—Part 1 & 2. *Cement International* vol. 3: 80–86 & vol. 4: 83–93.

Jasiczak J. & Łowińska-Kluge A. 1998. Slag cement CEM III/A (Cement hutniczy CEM III/A), *Materiały Budowlane* 2:17–19 *(in Polish)*.

Lang E. 2005. High slag blast furnace cement for high-performance concrete, *Global Slag Conference*, 14–15 November, Dusseldorf.

Locher F.W. 2006. *Cement—Principles of production and use,* Dusseldorf: Verlag Bau+Technik GmbH.

Małolepszy J. 1998. *Properties of concrete with blast furnace cement CEM III/A (Właściwości betonu z zastosowaniem cementu hutniczego CEM III/A)*, Proceedings from scientific-technical symposium in Chorula *(in Polish)*.

Ostrowski M. 2017. *Role of fly ash and granulated blast furnace slag in affect on properties of concrete mix and hardened concrete with low Portland clinker content (Rola popiołów lotnych i granulowanego żużla wielkopiecowego w kształtowaniu właściwości mieszanki betonowej i stwardniałego betonu o niskiej zawartości klinkieru portlandzkiego)*, PhD Thesis, Gliwice: Politechnika Śląska *(in Polish)*.

PN-B-19707:2013 *Cement. Special cements. Composition, specifications and conformity criteria (in Polish)*.

PN-EN 196-9:2010 *Methods of testing cement—Part 9: Heat of hydration—Semi-adiabatic method.*

PN-EN 197-1:2012 *Cement—Part 1: Composition, specifications and conformity criteria for common cements.*

prPN-B-06265:2016-10 *National supplement to PN-EN 206 Concrete—Specification, performance, production and conformity (in Polish)*.

Schneider M. & Romer M. & Tschudin M. & Bolio H., Sustainable cement production—present and future. *Cement and Concrete Research* 41: 642–650.

Sprung S. & Kropp J. 2001. *Cement and Concrete, Encyclopedia of Industrial Chemistry, Sixth Edition* (electronic release).

Torii K. & Sasatani T. & Kawamura M. 1998. Chloride penetration into concrete incorporating mineral admixtures in marine environment., *6th CANMET/ACI Int. Conf. Fly ash, silica fume, slag and natural pozzolans in Concrete* vol. 2: 701–716, Bangkok/Thailand.

Underground Infrastructure of Urban Areas 4 – Madryas et al. (Eds)
© 2018 Taylor & Francis Group, London, ISBN 978-1-138-55953-0

Geotechnical interaction in underground space – theory and practice

W. Bogusz & T. Godlewski
Building Research Institute, Warsaw, Poland

ABSTRACT: While greenfield sites at highly urbanized areas are often not available anymore, the continuous development of existing grey field sites is the only meaningful option. Every year, the underground space is being developed to a larger extent. An underground part of a structure can often stretch across the entire available land with excavation protected by retaining walls constructed directly next to existing buildings and infrastructures. One of the main challenges associated with underground constructions is their impact on existing buildings and other structures adjacent to the developed site. As these structures are often highly susceptible to excavation-induced ground movements, their behavior have to be considered in a design as one of the geotechnical-related limit states. The paper presents the analysis of the impact of new investments on existing structures as a way of assessing geotechnical risk inherent in the use of underground space, in limit state design framework of Eurocode 7.

1 INTRODUCTION

Design and construction of underground structures poses a significant risk to adjacent buildings and elements of infrastructure (Boscardin & Cording 1989; Wysokiński & Kotlicki 2002); Mitew-Czajewska & Siemińska-Lewandowska 2008). It is commonly recognized that a construction of such structure may have a significant impact on other structures located in its vicinity; especially, for deep excavations and shallow tunneling at highly urbanized areas, this is a subject of significant concern and a main geotechnical risk inherent in the execution of an underground project. Limitations imposed by the need to ensure the serviceability of neighboring structures, in some cases, may even be a major factor governing the choice of a construction method, specific design solutions, or organization of construction activities at the site.

2 RISKS ASSOCIATED WITH THE IMPACT OF UNDERGROUND CONSTRUCTIONS

As the subsoil is often composed of highly variable material provided by nature, and its behavior is often controlled by highly non-linear relationships concerning stress- and strain-dependence, the soil-structure interaction problem for underground structures is one of the most difficult issues to analyze. This complexity of the problem is further increased when considering the impact of the project on the surrounding area. The extent of the zone of influence, caused by the change of in-situ or groundwater conditions, may reach far beyond the area of a construction site, affecting other existing buildings and structures, as well as the interests of their owners and inhabitants.

In the Limit State Design framework, for any design, a verification of all relevant ultimate limit states (ULS) and serviceability limit states (SLS) have to be conducted. For most typical geotechnical structures, necessary analysis is limited to the structure itself and all its elements; however, when an underground structure (i.e. deep excavation, shallow tunnel, etc.) is considered, this also involves neigh boring structures in the expected influence zone. On the one hand, due to technical reasons, as excessive deformation of the soil due to construction activities may cause exceedance of SLS criteria, and finally may lead to ULS in the neighboring structures. On the other, due to social and administrative reasons, as respecting the interests of 3rd parties (i.e. neighbors) is expected by law and is a requirement for obtaining a building permit in Poland.

This paper does not explicitly cover the impact of tunneling. Their construction causes changes in the stress state in the subsoil, as well as it may result in ground loss, often expresses as the ratio of final soil volume loss measured as the trough of the surface soil-settlement profile to the cross-section area of the tunnel. The methods of assessing tunneling-induced deformations were summarized by Guglielmetti et al (2007). Although the assessment of maximum deformation due to tunneling differs from the procedure used for deep excavations, the basic principles when assessing the impact of tunneling activities on adjacent structures are similar.

Tunnels and structures, which require a construction of a deep excavation in urbanized areas, should be designed and executed with the limitation of the subsoil deformation in mind, in order to:

- avoid excessive strains and additional forces in the adjacent structures, which can threaten their bearing capacity—considered as ULS;
- avoid or limit the occurrence of damage or displacements to the adjacent structures, which can worsen the state or serviceability conditions of these structures in a noticeable way—considered as SLS.

The impact of deep excavations and other construction activities may vary at different stages of the execution phase. The main factors affecting their behavior in practice are:

- Unloading due to demolition of existing structures;
- Unloading due to deep excavation;
- Changes in water pressures due to dewatering;
- Deformation of retaining walls and the stress changes in retained soil;
- Loading from the new structure.

These factors may be limited on some projects, as well as they may include additional influences on the others. For relatively light-weight structures like Metro stations, the most critical influence will be due to the excavation and usually no additional loading will follow. While for a high-rise construction, excavation phase will be followed by incremental loading from the construction of the structure itself.

The verification of the impact on the neighboring structures is composed of following main steps:

- Assessing the extent of the zone of influence and identifying structures located within it;
- Assessing the impact of the construction works on the ground displacements within the influence zone;
- Investigating the type and the state of the structures in the influence zone, as well as the limiting values of their deformations;
- Assessing predicted deformation of adjacent structures due to ground displacement, and the impact of those deformations on the condition and serviceability of the structures;
- Verifying the limit states for adjacent structures;
- Documenting the current state and damages of the existing structures located in the active zone of influence, as well as those located in vigilance zone which are in poor technical condition;
- Design and preparation of the remediation measures, if necessary.
- Recommendations for the monitoring program for execution and maintenance phase.

It should be recognized that construction of underground structures is dominated by the actions and resistances caused by the interaction with the soil mass; therefore, the uncertainty inherent in the geotechnical conditions are transferred to these elements of the analysis as well. The understanding of the uncertainty of the parameters of the site is critical in geotechnical design (Simpson 2011); especially, when considering underground structures. In order to obtain a reliable prediction, appropriate calculation model has to be used, and proper parameters should be provided in the geotechnical investigation report.

Essex (1996) stated that the construction of underground structures provides a greater inherent risk of encountering unforeseen or differing ground conditions than other geotechnical activities. Furthermore, according to Simpson (2011), next to human errors, the most common causes of geotechnical failures are in fact ground conditions or water pressures significantly differing from design assumptions. To decrease a probability of failure due to human error, proper quality assurance should be provided. For structures of higher than usual consequences of failure (Geotechnical Category 3), this may include higher supervision and inspection levels, which may include additional verification and supervision by the third party (EN 1990: 2004). However, in regard to the scope of investigation, investors are often unwilling to provide or fund adequate geotechnical investigation programs. Often incorrectly considered by investors as separate elements of design, in practice, a geotechnical investigation report and an impact assessment analysis are prepared separately, without any coordination between them or other elements of geotechnical design. Correctly conducted site investigation can be very beneficial for obtaining accurate and reliable prediction of expected soil-structure interaction behavior. Baecher & Christian (2003) divided a site characterization program into three main phases: reconnaissance, preliminary, and detailed investigation, for verification and refining of the geotechnical ground model. With each subsequent stage, the uncertainties in the ground model are reduced. Unfortunately, clear presentation of the geotechnical ground model, its incremental updating as new information arise, and performing additional testing to remove the most significant uncertainties in the design of underground structures, is not a common practice among designers, despite the obvious benefits. At the same time, all too often, investors transfer all the responsibility for differing ground conditions to the designer or the contractor. Persistent use of the lowest-price criterion in selection of geotechnical investigators, designers, and contractors, complicates the proper project development. However, detailed insight into these issues is beyond the scope of this paper.

Designers, aware of the risks and uncertainties associated with the construction of underground structures, often tend to be over-conservative in their design assumptions and solutions. For structural part of the design, they prefer to increase reliability by providing an excessive margin of safety and sufficient robustness rather than decreasing inherent uncertainty; even though such decrease may be obtained by simply investing in additional ground investigation, as well as performing more advanced analysis.

The apparent cost and time savings at the preliminary (building permit) design stage, made on ground investigation or proper analysis, may lead to significant increase in costs at the execution stage, i.e. due to encountering differing ground conditions, underpinning existing structures, additional support for the retaining walls, or the need for repair of damaged buildings after the construction. Especially the latter case is inconvenient for investors, as it may be associated with lengthy litigation process; its cost is not only expressed in monetary value, but also in public perception of the company and its good name that may be tarnished. When the information on the damage to existing structures, due to previous investments, becomes a public knowledge, the investor can expect a very strong opposition for future investments.

For construction of deep excavations, especially in close proximity to existing structures, diaphragm walls are usually preferred. To increase safety level and decrease the impact on existing structures, the increase of the thickness of the wall, additional supports, and limiting the width of panels can be considered. Diaphragm walls allow for relatively safe construction of a retaining wall at a small distance from foundations of existing buildings. A potential failure of a neighboring foundation during construction of a diaphragm wall, as investigated i.e. by Choy et al (2007), is not considered here, although such possibility should be taken into account in a design by verifying the stability of a diaphragm wall panel during its excavation. In general, excavating closer than 0.50 m from existing spread foundations or retaining walls and 2.0 m from existing piles or tunnels should be avoided.

Similarly, the installation-induced displacements are just briefly considered in this paper, but the impact of the construction procedure of a retaining wall should not be ignored. L'Amante et al (2012) concluded that this displacement is small, but it is not negligible in comparison to the one induced by the excavation.

Finally, it is important to account for potential perception of the construction activities, as the impact of underground construction on adjacent structures is not only limited to technical aspects. Although not often considered in the design, the choice of retaining wall type should include psychological factors (Wysokiński & Kotlicki 2002). The construction of a deep excavation may impact inhabitants of adjacent buildings by imposing:

- Fear of construction of the deep excavation, often in full view from neighboring buildings;
- Inconveniences associated with the noise and the vibrations due to construction activities.

In some cases, remediation measures used by contractors have more of a psychological impact, reducing the subjective perception of risk rather than its probability of occurrence.

3 EXTENT OF THE INFLUENCE ZONE

The influence zone of the underground construction can be defined as an area in which noticeable displacements of the ground can occur as a result of excavation and construction, consequently affecting other structures located within it. The level of influence affecting specific structures will vary, depending on the distance of the considered structure from the excavation.

When considering the extent of the influence of underground construction activities on neighboring area, the exact mechanism of soil-structure interaction is often complex and it depends on a combination of various factors. Therefore, empirical estimation based on the previous practice and experience is often preferred as the first approximation. According to ITB Recommendations (Wysokiński & Kotlicki 2002), the assessment of the extent of the influence zone is not necessary only when the minimum distance from the nearest structure is larger than 4 times the excavation depth ($4H_{excav}$), without dewatering.

Generally, two main influence zones can be distinguished. The first one is the active zone of influence, where excavation-induced displacements can potentially threaten the bearing capacity of neighboring structures. The extent of active zone is based on the potential failure wedge in the soil due to failure or extensive deformation of the supporting structure. The second one is the vigilance zone of influence, where measurable displacements may occur, and the serviceability of adjacent structures is of the most concern. The extent of these zones of influence will depend on a number of factors, but most notably, on the dominant type of the soil and the depth of the excavation. The ranges of the influence zone suggested for the use in Poland, based on the experiences from the construction of Metro line I and deep excavations in Warsaw, were presented in ITB recommendations (Wysokiński & Kotlicki 2002), and they are presented in Figure 1 and Table 1. These values are valid for relatively homogeneous ground conditions, which often is not the case encountered in practical geotechnical design. Usually, in heterogeneous conditions, the more conservative assumptions should be made.

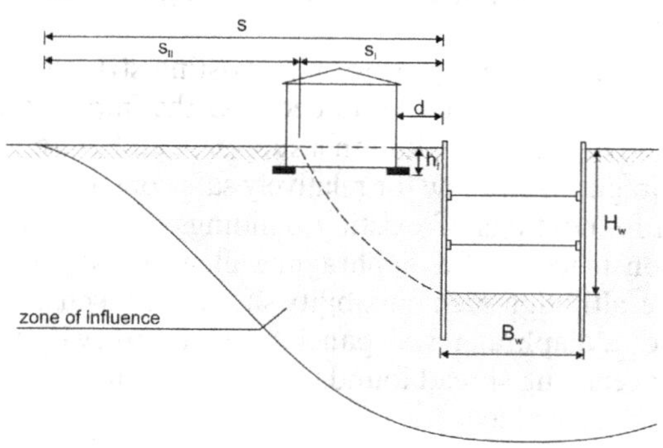

Legend:
H_w i B_w – depth and width of excavation
S – range of influence zone
S_I – range of active zone influence
S_{II} – range of vigilance zone influence
d – distance of the structure from the excavation
h_f – foundation depth

Figure 1. Idealized schematic of the influence zone for a deep excavation near an existing structure—the extent of the influence zone is dependent on ground conditions.

Table 1. The extent of the area of influence of deep excavations (Wysokiński & Kotlicki 2002).

	Zone of influence	
Soil type	S_I	S_{II}
Sands	0,50 H_w	2,0 H_w
Glacial tills	0,75 H_w	2,5 H_w
Clays	1,00 H_w	$3 \div 4$ H_w

The ITB Recommendations (Wysokiński & Kotlicki 2002) stipulated that the extent of the zone of influence, estimated based on the Table 1, can be reduced by 20% if no dewatering outside the excavation is anticipated. Furthermore, for excavations larger than 60 m in both directions, as well as in the presence of relatively soft soils, the expected range of influence should be increased.

Similar division into two zones has been recommended by ITA-AITES (2014). When no other factors are considered (especially ground conditions), they recommended choosing the extent equal to the higher value out of:

- For active zone: The excavation depth (*H*) or 50 m;
- For vigilance zone: Twice the excavation depth (*2H*) or 100 m.

This approach can be assumed as very conservative when deep excavations and shallow tunnels are considered.

Usually, the extent of the area of influence is derived in reference to the deformation of ground surface. This approach may not be directly applicable for assessing the impact on structures with deep foundations or tunnels. Furthermore, usually greenfield conditions are assumed in the analysis. While this assumption may be regarded as conservative for prediction of the zone of influence, the presence of stiff structures may limit the extent of the impact on them, while increasing its magnitude. This may result in less favorable differential settlements.

Presented values are empirically-based and should be regarded as appropriate for over-consolidated soils commonly encountered in Warsaw and some other regions of Poland. The extent of the area affected by the excavation, especially in case of failure occurrence, can be significantly larger, i.e. for a 30 m deep excavation in soft clays and at failure, Simpson et al (2009) reported that substantial ground movements occurring up to 70 m from the excavation. More accurate prediction of the extent of the influence zone can be obtained from the analysis using numerical methods, i.e. finite element method (FEM).

The extent of the zone of influence as well as the location of the buildings within it should be presented at least in a layout map in the geotechnical design part of the preliminary design (building permit design).

4 SERVICEABILITY LIMIT STATE VERIFICATION

The impact of underground construction activities on adjacent structures has to be recognized as one of the limit states for consideration. According to Eurocode 7 (2004), the movement of a retaining structure that may cause collapse or affect the appearance or efficient use of the nearby structures or services, shall be considered in geotechnical design. Limiting values for the allowable displacements of walls and the ground adjacent to them shall be established while taking into account the tolerance to displacements of supported structures and services. Although not explicitly mentioned, this also relates to all effects of actions related to the construction activities, i.e. uplift due to unloading of over consolidated subsoil or subsidence induced by dewatering. Impact of additional displacements on the technical condition of structures should be analyzed for all relevant structures located within the influence zone. It can be omitted for structures located in the vigilance zone for which predicted displacements are lower than 5 mm, unless their technical state is very poor or additional serviceability criteria have to be considered.

According to Eurocode 7 (2004), the design may be justified by checking that the estimated displacements do not exceed the limiting values. The basic expression guiding the verification of SLS can be used:

$$E_d \leq C_d \tag{1}$$

where E_d = design value of an effect of the actions; C_d = limiting design value of the effect of an action.

Depending on a local practice and a type of the structure under consideration, different parameters may be considered for verification, as well as different factors can influence the choice of SLS limiting value (i.e. serviceability, damage to structural or decorative elements, water-tightness, etc.). In Poland, the verification of the SLS can be conducted in a simplified manner based on comparison of maximum predicted vertical displacement and the limiting value of this deformation. However, Boscardin & Cording (1989) recognized that excavation-induced displacements have greater impact on structures than settlements due to their self-weight; they related the damage potential of masonry buildings to limiting values of tensile strain, as the main contributing factor.

As excavation-induced SLS will usually precede the occurrence of ULS, meeting SLS criteria makes verification of the ULS not necessary, when it requires much higher displacements to occur.

Avoidance of this risk affecting serviceability may not always be justified financially when the cost of remediation measures would be significantly higher than the cost of future repairs. With the agreement of the owner of a neighboring structure, a temporary worsening of serviceability or the condition of the structure may be permitted, during the construction phase, but only if the structure can be restored to the previous state afterwards. In such case, sufficient margin of safety in regard to ULS should be provided.

5 DISPLACEMENT PREDICTION

Various calculation models can be used for prediction of deformations due to the construction of underground structures. These models can be divided into three main categories: analytical, semi-empirical, and numerical models (EN 1997: 2004). In Poland, the latter two types are preferred, as so far they offered the most reliable results, validated by the comparable experience. As all calculation models are idealizations of expected soil-structure interaction, they involve varying levels of simplification and resulting model uncertainty. Due to that fact, simpler calculation methods are often more conservative and robust, to ensure sufficient reliability of displacement prediction. When considering the choice of the calculation method, the use of the most appropriate model is recommended. For relatively simple cases, a balance between increased accuracy of prediction and the increase in the numbers of required parameters may favor simplified semi-empirical models, while assuring sufficient reliability. These models describe the behavior of the ground in a simplified manner, and they are calibrated based on the behavior observed in the past (i.e. Figure 2). A method most commonly used in Poland for displacement prediction is based on semi-empirical calculation model presented in ITB recommendations (Wysokiński & Kotlicki 2002). As more complex projects are often considered, especially, when other underground structures already exist at the site and in the presence of difficult ground conditions, these simple semi-empirical models should be considered only as preliminary evaluation (i.e. for obtaining of a building permit) and the analysis with the use of more advanced numerical methods should follow at later stages of the design.

According to Eurocode 7 (EN 1997: 2004), a more detailed investigation, including displacement calculations, shall be undertaken where nearby structures and services are unusually

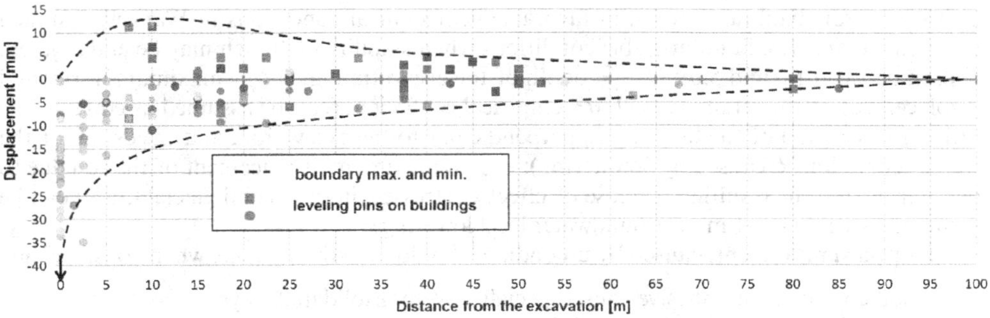

Figure 2. Example of displacements of leveling pins on selected buildings along part of the II metro line in Warsaw.

sensitive to displacement, and where comparable experience is not well established. With the use of sufficiently advanced constitutive model, FEM analysis may provide more accurate prediction of displacements, especially, those including unloading of the overconsolidated soils, complex stratification or geometry of the structure, as well as uncommon types or distribution of supports for retaining walls.

When reliable prediction of the behavior of adjacent structures is difficult, the observational method can be used (Peck 1969, EN-1997-1: 2004).

5.1 *ITB Recommendations (semi-empirical prediction model)*

The semi-empirical method presented in the Recommendations of ITB No. 376/2002 (Wysokiński & Kotlicki 2002) became very popular due to lack of other normative requirements, as well as its simplicity, ease-of-use, and comprehensive approach to the problem.

The evaluation of both, the extent of the zone of influence and the prediction of deformation, is based on simplified assumption that retaining structure is sufficiently stiff with the use of common supporting measures, and that the results of the analysis are mainly affected by the depth of the excavation and dominant ground conditions. If necessary, additional factors, such as the area of the excavation, dewatering, the extent of ground anchors, or the time of the excavation, might have to be considered.

The resulting lower- and upper-bound prediction should be considered as a possible range of displacement values that may occur as a result of an underground construction; these values may generally be regarded as conservative.

As excavation-induced horizontal strains had not been a common cause of damages to buildings in Poland, the semi-empirical method presented in ITB Recommendations (Wysokiński & Kotlicki 2002) based its estimation on vertical displacements only. Although simplified, this approach is further justified by the ease-of-use expected by geotechnical designers, as well as possibility of direct reference to the values being measured at the site during the construction. This does not preclude the use of other criteria, which may be justified for some structures and situations.

According to these recommendations, downward movement of the ground should be estimated for every analyzed case. Maximum downward vertical displacement can be evaluated as the sum of displacements caused by the construction of a retaining wall and its horizontal deformation during construction:

$$\max v_0^{(-)} = v_i + v_u + v_w = \alpha \cdot \left(H_{excav}\right)^{0.5} + 0.75 \cdot \max u_k + \theta \cdot v_{w;\max} \tag{2}$$

where $\max v_0^{(-)}$ = maximum downward vertical displacement; v_i = maximum displacement caused by the construction of a retaining wall; v_u = maximum displacement due to horizontal deformation of the wall; v_w = maximum displacement of the building due to dewatering (usually equal to 0 if no dewatering outside the excavation is expected); α = empirical coefficient (for diaphragm walls, in favorable geotechnical conditions in Warsaw, assumed between 1.0 and 1.3; in unfavorable conditions when stability of trench may be endangered, chosen between 1.3 and 5.0); H_{excav} = depth of the excavation; max u_k = maximum horizontal displacement of the retaining wall; θ = empirical coefficient (can be assumed as equal to the ratio of dimension of the building L, perpendicular to the excavation, and the extent of the water depression R, caused by dewatering); $v_{w;max}$ = maximum displacement of the soil due to dewatering and the resulting increase of effective stresses (in sands and glacial tills, it can be assumed as 1 mm per 1 m of groundwater level lowering).

An upward movement, caused by unloading, should be estimated, i.e. when considering:

- Subsoil composes of cohesive soils, especially, overconsolidated clays;
- Construction of relatively light building (i.e. Metro stations), when loads from the structure are lower than the self-weight of the excavated soil;
- Unusually long construction period for the underground part of the structure.

Maximum upward vertical displacement behind the wall can be evaluated based on the expected value of the maximum uplift at the bottom of the excavation:

$$\max v_0^{(+)} = \eta \cdot v_{\max} \tag{3}$$

where max $v_0^{(+)}$ = maximum upward vertical displacement; η = empirical coefficient (0.3 for minimum embedment of the retaining wall of 3 m below the bottom of the excavation; 0.6 for other walls); v_{max} = expected value of the maximum uplift at the bottom of the excavation due to unloading.

Both, a maximum horizontal deformation and a maximum uplift at the bottom of the excavation have to be assumed or predicted beforehand, i.e. based on comparable experience. The calculation of the maximum horizontal displacement of the wall should include displacements at different stages, as well as all deformation, i.e. resulting from compressive stresses in struts, thermal actions, or shrinkage of concrete. For soldier-pile wall, this value should include the deformation of soil behind the lagging.

Based on calculated maximum and minimum vertical displacements, as well as predicted extent of the influence zones, predicted lower- and upper-bound distribution of ground level deformations can be estimated. It is assumed that maximum values occur near the retaining wall, they are equal to zero at the end of the entire influence zone, and between active and vigilance zone they are equal to 50% of the maximum values; in-between, the values can be interpolated linearly.

In the case of the structures with spread foundations, their predicted deformation may be assumed as equivalent to the displacement of the ground at the location and the depth of the foundation. In most cases, assuming a displacement equal to the ground surface displacement may be considered as a conservative approach. For buildings founded at the depth lower than or equal to 2.5 m below ground level, the predicted displacement v can be assumed equal to the displacement of the ground surface. For greater depth, it can be adjusted as follows:

$$v = v_0 \cdot \frac{H_{excav} - h_f}{H_{excav}} \tag{4}$$

where v_0 = predicted displacement of the ground surface at specific distance from the excavation; h_f = foundation depth.

5.2 *FEM analysis (numerical prediction model)*

The use of numerical method, allowing for more realistic soil-structure interaction, compatibility of displacements, as well as to account for the stiffness of the structure and the soil, may provide more realistic prediction of displacements. In some cases, when a complex spatial distribution of underground structures and soil layers is present, usually outside the range of applicability of simpler calculation models, performing spatial (3D) FEM analysis is often the only viable solution to obtain a reliable prediction. The increasing availability of software used for advanced numerical analysis provides an opportunity for cost optimization by obtaining more accurate prediction of soil-structure interaction, especially, in the underground space.

The use of advanced numerical methods, like FEM, requires appropriate input parameters, good understanding of soil mechanics, and their implications for the analyzed problem. The analysis of underground structures often requires the use of stress- and strain-dependent constitutive models, i.e. Hardening Soil with small strain stiffness (Benz 2007), as deeper strata subjected to lower strains and higher stresses exhibit much higher stiffness (Godlewski et al 2015); this significantly affects the results of prediction when impact on adjacent structures is considered. Together with the use of complex 3D models, such analysis allows to account for:

- The complexity of geotechnical conditions—including complex soil layering (i.e. Figure 3a);
- The complexity of newly designed structure, the sequence and the schedule of the construction (i.e. Figure 3b);
- The structural resistance and the stiffness of structural elements of existing structures;
- Recreating the initial state—including previously constructed structures and stress changes in the soil;
- The expectations of other stakeholders in regard to the reliability of prediction and identification of the behavior of adjacent structures during construction.

The use of 3D models has an advantage when dealing with complex geometries of excavation pits. This allows considering the constraining effect of stiff corner bracing on the deformations (Orazalin et al 2015); an effect often observed in practice. Furthermore, the distribution and installation sequence of supports, i.e. struts, anchors, reinforced concrete slabs, that can be taken into account can have a significant impact on the distribution of ground surface displacements. Zhang et al (2015) stated that the initial cantilever deflection at the top of the retaining wall contributes significantly to the final wall movement, and it has a visible effect on the ground settlement profile.

Nevertheless, there are some complications involved with the use of advanced numerical methods, as well, of which a designer should be aware. Primarily, the parameters required for advanced non-linear constitutive models are difficult to determine from routine soil testing programs (Osman & Bolton 2006). Secondly, the use of complex numerical calculations may give a false sense of security (Simpson 2011). Therefore, a parametric or sensitivity analysis may be necessary to establish a range of possible displacement values as well as the impact of the change in input parameters, especially, when significant uncertainties in regard to the soil parameters are expected.

a) b)

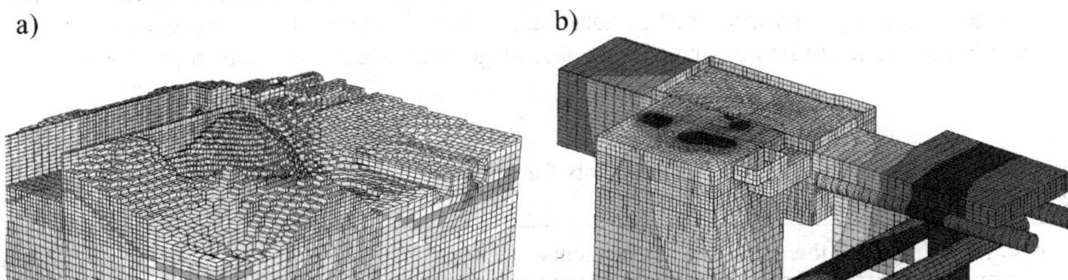

Figure 3. a) The variation of the upper level of Pleistocene strata near an existing Metro station. b) Map of displacements for complex spatial distribution of underground structures due to the construction of deep excavation and additional loading from new high-rise building.

6 SERVICEABILITY LIMITING CRITERIA

The limiting criteria for verification of serviceability limit state should be chosen on a case-by-case basis. A number of factors can affect the choice of these values, namely: type of the structure (i.e. a building, underground installations, or a tunnel), type of foundation and load-bearing structural elements, the condition and fatigue of these elements, possibility of ductile or brittle failure, consequences of damage, possibility of repair, or even the value of a structure to the society (i.e. monuments, critical infrastructure). Some default but often conservative values are given by various authors (i.e. Boscardin & Cording 1989; Wysokiński & Kotlicki 2002).

Consideration of serviceability conditions of existing structures is especially important for elements of critical infrastructures, i.e. existing tunnels, metro lines, etc. Such structures often have their own strict serviceability criteria imposed by authorities responsible for their maintenance and operation. When considering a complex layout of underground structures, only advanced methods of analysis may provide sufficiently detailed information of the system behavior and resulting impact on existing elements.

6.1 *Buildings and typical structures*

Detailed investigation concerning the type of a structure and its load-bearing elements, as well as documentation of existing damages, should be performed for all buildings located in the active zone of influence. This should be also done for building in the vigilance zone which are in poor technical condition or with preexisting visible cracks. This is especially important for buildings constructed before year 1930, which often do not have structural elements improving their spatial rigidity (Wysokiński & Kotlicki 2002). For other buildings, general assessment may be sufficient.

The limiting value for verification of SLS according to Equation 1, for buildings in good or average technical condition, can be chosen from Table 2. The choice of limiting value of vertical displacement for serviceability verification should be based on the results of technical assessment of a building, depending on following factors:

- Type of the building and its load-bearing structural elements;
- Current technical condition and preexisting damage (i.e. cracks);
- Susceptibility to differential deformations, as well as their impact on bearing capacity and serviceability of the structure;
- Comparable experience with similar buildings subjected to excavation-induced deformations.

Location of the building in relation to the excavation may also be considered in the analysis, as buildings located perpendicular to the excavation are usually more susceptible to deformations than parallel ones (Wysokiński & Kotlicki 2002). This issue has been investigated in more detail by Boscardin & Cording (1989).

When information on the foundations of the buildings is not available, either the most conservative assumptions should be undertaken, or the on-site verification (i.e. by excavating the foundation) should be performed as a part of geotechnical investigation program.

Table 2. The limiting values of displacements for buildings (adopted after Wysokiński & Kotlicki 2002).

Types of buildings and their load-bearing elements	v_{SLS} [mm]	v_{ULS} [mm]
Masonry buildings without roof and floor bands, with wooden or steel-framing floors	5–7	15–18
Masonry buildings with suspended beam-and-slab or reinforced concrete floors, or buildings constructed with precast concrete elements	7–9	20–25
Cast-in-place concrete or steel buildings	9–11	25–35

Table 3. The limiting values of displacements for installations (adopted after Wysokiński et al 2011).

Types of installations	v_d [mm]
Single cables	200
Cable bands (i.e. electrical, telecommunication)	150
Water pipes Φ 200 mm	100
Natural gas pipes Φ 100 mm	150
Natural gas pipes Φ 400 mm	50
Sewage pipes	10 to 25

6.2 *Installations and utilities*

Underground installations and utilities may be subjected to the ground movement caused by excavation. To avoid the disruption in their service, their vertical displacements should be limited as well. Examples of limiting values are provided in Table 3.

6.3 *Railway and metro lines*

Limiting values of deformations for railway and metro lines are usually set by their owners, based on technical requirements necessary to maintain their serviceability.

According to the ID-3 specification of Polish railways (PKP 2009), generally, the allowable differential settlement of a railway track should be limited to 4 mm/year at a distance of 30 m or 10 mm/year at a distance of 200 m if no other criteria were specified. Usually, a local railway authority responsible for operation of a specific track should be consulted concerning the specific value. Should no other criteria be required, a value of differential settlement of 4 mm per 30 m should be used, as the time and the area of underground construction activities are often limited.

Limiting values for various parameters of the railway tracks of the Metro line are provided by Metro authorities in regard to the measurement method and the maximum speed allowed at a specific location (Metro Warszawskie 2014). All relevant criteria should be verified to ensure serviceability of the line. This often requires calculating a distribution of displacements along the tracks with the use of advanced numerical methods and 3D geometrical model.

7 MONITORING

Monitoring during the construction of the underground structure is important not only for the purpose of verification of design assumptions, but also to safe-guard oneself from unjustified claims of 3rd parties. Monitoring should involve taking measurements and visual inspections of adjacent structures and the excavation.

ITB (Wysokiński & Kotlicki 2002) and ITA-AITES (2014) recommendations provide the general guidance on the extent of the area that should be subjected to monitoring. Based on analysis performed in the design, this area may be limited to the influence zone and the structures located within it. All the structures located within the active zone of influence, as well as those most susceptible to the deformation of the subsoil in the vigilance zone, should be monitored the most extensively. Other structures, located in the vigilance zone, are of a lower priority. However, they also should be monitored regularly, and their observations should not be limited only to the parts located within the zone; for a rectangular structure without expansion joints, at least 4 leveling pins should be placed at the corners to allow assessment of the differential deformation and the tilt of the structure (Wysokiński & Kotlicki 2002).

Monitoring program, involving the performance of the new as well as adjacent structures, should begin prior to the start of construction works in order to establish background variations in measured parameters. Then, monitoring should be continued after the end of construction works for a sufficient period of time allowing stabilization of measurements.

A geotechnical design of an embedded retaining wall should include estimation of its displacements, both the most probable value expected to occur, based on the most probable characteristic values of soil parameters, as well as the limiting value. The limiting values for the lateral wall movement should be based on the allowable SLS displacements of adjacent structures.

8 CONCLUSIONS

The paper presented the general description of the static soil-structure interaction problem when considering underground structures and their interaction in the underground space. When serviceability of adjacent structures has to be considered, a reliable prediction of the behavior as well as the quantitative assessment of the deformation is required. The geotechnical analysis of the impact on adjacent structures is a necessary element of a design involving underground construction works. The analysis of the predicted behavior of adjacent structures and their serviceability does not differ from other limit states in Eurocode 7 (EN 1997: 2004). However, the need to limit deformations of neighboring structures may considerably affect design decisions.

The increasing complexity of construction projects and the availability of more advanced prediction tools make it reasonable to conduct more advanced analysis. Prediction of the extent of influence zone as well as the displacements, offered by simplified semi-empirical calculation models, can be regarded as safe and conservative within the boundaries of their applicability. When complex soil-structure interaction problems are considered (Geotechnical Category 3 according to Eurocode 7), the complementary use of more advanced prediction models and calculation methods should be considered by designers and investors. The choice of appropriate calculation model should take into account the risk profile of the investment and possible consequences of failure. Nowadays, some institutions require more advanced analysis before allowing a construction near their area of operation, i.e. near existing Metro stations and lines in Warsaw (Metro Warszawskie 2014). Such detailed analysis may not always be possible at the preliminary design stage; however, this does not remove the responsibility of conducting such analysis, from the investor and the designer, at a later stage, even after obtaining the building permit. Information obtained from the analysis might be valuable for evaluating the most probable behavior of the structure and identifying and assessing possible risks. Such knowledge is necessary to effectively manage the geotechnical risk throughout the entire project.

Conducting advanced analysis and increasing the reliability of the prediction may significantly reduce the costs suffered at the execution stage. These costs may be foreseen, i.e. the cost of monitoring and the strengthening of neighboring structures, as well as unforeseen, i.e. the cost of repair and compensation. Without sufficient documentation preceding the construction, as well as monitoring results obtained during the execution, the party responsible for managing the geotechnical risk is always at a disadvantage in the litigation process, should the neighbors submit a claim to the investor. In addition to suffering additional expenses, this may negatively impact the reputation of the investor, even if the contractor was contractually responsible for the lack of foresight and will be affected financially. The possible damage to the reputation of the investor, caused by ignoring the risk, is often not taken into consideration in the decision making process. However, it may result in an adverse attitude of authorities and other people, during work on future projects.

REFERENCES

Baecher, G. & Christian, J. 2003. *Reliability and Statistics in Geotechnical Engineering*, Wiley & Sons.

Benz, T. 2007. *Small-strain stiffness of soils and its numerical consequences*, PhD Thesis, Universitat Stuttgart, Stuttgart, Germany.

Boscardin, M.D. & Cording, E.J. 1989. Building response to excavation-induced settlement, *ASCE J. of Geotech. Eng.*, Vol. 115, No. 1, 1–21.
Choy, C.K., Standing, J.R. & Mair, R.J. 2007. Stability of a loaded pile adjacent to a slurry-supported trench. *Geotechnique* 57, No. 10, 807–819.
EN 1990: 2004 *Eurocode—Basis of design.*
EN 1997-1: 2004. *Eurocode 7 – Geotechnical design—Part 1: General rules.*
Essex, R.J. 1996. Means of Avoiding and Resolving Disputes During Construction, *Tunneling and Underground Space Technology*, Vol. 11, No. 1, pp. 27–31.
Godlewski, T. Szczepański, T. & Bogusz W. 2015. Determination of soil stiffness parameters using in-situ seismic methods—methodological aspects. *Inżynieria Morska i Geotechnika*, Vol. 36, 371–376.
Guglielmetti, V. Grasso, P. Mahtab, A. & Shulin, X. 2007. *Mechanized tunneling in urban areas. Design Methodology and construction control.* Taylor & Francis, London.
ITA-AITES Report. 2014. ITAtech *Guidelines on Monitoring Frequencies in Urban Tunneling*. ITAtech Activity Group MONITORING, 20975 ITA Report No 3.
L'Amante, D. Flora, A. Russo, G. & Viggiani, C. 2012. Displacements induced by the installation of diaphragm panels. *Acta Geotechnica* 7, 203–218.
Metro Warszawskie, 2014. Technical requirements for design and construction of investments which may affect the metro structures, ITB & Metroprojekt. Warsaw.
Mitew-Czajewska, M. & Siemińska-Lewandowska, A. 2008. The effect of deep excavation on surrounding ground and nearby structures, *Proc. of the 6th Int. Symp. on Geotech. Aspects of Underground. Construction in Soft Ground*, Shanghai.
Orazalin, Z.Y. Whittle, A.J. & Olsen, M.B. 2015. Three-dimensional analyses of excavation support system for the Stata Center Basement on the MIT Campus. *ASCE J. of Geotech. and Geoenv. Eng.*, Vol. 141, Issue 7.
Osman, A.S. & Bolton, M.D. 2006. Ground movement predictions for braced excavations in undrained clay, *Journal of Geotech. and Geoenv. Eng.*, Vol. 132, No. 4, ASCE.
Peck, R.B. 1969. Advantages and limitations of the observational method in applied soil mechanics. *Geotechnique* 19, No. 2, 171–187.
PKP, 2009. *Technical specification on the maintenance of the track bed ID-3*, PKP, Warsaw (in Polish).
Simpson, B. 2011. Reliability in geotechnical design—some fundamentals, *Proc. of 3rd Int. Symp. On Geotech. Safety and Risk*, Munich.
Simpson, B. Nicholson, D. Banfi, M. Grose, B. & Davies, R. 2009. Collapse of the Nicoll Highway excavation, Singapore. *Proc. of 4th Int. Forensic Eng. Conf.*, ICE, London.
Wysokiński, L. & Kotlicki, W. 2002. Protection of buildings adjacent to deep excavations. ITB Recommendations No. 376/2002, ITB, Warsaw (in Polish).
Wysokiński, L. Kotlicki, W. & Godlewski, T. 2011. *Geotechnical design according to Eurocode 7*. ITB, Warsaw (in Polish).
Zhang, D.M. Phoon, K.K. Huang, H.W. & Hu, Q.F. 2015. Characterization of model uncertainty for cantilever deflections in undrained clay. *ASCE J. Geotech. Geoenviron. Eng.*, 141.

Underground Infrastructure of Urban Areas 4 – Madryas et al. (Eds)
 ISBN 978-1-138-55953-0

Comparison of global liner design codes

B. Falter
Department of Civil Engineering, University of Applied Sciences Munster, Germany

ABSTRACT: The German design code DWA-A 143-2:2015 differentiates between four Host Pipe States. In some foreign standards (e.g. ASTM F 1216 and SRM) two states are defined: Sewer partially deteriorated (structurally safe) or fully deteriorated (deteriorated pipe). More important is a different theoretical basis, the application of imperfections and the need of regarding liner stresses.

Experiments are helpful to understand stability and break failure modes and to create appropriate imperfections. An overview on numerous liner buckling results is presented and compared with the theoretical critical water pressures in a large scale of liner slenderness r_L/t_L.

1 INTRODUCTION

The paper reports about common aspects and differences between global codes.

a. The background of Host Pipe States I and II (partially deteriorated pipe) is the liner in the host pipe subjected to water pressure, described either by an unsupported pipe applying the support factor $K = 4$ resp. 7 or by a rigidly bedded pipe applying reduction factors regarding three different imperfections.
b. The background of States III and IIIa (fully deteriorated pipe) is the liner, described either by a bedded liner-pipe-soil system or an elastically bedded pipe.

Abbreviations used by DWA-A 143-2 are summarized in Table 1.

Table 1. Definitions used in the German code DWA-A 143-2.

Definition	A 143-2	Definition	A 143-2
Wall thickness of the old pipe	t	Flexural, compression, shear strength	σ_{fl}, σ_c, τ_u
Wall thickness of the liner	t_L	Partial safety factor of loading	γ_F
ratio r/t (slenderness)	r_L/t_L	partial safety factor of resistance	γ_M
Local imperfection w_v, ovality	ω_v	Characteristic value, design value	index k, d
$w_{GR,v}$ (four hinge deformation)	$\omega_{GR,v}$	e.g. water table above invert	$p_{a,d} = p_{a,k} \cdot \gamma_F$
annular gap w_s (related to the radius r_L)	ω_s	e.g. long-term liner E-modulus	$E_{L,d} = E_{L,k}/\gamma_M$
Shrinkage of grout, mortar	ω_M	Critical water table	crit p_a
		Critical vertical soil pressure	crit q_v

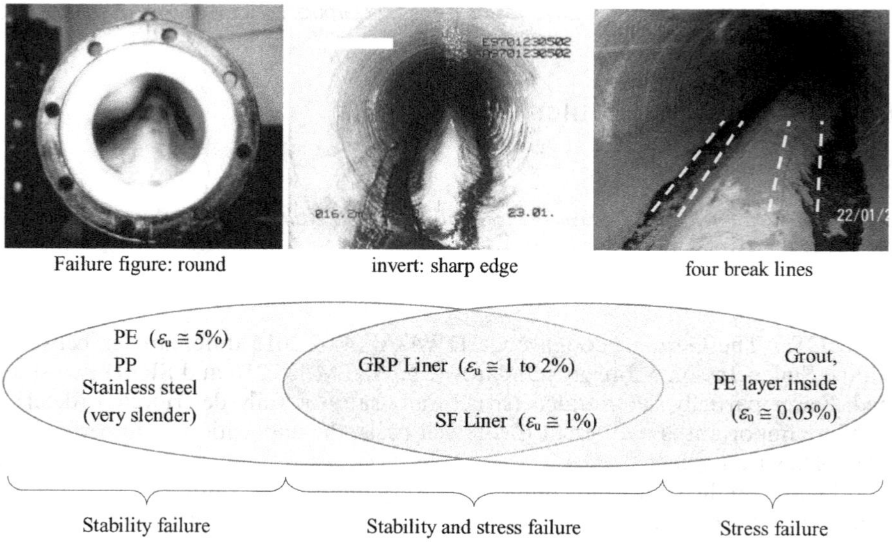

Figure 1. Relevant failure modes for different liner material (PE, GRP, grout, etc).

2 FAILURE MODES OF LININGS

Figure 1 shows three types of liners subjected to water pressure after failure. From left to right: polyethylene (PE), glass reinforced pipe (GRP) and grout.

In all cases, failure was caused from the liner wall moving from the invert toward the center of the pipe. However, each failure mode is quite different from the other. The PE liner material fails elastically, even in places of maximum deformation and up to the pipe crown. Though the mortar liner appears to buckle in a similar fashion, a closer examination shows four break lines in the longitudinal direction and a kinematic appearance is the result. A thin-walled GRP liner should buckle elastically but, after a certain amount of deformation, it breaks in the line of maximum stress.

Figure 1 shows that different failure criteria dependent on the lining material should be established. A design code must refer to both the proof of buckling *and* the proof of stresses. Cured-in-place pipe (CIPP) lining material shows both behaviors, thus both procedures are required in the code DWA-A 143-2.

3 A BRIEF LOOK AT THE DESIGN CODE DWA-A 143-2 (2015)

3.1 *Host pipe states*

Host Pipe States I to IIIa can be defined as shown in Table 2. This shows that old pipes in both states II and III have four cracks and are unstable without soil support. The difference between states II and III can be seen by forecasting future ground stresses in the vicinity of the rehabilitated sewer. If there will be no excavation, no settlements and minor traffic load (e.g. sufficient cover height) the system remains undisturbed and Host Pipe State II can be assumed.

The category HPS IIIa may be applied as well if the old pipe the pipe might possibly corrode and no compressive strength remains for long-term.

3.2 *Stability proves procedures*

The Figures 2 to 5 shows an overview of the liner design steps. The code DWA-A 143-2 contains several of tables for reduction factors κ and section force factors n and m to allow calculations without a computer.

Table 2. Definition of the Host Pipe States I to IIIa.

Host Pipe State (HPS) Old pipe:	I (no cracks)	II (4 longitudinal cracks)	III (4 longitudinal cracks)	IIIa (fully corroded, fragments etc.)
Sufficient flexural strength σ_{fl}	+	–	–	–
Old pipe-soil system undisturbed	+	+	–	–
Sufficient compression strength σ_c	+	+	+	–

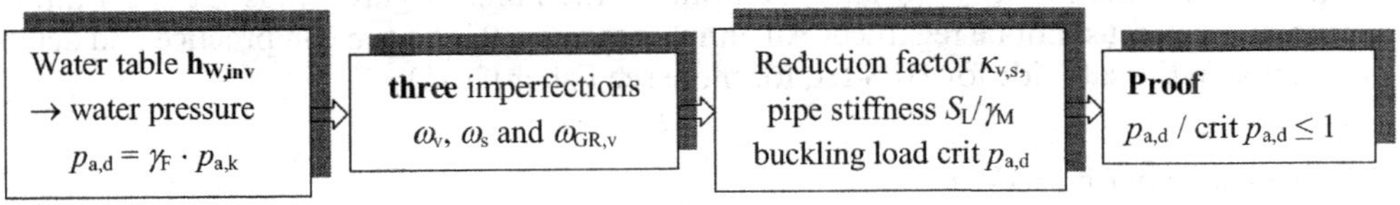

Figure 2. Stability proof—water pressure ($\kappa_{v,s}$ valid for all values of E_L).

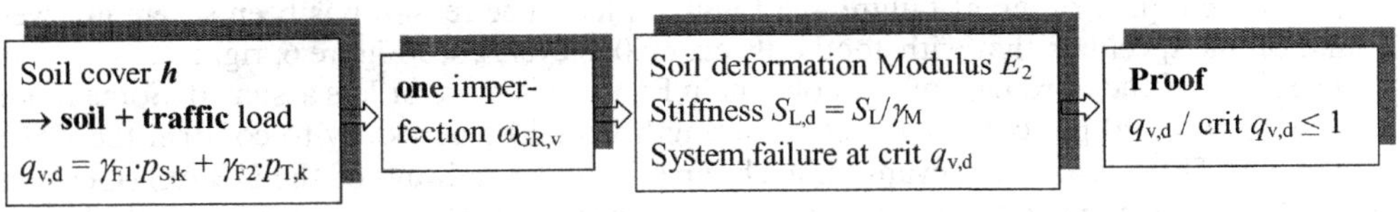

Figure 3. Stability proof—soil and traffic loads (crit $q_{v,d}$: exactly valid for $E_{L,d}$ = 1,000 and 4,400 MPa, E_2 = 3 and 8 MPa, K_2' = 0.2, e_G/t = 0.25—thus manual calculation is restricted to synthetic and glass fiber reinforced resin liners.).

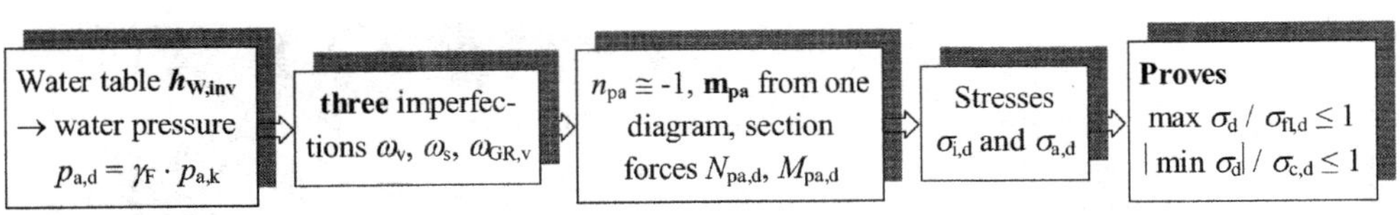

Figure 4. Stress proves—water pressure ($\kappa_{v,s}$ valid for all values of E_L).

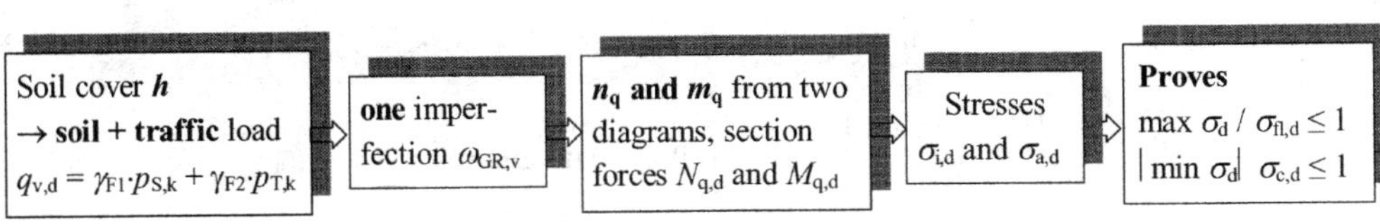

Figure 5. Stress proves—soil and traffic loads (n, m: exactly valid for E_2 = 3 and 8 MPa, K_2' = 0.2, e_G/t = 0.25, $E_{L,d}$ = 1,400 and 6,000 MPa. Please note: For HPS III γ_M = 1 is prescribed.).

3.3 *Stress proves procedures*

The procedures are based on parameters that have been evaluated by Finite Element analyses for circular and oval profiles (B:H = 1:1.5) of all diameters. They are exactly valid for needle felt liners with E_L = 1,400 MPa and glass fiber reinforced liners with E_L = 6,000 MPa—for similar parameters they can be used as well. For greater deviations a computer aided design must be performed (Falter, 2017).

Loadings are: The water pressure p_a perpendicular to the liner surface and the vertical loading $q_v = 0.75 \cdot p_S + p_T$ from soil cover and traffic needed only for HPS III and IIIa. For the calculation of the section forces dimensionless factors (n, m) are multiplied with the loading and the liner radius r_L resp. r_L^2, cf. the calculation of a beam of length l subjected to uniform loading q:

$$M = m \cdot q \cdot l^2$$

where the coefficient m is 0.125 (= 1/8).

In HPS III the two imperfections ω_v and ω_s are set zero. For the vertical pressure q_v the hinge ring deformation $\omega_{GR,v}$ (ovality) is affine to the failure figure—thus the local initial deformation ω_v must not be regarded. An annular gap $\omega_s = 0$ is possible in practice and delivers results on the safe side for HPS III, for more see Falter (2015).

3.4 *Fatigue proof procedure*

To get information on fatigue strength of GRP linings several probes were subjected to cyclic loading. The upper stress σ_u was 100 per cent, 70 per cent, 50 per cent, 40 per cent and 30 per cent of the static strength, the lower stress was $0.3 \cdot \sigma_u$. Two specimens subjected to the lowest stress showed quite different failure, see Figure 6, left. The reason has been severe grooves found on the specimen that withstood only ca. 60,000 cycles, see Figure 6, right.

The slope of the third part of the polygon in Figure 6 is more or less assumed. Some more tests with 30 to 40 per cent of the static strength would be necessary to confirm the slope of the line. These tests are relevant for each pipeline under railways of the DB AG (German Rail AG) that is planned to be rehabilitated by a liner. Under streets with a cover height less than 1.5 m and traffic by heavy trucks the fatigue test is obligatory in the code DWA-A 143-2.

The fatigue proof has to be performed by comparing the liner stresses calculated from traffic loads with the double amplitude $2\sigma_A = \sigma_u - \sigma_l$.

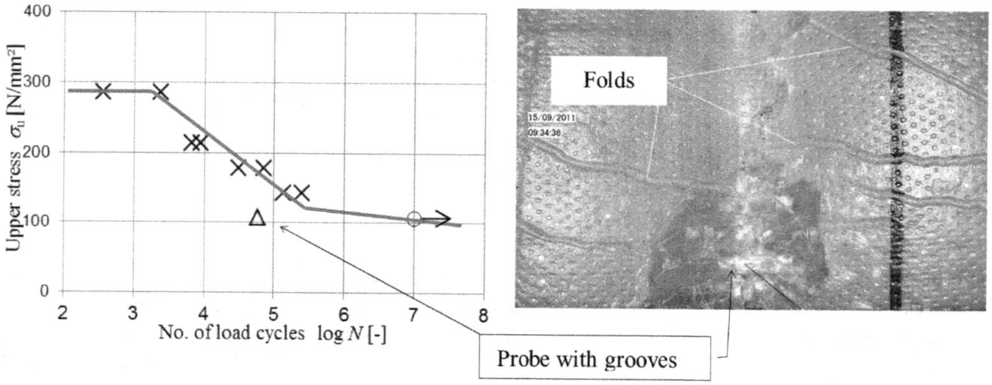

Figure 6. Tests on GRP liners with max N = 10^7 cycles (Wöhler curves), right Figure: Fingerhut (2015).

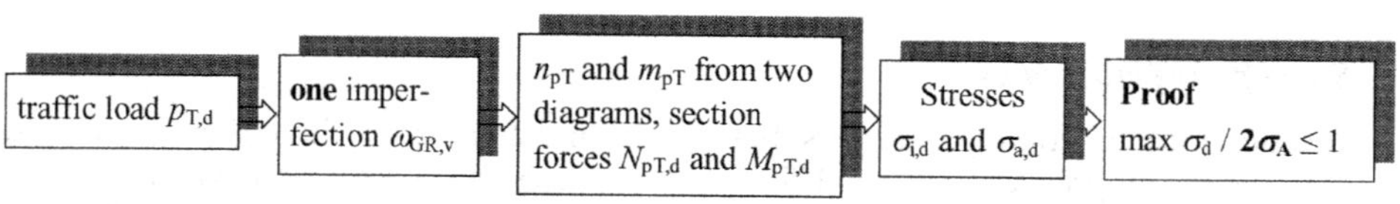

Figure 7. Fatigue stress proof—traffic loads.

4 COMPARISON OF CODES

The comparison in the following tables refers to the ASTM and DWA codes. In Table 3 the main topics of a Host Pipe State I and II design are summarized.

The critical water table prescribed by the ASTM code is derived from the model of the free ring with a support factor K. The factor is chosen as $K = 7$ in US, in UK: $K = 7$ for good and $K = 4$ for poor installation. S_L is the long-term liner pipe stiffness.

$$\text{crit } P = K \cdot 3 \cdot S_L \tag{1}$$

Introducing the pipe stiffness S_L into Equation (1) yields an equation where the slenderness r_L/t_L of the liner has the exponent 3, see Equation (2)

$$\text{crit } P = K \cdot 3 \cdot E_L \cdot (t_L/r_L)^3/12/(1-\mu^2) \tag{2}$$

The DWA approach starts with the Glock (1977) formula of a perfectly round and rigidly encased ring but reduced by a factor $\kappa_{v,s}$ for imperfections ω_v, $\omega_{GR,v}$ and ω_s:

$$\text{crit } p_a = \kappa_{v,s} \cdot \alpha_D \cdot S_L \tag{3}$$

where $\alpha_D = 2.62 \cdot (r_L/t_L)^{0.8}$ is the coefficient of failure by *snap through*.

The name is derived from elastic material that fails suddenly from the bottom to the pipe crown, see Figure 1. Liner material with creep behavior will move slowly depending on the water level and the loading history.

$$\text{crit } p_a = \kappa_{v,s} \cdot 2.62 \cdot E_L \cdot (t_L/r_L)^{2.2}/12/(1-\mu^2) \tag{4}$$

Table 3. Host Pipe States I and II (partially deteriorated), CIPP.

	ASTM F 1216	DWA-A 143-2
1. Stability proof	Unsupported ring,	Rigidly encased ring, Glock (1977)[1,2]
Theoretical **buckling model**	enhancement factor $K = 7$	reductions for imperfections
loading	Water table P	Water table p_a
Safety factor(s)	Global factor $N = 2$	Partial factors $\gamma_F = 1.5$, $\gamma_M = 1.35$
Soil parameters	– (Not required)	Required for HPS II and weak soil
Imperfections due to the	Ovality Δ [3)]	Ovality $\omega_{GR,v}$ [3]
old pipe and liner	–	local imperfection ω_v, annular gap ω_s
Reduction factor(s)	$C = f_1(\Delta)$ where C is *independent* of r_L/t_L	$\kappa_{v,s}(r_L/t_L) = f_2(\omega_{GR,v}, \omega_s, \omega_v)$[1] where $\kappa_{v,s}$ is *depending* on r_L/t_L
Critical water table	crit $P = K \cdot 3 \cdot S_L$	crit $p_a = \kappa_{v,s} \cdot \alpha_D \cdot S_L$
	Eq. (X1.1)	where α_D = snap through factor[1]
	Eq. (X1.2)	Tables with m_{pa} for M_{pa} and n_{pa} for N_{pa}
	relevant for big values Δ	$\rightarrow \sigma \cong N_{pa} / t_L \pm 6M_{pa} / t_L^2$
2. Stress proof		
Profiles different from circle		
Oval profile $B{:}H = 1{:}1.5$, buckling	–	Substitute circle with radius $0.6 \cdot H - t_L/2$[1) 2)]
Stress proof	–	see 2. with *separate* factors m_{pa} and n_{pa}
Other profiles, buckling and stress proof	–	Non-linear FE-Analysis with adequate imperfections

[1]Evaluation by experiments, see Figs. 10 & 11. [2]Experiments on these items see chapter 5. [3]Ovalities Δ and $\omega_{GR,v}$ are defined equal.

The dimensionless expression crit p_a / E_L can be drawn in a double logarithm chart as lines with different slopes (slope 3 resp. 2.2). Equation (2) delivers a straight line. As the reduction factor $\kappa_{v,s}$ in Equation (4) depends itself on r_L/t_L the line is slightly curved, see Figure 10.

Table 4 shows the main topics of a Host Pipe State III and IIIa design. It is obvious that some more parameters are necessary in the DWA code, especially those related to the soil and the old pipe.

The soil loads are reduced by the concentration factor $\lambda_R = 0.75$ what is an average value for flexible pipes—please note that the pipe cracked four times behaves now like a flexible pipe. According to DWA-A 143-2 the soil pressure can be reduced by lower values ($\lambda_R < 0.75$) for deeply buried sewers applying *Silo Theory*. The soil pressures from traffic loads have been measured in the IKT Test Stand for the new code EN 1991-2 (double axis, each 30 t). The tests have been performed on different pipe material, pavement and cover height (Falter & Wolters, 2008).

In the ASTM code the critical vertical pressure is calculated by the well-known equation of an elastically bedded ring. In the DWA code three methods are presented to evaluate the critical vertical pressure of soil and traffic loads.

In the following the method 2 of Table 4 is discussed more deeply. The critical load of the old pipe-soil system is evaluated on the quarter system in Figure 8 by elementary equilibrium conditions if t_L is set zero.

Table 4. Host Pipe States III and IIIa (fully deteriorated, gravity pipe), CIPP.

	ASTM F 1216	DWA-A 143-2
1. Stability proof		
Theoretical **buckling model**	Elastically bedded ring	Liner-old pipe-soil system
Loading: Soil pressure	wHR_W	0.75 p_S or $\kappa \cdot \gamma_S \cdot h^2$
traffic load	W_s due to Heavy Truck	p_T due to EN 1991-2[1]
water table	P	p_a
Interaction of loadings	$q_t = \Sigma(P + wHR_W + W_s)$	loadings q_v and p_a weighted
Safety factor(s)	Global factor $N = 2$	Partial factors $\gamma_F = 1^3$ resp. 1.35, $\gamma_M = 1.35$
Soil parameters:		SG 1 to 4
Soil groups		E_2 (Index 2: springline zone)
Modulus of elasticity	E_S	$K_2 \le 0.5$
lateral pressure coefficient	– (K_2 is set 1)	K_{pgh}
passive soil reaction	–	
Imperfections due to the old pipe	Ovality Δ	Ovality $\omega_{GR,v}$[1] imperfections $\omega_v = \omega_s = 0$ load interaction: $\omega_s = 0$ resp. 0.5
Old pipe wall thickness	–	t [4)]
state of springline region	–	hinge eccentricity e_G[1,4]
compressive strength	–	σ_c [1)]
Reduction factor	$C = f_1(\Delta)$ where C is *independent* of r_L/t_L	$\kappa_{GR,v} = f_2(\omega_{GR,v}, r_L/t_L)$ $\kappa_v = 1$, $\kappa_s = 1$
Critical vertical load	crit $q_t = K \cdot 3 \cdot S_L$ Eq. (X1.3)	Methods for crit q_v: 1. Direct tables 2. Tables for old pipe without liner[4] 3. Maximum of load-deflection curve[1]
	–	Factors m_q for M_q, n_q for N_q
2. Stress proof		$\rightarrow \sigma \cong N_q / t_L \pm 6M_q / t_L^2$
Profiles different from circle		
Oval $B{:}H = 1{:}1.5$, buckling	–	Substitute circle with the crown radius r_c[1]
Stress proof	–	see Method 2. same factors as circular liner
Other profiles, buckling and stress proof	–	e.g. non-linear FE-Analysis (if no appropriate substitute radius is available)

[1]Experimental evaluation [2]Factor κ from silo theory (adopted from code DWA-A 161, pipe jacking). [3]$\gamma_F = 1$ prescribed for deformation controlled loadings like soil pressure and for p_T due to EN 1991-2. [4]On the safe side, to be treated more deeply in the following.

It can be seen from Figure 8 that the lateral soil pressure coefficient K_2 is important in the stability analysis as the lateral pressure q_h stabilizes the system. In the German code A 127 for newly laid pipelines the coefficient K_2 is small (0.1 to 0.2) for weak soil and 0.3 to 0.4 for sand or gravel.

Furthermore the lateral bedding support q_h* is restricted by plastic soil limits (passive soil reaction). Applying the full triangular support q_h* is not on the safe side if an old sewer in soft soil has to be rehabilitated.

In Table 4 the pipe wall thickness t and the eccentricity e_G of the assumed hinges in the cracks are important parameters for a Host Pipe State III design. This has been shown by the experimental configuration in Figure 9. The tests have been performed on new concrete pipe and 100 years old pipe fragments.

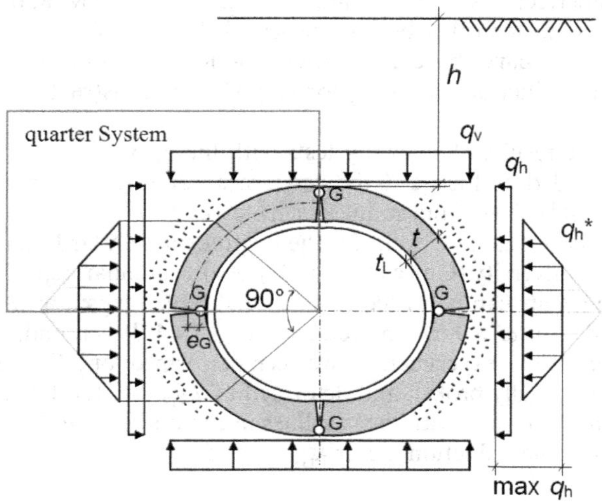

Figure 8. Typically cracked old pipe with liner, Host Pipe State III, double symmetric system, vertical, horizontal soil pressures q_v, q_h and bedding reaction q_h*, G = eccentric hinge, e_G = eccentricity of hinges.

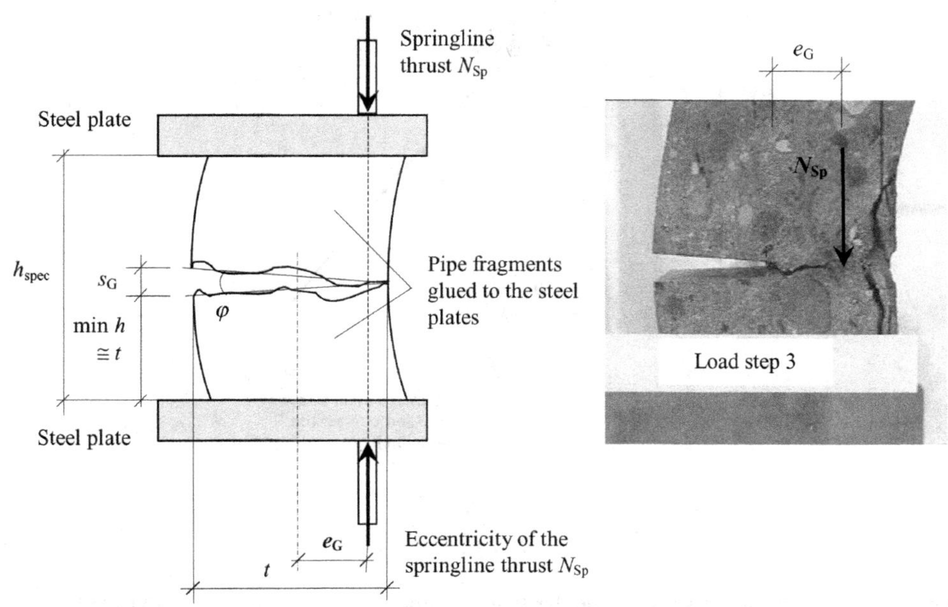

Figure 9. Test configuration with eccentrically loaded fragments of the old pipe springline.

The angle φ between the upper and the lower fragment was chosen to generate ovality values $\omega_{GR,v} = 3\%$, 6% and 9% (Steffens et al., 2002). It is seen from the photo in Figure 9 that increasing thrust N_{Sp} causes spalling and decrease of eccentricity e_G.

The parameters t and e_G/t have been found for the DWA code to be important for a *safe and economic* liner design. Poor springline zones are described by $e_G/t = 0.25$, for a good situation $e_G/t = 0.45$ can be chosen.

5 EXPRIMENTAL EVALUATION

Figures 10 and 11 show the ASTM and DWA critical pressures crit p_a related to the E-modulus. The dotted DWA line is not relevant as no imperfections are applied.

For standard imperfections $\omega_v = 2\%$ and $\omega_s = 0.5\%$ the ASTM and DWA curves seem to be not very far from each other but, as the scale is logarithmic the difference should not be neglected. The DWA curve behaves somewhat non-linear visible for $r_L/t_L > 70$ caused by a rapidly decreasing reduction factor $\kappa_{v,s}$ for very slender constructions: $r_L/t_L = 10$ to 100, $\kappa_{v,s} = 0.48$ to 0.11.

In both figures the results of buckling tests with linings subjected to water pressure on their outside are included. In Figure 10 the slenderness range r_L/t_L is larger to show results of spiral wound liner tests (uPVC) and stainless steel tests.

Figure 11 shows more detailed results for the Insituform Standard Liner long-term tests at the Louisiana Tech University (Guice et al., 1994). As the annular gap of each specimen had been measured the results can be sorted to small and larger gaps: for $\omega_s \leq 0.6\%$ the average value of p_{test}/E is 1.49, for $\omega_s > 0.6\%$ an average of $p_{test}/E = 1.28$ is found.

The stainless steel sleeves experiments have been performed for a Host Pipe State II condition (old pipe longitudinally broken and deformed by $\omega_{GR,v} \cong 3\%$) (Falter & Wolters, 2008). In order to compare the results with the two lines in Figure 10 for $\omega_{GR,v} = 0$ the test load was divided by an appropriate reduction factor $\kappa_{GR,v} \cong 0.75$.

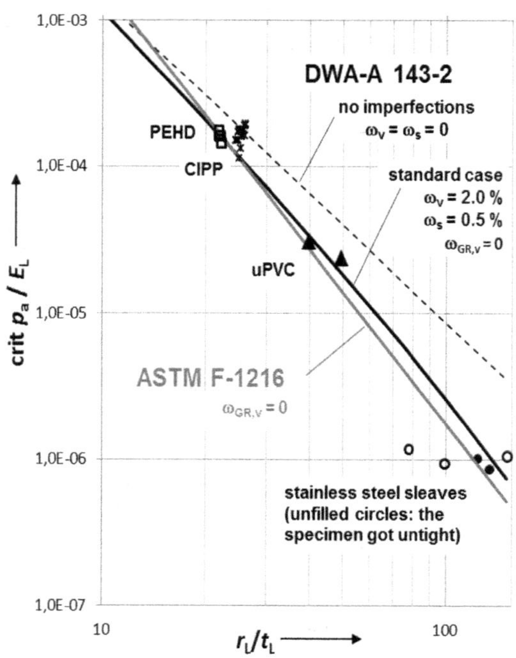

Figure 10. Experimental results for liner buckling Equations (2) and (4) for $r_L/t_L = 10$ to 150.

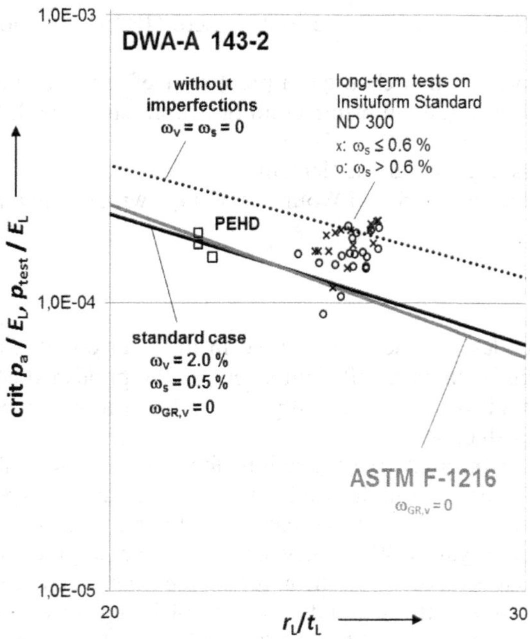

Figure 11. Liner buckling equations for $r_L/t_L = 20$ to 30 and experiments (detail of Figure 10).

The tests on oval PE linings have been reported in (Falter et al., 2008). The oval form can be transformed to a circular liner by the substitute radius $r_E = 0.6 \cdot H - t_L/2$ proved by Finite Element Analyses. To regard creep effects and high compression stresses of PE the material parameters had to be adjusted.

From Figures 10 and 11 it can be summarized:

- A fairly good agreement is found between DWA and ASTM in the region of $r_L/t_L \cong 20$ to 30.
- For liners with initial deformations (ovality) the ASTM line is reduced by a factor *C independent* of the slenderness r_L/t_L where an increasing slenderness is treated by *a more severe reduction* in the DWA code.
- The use of the DWA Equation (3) *without imperfections* ω_v and ω_s would not be on the safe side.
- Experiments in a wide slenderness range show a good coincidence with theory, Equation (3).

6 RESEARCH NEEDS

Although a number of research papers have been published since the beginning of liner design, some additional examinations should be performed in the near future. These examinations include:

- Long-term strength—reduction factors for strength and E modulus $A_{1\sigma}$ and A_{1E} are *not the same* but used in practice (Fingerhut, 2015)
- Fatigue strength ($2\sigma_A$) of linings (see remarks to Figure 6)
- Non-linear material behavior, e.g. reduction for PE liners to regard the dependency of E on σ
- Assessment of soil parameters from inside of pipeline, e.g. by the MAC test applied in France and Germany (Thépot, 2004)

- Treatment of very soft soil, e.g. $E_S < 5$ MPa, where HPS II is no longer valid (change to HPS III)
- Calibration of theoretical models used in practice: Definition of minimum requirements for computer analyses, e.g. equilibrium condition, consistence with limit states crit p_a and crit q_v for combined loadings
- Longitudinal effects (e.g. laterals, settlements)
- Composite cross section, e.g. Spiral Wound PVC Pipe with grouted annulus.

7 CONCLUSIONS

The comparison of codes for liner design used in different countries shows that standard CIPP liners are dimensioned with sufficient safety by the published equations but there are limitations of the codes if no stress proof is prescribed. For intact pipes the water pressure is the relevant loading in all codes.

For very slender constructions the reductions for imperfections—if regarded at all—are severe. Care must be taken if formulas that have been checked by experiments in the region of $r_L/t_L \cong 20$ are applied for thinner liners like $r_L/t_L \cong 40$ or even more.

In any case the annular gap width w_s between liner and host pipe must be kept as small as possible in projects as it is a crucial criterion for slender liners. In accessible sewers it is possible to measure w_s what is a strong quality criterion of lining works.

The number of field and material parameters to be investigated is different in the codes. For example, for Host Pipe State III the wall thickness, the state of the old pipe compression zones and the old pipe strength is not regarded in ASTM code. To achieve economic *and* safe linings geometric and physical parameters of the old pipe should be known before starting the rehabilitation work.

ACKNOWLEDGEMENTS

The author wants to thank the following for their funding support: BMBF (Project ASSUR) & MUNLV (Project MIBAK), StEB Cologn (Sewerage Authorities), the University of Applied Sciences Muenster, Department of Civil Engineering.

I would like to thank for their cooperation: ITTRC meetings initiated by Ray Sterling et al., IFES (Institute for Experimental Statics at the University of Applied Sciences, Bremen), IKT (Gelsenkirchen), DWA (Hennef), Laboratory for Material Sciences (Muenster), TBU (Greven), and many young engineers in my research group.

This paper was previously published in Trenchless Australasia magazine and was originally presented as part of the conference program for No-Dig Down Under 2015, the biennial conference of the Australasian Society for Trenchless Technology (ASTT). For further information visit www.trenchless-australasia.com or contact Editor Nick Lovering on nlovering@ gs-press. com.au.

REFERENCES

ASTM-Standard F 1216-09. Standard Practice for Rehabilitation of Existing Pipelines and Conduits by the Inversion and Curing of a Resin-Impregnated Tube.

DWA-A 143-2 (2015). Sanierung von Entwässerungssystemen außerhalb von Gebäuden. Teil 2: Statische Berechnung zur Sanierung von Abwasserleitungen und -kanälen mit Lining- und Montageverfahren (Static Calculation for the Rehabilitation of Drains and Sewers Using Lining and Assembly Procedures). Hennef.

Falter, B., Eilers, J., Müller-Rochholz, J. & Gutermann, M. (2008). Buckling experiments on polyethylene liners with egg shaped cross-sections. Geosynthetics International, Volume 15, No. 2, pp. 152–164.

Falter, B. & Wolters, M. (2008). Außendruckprüfungen an Quick-Lock-Stahlmanschetten DN 200 bis DN 600 für Altrohrzustand II (Outside pressure tests on Quick-Lock Steel Sleeves ND 200 to ND 600 for Host Pipe State II). Report for Uhrig Kanaltechnik GmH, Geisingen (not published).

Falter, B. & Wolters, M. (2008). Mindestüberdeckung und Belastungsansätze für flach überdeckte Abwasserkanäle (MIBAK) (minimum cover height and loading for sewers with shallow cover). Co-operative Research Proj. IV-9-042 3E1, IV-7-042 3E1 0010. Funded by MUNLV. Final Report.

Falter, B. & Wagner, V. (2015). Linerdimensionierung nach DWA-A 143-2:2015 (Liner dimensioning according to DWA-A 143-2:2015). 3R, No. 1–2, pp. 36–46.

Falter, B. (2017). Software LinerB, Version 8.7 – Computer aided calculation for Host Pipe States I to IIIa with load combinations (interaction), in German and English.

Fingerhut, S. (2015). Experimentelle Ermittlung der charakteristischen Materialparameter von GFK-Linern aus der Baupraxis—Scherfestigkeit, Dauerschwingbreite (Ermüdung) und Langzeit-Biegezugfestigkeit (Experimental Evaluation of Characteristic Material Properties of GRP Linings—Shear Strength, Fatigue and Long-Term Flexural Strength). Masterthesis, Univ. of Appl. Sc. Münster.

Glock, D. (1977). Überkritisches Verhalten eines starr ummantelten Kreisrohres bei Wasserdruck von außen und Temperaturdehnung (Post-Critical Behavior of Rigidly Encased Circular Pipe Subjected to Outside Water Pressure and Temperature Rise). Stahlbau, Volume 46, pp. 212–217.

Guice, L.K., Straughan, T., Norris, C.R. and Bennett, R.D. (1994). Long-Term Structural Behavior of Pipeline Rehabilitation Systems. TTC Technical Report #302, Louisiana Tech University Ruston, Louisiana, USA.

Steffens, K. (ed.), Falter, B., Grunwald, G. & Harder, H. (2002). Abwasserkanäle und -leitungen, Statik bei der Substanzerhaltung und Renovierung (ASSUR) (Drains and Sewers, Statics in Maintanance and Rehabilitation). Cooperative Research Project 01RA 9803/8, funded by BMB+F. Final Report: Inst. for Experimental Statics, Univ. of Appl. Sc. Bremen.

Thépot, O. (2004). Prise en compte des caractéristiques en petites déformations des sols dans l'étude du comportement des collecteurs enterrés. Thèse de doctorat.

WRc/WAA 4th ed. Sewerage Rehabilitation Manual (SRM), UK Water Research Centre/Water Authorities Association.

Underground Infrastructure of Urban Areas 4 – Madryas et al. (Eds)
© 2018 Taylor & Francis Group, London, ISBN 978-1-138-55953-0

The influence of bedding conditions on the safety state of sewage conduits

Z.A. Fyall
Faculty of Civil Engineering, Wroclaw University of Science and Technology, Wroclaw, Poland

ABSTRACT: In the case of linear structures such as sewers, an important issue is to determine the soil-water conditions along the route of the designed conduit. The exact definition of these conditions requires costly geotechnical investigations. To reduce the cost of such research, test holes are made over relatively long distances of usually about 100 m. This may lead to an incorrect determination of the bedding conditions of a conduit, which is hazardous for its safety. Another threat to safety is a change in conduit bedding conditions, which is caused by the negligent compaction of backfilling in the area of a sewer's foundations. The paper presents the results of analysis of the impact of changes in soil-water conditions and also the impact of assembly errors related to the provision of required conduit bedding conditions on the state of their safety. The safety state of conduits made of both flexible and rigid pipes laid in trenches with supported walls was analysed.

1 INTRODUCTION

The load-bearing capacity of a sewage conduit depends on, among other things, its bedding conditions. The most important for the safety of a conduit's construction is the precise determination of the parameters of soil along the conduit's route. For linear objects such as sewage conduits, it is necessary to assume the maximum distances between the test holes that enable the soil-water conditions to be examined. The smaller the distance between the holes, the more accurate the examinations are. However, this would generate big costs.

In practice, test holes are made with distances of no more than 100 m. There are unfortunately many cases where soil conditions are so variable that their identification at such distances can lead to errors. Such errors endanger the safety of a conduit along sections where there is weak soil. The article attempts to assess how serious this hazard is and how it influences a conduit's foundations.

Important for the safety of a sewage conduit is the exact fulfilment of the requirements for the method of executing the bedding layer, backfill and also the method of removing excavation protection, which should all be assumed in static-strength calculations. Negligent execution of these works leads to a change in the angle of a conduit's support and a change in the deformation modulus of the backfill soil. This can result in a reduction of the load-bearing capacity of a conduit and threaten its safety. The impact of executive errors in this area will be analysed in the paper.

2 GENERAL ASSUMPTIONS FOR THE CALCULATIONS

The paper presents computational analysis, which was carried out on the basis of German guidelines [1] and standard [2], of two conduits. The probability of their failure due to inappropriate determination of soil-water conditions and also inappropriate compaction of soil in the bedding zone of the examined conduits was assessed. The state of safety was analysed for both a flexible conduit made of high-density polyethylene (PEHD) pipes and a rigid conduit made of stoneware pipes. Both pipelines were laid in an open trench.

The following assumptions were made considering the flexible conduit: native soil from group G1 (according to [1]), a weak soil lens from group G3, backfill from group G1, a volume weight of soil $\gamma_B = 20$ kN/m^3, the nominal diameter of the conduit equal to 600 mm, conduit made of PEHD, depth of the conduit's cover h = 6 m, width of the trench b = 1.4 m and the trench protected using steel sheet piling.

In turn, the following was assumed for the rigid conduit: native soil from group G1 (according to [1]), a weak soil lens from group G3, backfill from group G1, a volume weight of soil $\gamma_B = 20$ kN/m^3, the nominal diameter of the conduit equal to 200 mm, the outer diameter of the conduit equal to $d_a = 0.242$ m, conduit made of stoneware, depth of the conduit's cover equal to h = 4 m, width of the trench equal to b = 1.4 m and the trench protected using steel sheet piling.

3 THE EFFECT OF ACCURACY IN THE DETERMINATION OF SOIL-WATER CONDITIONS ON THE SAFETY STATE OF A SEWAGE CONDUIT

The construction of a sewage conduit is designed whilst assuming soil-water conditions that are based on the results of geotechnical investigations. An optimally designed construction is one in which the safety coefficient is minimal. However, it must still comply with the current state of technical knowledge or the requirements defined in the relevant law regulations. Standard soil-water tests, with a distance between test holes of about 100 m, are sometimes carried out along the pipeline route where a weak soil lens can be present between the test holes. Laying pipes that are designed for standard soil-water conditions along such sections may lead to conduit failure. In order to assess the probability of such failure, an appropriate computational analysis was performed. The state of safety was analysed for both a flexible conduit made of PEHD pipes and a rigid conduit made of stoneware pipes, for which relative vertical deformations and the safety coefficient according to [1] and [2] were calculated. The bedding angle of the flexible conduit was assumed as $2\alpha = 180°$ and for the rigid conduit as $2\alpha = 120°$.

The distribution of loads acting on the conduit is shown in Figure 1 [1].

The value of loads acting on the conduit was determined on the basis of the commonly used German guidelines [1]. The vertical resultant load of the conduit was determined using relation (1):

$$q_v = \lambda_{RG} \cdot p_E + p_V \tag{1}$$

where λ_{RG} = load concentration factor, which is determined from formula (2); p_E = soil stresses due to the ground load and surface load that were determined using formula (3); p_V = soil stresses due to traffic load, which were assumed on the basis of the diagrams included in [1],

$$\lambda_{RG} = \frac{\lambda_R - 1}{3}\frac{b}{d_a} + \frac{4 - \lambda_R}{3} \tag{2}$$

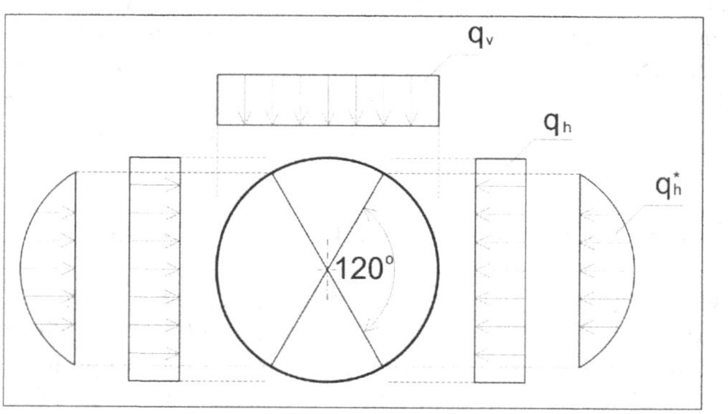

Figure 1. The distribution of loads acting on the conduit [1]: q_v – vertical load, q_h – horizontal load, q_h* – horizontal load considered as the bedding reaction pressure resulting from flexible pipe deformation.

where b = width of a trench measured at a level of the conduit's crown; d_a = outer diameter of the conduit; λ_R = load concentration factor, which is determined from formula (5)

$$p_E = \kappa \cdot \gamma_B \cdot h \tag{3}$$

where κ = load reduction factor according to the silo theory, which is determined from dependence (4); γ_B = volume weight of soil; h = depth of the conduit's cover,

$$\kappa = \frac{1 - e^{-2\frac{h}{b}K_1 \cdot \tan\delta}}{2\frac{h}{b}K_1 \cdot \tan\delta} \tag{4}$$

where K_1 = ground pressure ratio in the backfill zone; δ – angle of wall friction.

$$\lambda_R = \frac{\max\lambda\ V_s + a'\frac{4K_2K'}{3}\frac{\max\lambda - 1}{a' - 0.25}}{V_s + a'\frac{3 + K_2K'}{3}\frac{\max\lambda - 1}{a' - 0.25}} \tag{5}$$

where K_2 = ground pressure ratio in the backfill zone, which depends on the soil group and pipe-soil system stiffness; a' = effective relative outreach of the pipe's cross-section, which is determined using formula (6); K' = modulus for deformations, which is determined using formula (7); maxλ = maximum load concentration factor, which is determined using formula (8); V_s = stiffness ratio, which is determined using formula (9)

$$a' = a\frac{E_1}{E_2} \geq 0.26 \tag{6}$$

where a = relative outreach of the pipe's cross-section—for circular cross-sections a = 1; E_1 = modulus of resilience in the soil zone E_1 according to Figure 2; E_2 = modulus of resilience in the soil zone E_2 according to Figure 2,

The designation of zones for which the modulus of resilience in soil was assumed in the calculations is presented in Figure 2 [1].

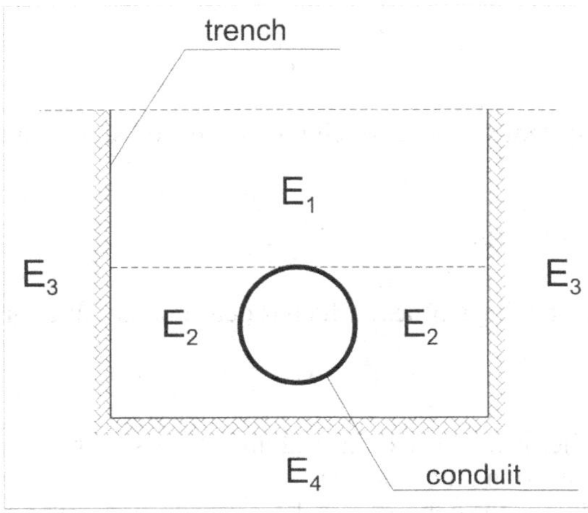

Figure 2. The designation of modules of resilience in soil for various zones [1]: E_1 – modulus of resilience in the soil zone above a conduit, E_2 – modulus of resilience in the soil zone on both sides of a conduit, E_3 – modulus of resilience in the native soil zone, E_4 – modulus of resilience in the soil zone below a conduit.

$$K' = -\frac{c_{v,qh} + \frac{c_{h,qh}}{c_{h,qv}} c_{h,qh^*} K^*}{c_{v,qv} + c_{h,qh^*} K^*} \tag{7}$$

where $c_{v,qv}$, c_{v,qh^*} = pipe deformation coefficients; K^* = coefficient for the bedding reaction pressure, which is determined using formula (13),

$$\max \lambda = 1 + \frac{\frac{h}{d_a}}{\frac{3.5}{a'} + \frac{2.2}{\frac{E_4}{E_1}(a' - 0.25)} + \left[\frac{0.62}{a'} + \frac{1.6}{\frac{E_4}{E_1}(a' - 0.25)}\right]\frac{h}{d_a}} \tag{8}$$

where E_4 = modulus of resilience in the soil zone located below the conduit,

$$V_s = \frac{8S_0}{|c_v^*| S_{Bv}} \tag{9}$$

where S_{Bv} = vertical bedding stiffness, which is determined using formula (10); S_o = circumferential pipe stiffness, which is determined using formula (11); c_v^* = pipe deformation coefficient, which is determined using formula (12),

$$S_{Bv} = \frac{E_2}{a} \tag{10}$$

$$S_o = \frac{E_R \cdot s^3}{12 \cdot d_m^3} \tag{11}$$

where s = pipe wall thickness; d_m = mean diameter of the pipe; E_R = short-term elasticity modulus of pipe material,

$$c_v^* = c_{v,qv} + c_{v,qh^*} \cdot K^* \tag{12}$$

$$K^* = \frac{c_{h,qv}}{V_{RB} - c_{h,qh^*}} \tag{13}$$

where V_{RB} = pipe-soil system stiffness, which is determined using formula (14):

$$V_{RB} = \frac{8 \cdot S_o}{S_{Bh}} \tag{14}$$

where S_{Bh} = horizontal bedding stiffness, which is determined using formula (15):

$$S_{Bh} = 0.6 \cdot \zeta \cdot E_2 \tag{15}$$

where ζ = correction factor for horizontal bedding stiffness, which was determined on the basis of a diagram included in guidelines [1],

The horizontal load q_h of the conduit was determined using dependence (16):

$$q_h = K_2 \cdot \left(\lambda_B \cdot p_E \cdot + \gamma_B \frac{d_a}{2} \right) \tag{16}$$

where λ_B = side load concentration factor in the zone next to the pipe, which is determined using formula (17):

$$\lambda_B = \frac{4 - \lambda_R}{3} \tag{17}$$

The bedding reaction pressure was determined using formula (18):

$$q_h^* = \frac{c_{h,qv} \cdot q_v + c_{h,qh} \cdot q_h}{V_{RB} - c_{h,qh^*}} \tag{18}$$

Using the above formulas, the vertical and horizontal loads acting on both conduits were calculated. The bedding reaction pressure was only considered in the case of the flexible conduit. The summary of loads regarding the flexible conduit is presented in Table 1.

The summary of loads acting on the rigid conduit is presented in Table 2.

The vertical deformation of the pipe was determined from dependence (19):

$$\Delta d_v = \frac{d_m}{8S_o}(c_{v,qv}q_v + c_{v,qh}q_h + c_{v,qh^*}q_h^*) \tag{19}$$

The permissible vertical deformation of the conduit cannot exceed a specific value. According to guidelines [1], which were used to determine the loads and deformations of the conduits, the long-term deformations cannot be greater than 6% for all types of conduits. In turn, according to standard [2], the long-term deformations of flexible conduits (made of thermoplastic materials) should not be greater than 15%.

Guidelines [1] recommend the use of the safety coefficient γ_{wym} with a value that depends on the safety class of the conduit. For conduits that are designed for the case in which the occurrence of failure would not cause a significant hazard for the safety state of a conduit, this coefficient should be assumed as 1.8. Standard [2] does not specify the value of the required safety coefficient.

The relative vertical deformations and safety coefficient according to [1] and [2] were calculated in order to analyse the state of safety for the flexible conduit made of PEHD pipes. A summary of the results of the conducted calculations is presented in Table 3.

Table 1. Summary of loads acting on the flexible conduit.

Group of soil at the level of the conduit's bedding	G1	G3
Vertical load of the conduit q_v [kN/m^2]	216.0	186.0
Horizontal load of the conduit q_h [kN/m^2]	47.6	16.9
Bedding reaction pressure q_h* [kN/m^2]	146.1	125.0

Table 2 Summary of loads acting on the rigid conduit.

Group of soil at the level of the conduit's bedding	G1	G3
Vertical load of the conduit q_v [kN/m^2]	192.8	160.8

Table 3. Summary of the calculation results of the flexible conduit.

Group of soil at the level of the conduit's bedding	G1	G3
Relative vertical deformation of the conduit [%]	7.3	10
Safety coefficient according to guidelines [1]	0.82	0.6
Safety coefficient according to standard [2]	2.05	1.5

Table 4. Summary of the calculation results for the rigid conduit.

Group of soil at the level of the conduit's bedding	G1	G3
Vertical load of the conduit [kN/m]	39.1	79.4
Declared destructive force F_N [kN/m]	40	40
Safety coefficient γ	2.2	1.1

The results of the conducted calculations show that a change in the conduit bedding conditions, which is caused by a change in the group of soil at the level of the conduit's bedding, results in a reduction of the conduit's safety coefficient of approximately 35%.

In the case of rigid pipes, the basic design criterion is the load-bearing capacity of the pipe, which is declared by the manufacturer whilst specifying the size of the crown compressive force (destructive force) F_N. The safety coefficient γ for the rigid pipe with the declared destructive force is determined using formula (20) [1]:

$$\gamma = \frac{F_N}{F_c} EZ \tag{20}$$

where EZ = installation figure that depends on the bedding angle of the conduit; F_c – total vertical load of the conduit, which is determined using formula (21):

$$F_c = q_v \cdot d_a \tag{21}$$

According to guidelines [1], the safety coefficient for a typical case cannot be less than 2.2. A summary of the computational results for the rigid pipe (with an assumption of the support angle of $2\alpha = 120°$) is shown in Table 4.

The results of the conducted calculations show that the change in the rigid conduit bedding conditions, which was caused by the change in the group of soil at the level of the conduit's bedding, results in a decrease in the conduit's safety coefficient of about 100%.

4 THE EFFECT OF THE ACCURACY OF BACKFILL COMPACTION IN THE ZONE OF THE CONDUIT BEDDING ON THE SAFETY STATE OF THE CONDUIT

The size and distribution of the loads acting on a sewage conduit depend on the quality of soil compaction in the bedding zone. The quality of this compaction determines the bedding angle, which is illustrated in Figure 3. The value of this angle influences the value of deformation coefficients: $c_{v,qv}$, $c_{v,qh}$, $c_{v,qh*}$, $c_{h,qv}$, $c_{h,qh}$ and $c_{h,qh*}$.

In the case of flexible pipes, a very accurate soil compaction in the bedding zone enables the bedding angle 2α to be assumed as 180°. In the case of inaccurate compaction, the value of 2α should be assumed as 120°. Computational analysis was performed in order to estimate the influence of the bedding angle on the safety state of the conduit. The analysis involved a flexible and rigid conduit, with their characteristics specified in point 2. However, both conduits were placed in G1 group soil. The loads and deformations were calculated using the dependencies as in point 2. The results of calculations are summarized in Tables 3 and 4.

Obtaining an accurate compaction of the backfill in the bedding zone is relatively laborious and requires the use of soils that compact well. Soils that have a value of uniformity coefficient U greater than 5, can be considered as soils that compact well. Moreover, the humidity of these soils should be close to the optimum value. In certain situations, it may be worthwhile considering the use of a pipe with a greater circumferential stiffness, for which a very precise compaction of backfill in the bedding zone will not be necessary.

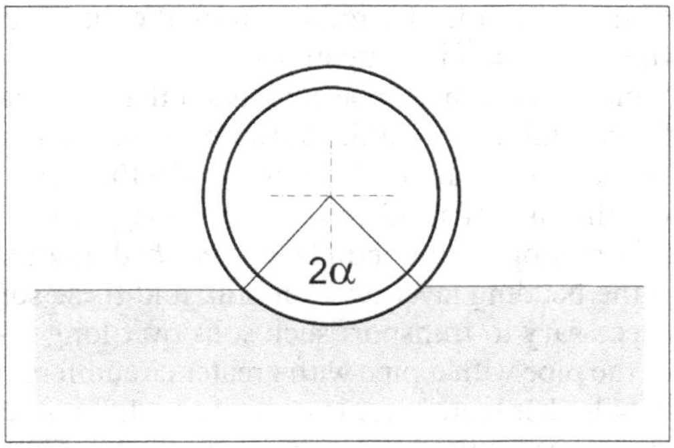

Figure 3. The bedding angle 2α.

Table 5. Summary of the calculation results for the flexible conduit.

Bedding angle	120°	180°
Relative vertical deformation of the conduit [%]	8.1	7.3
Safety coefficient according to guidelines [2]	0.74	0.82
Safety coefficient according to [1]	1.85	2.05

Table 6. Summary of the calculation results for the rigid conduit.

Bedding angle	60°	120°
Installation figure EZ	1.59	2.18
Total vertical load of the conduit [kN/m]	38.7	39.1
Declared destructive force F_N [kN/m]	40	40
Safety factor γ	1.64	2.2

In the case of the flexible conduit, the relative deformations and safety coefficient were calculated for a bedding angle equal to 120° (unsatisfactory quality of bedding compaction) and 180° (high quality of bedding compaction). A summary of the calculation results for flexible pipes is shown in Table 5.

The results of the conducted calculations show that the change in the bedding angle of the flexible conduit 2α from 180° to 120° results in a reduction of the safety coefficient of about 10%.

In the case of the rigid conduit, the relative deformations and safety coefficient were calculated for a bedding angle equal to 60° and 120°. A summary of the calculation results for the rigid pipe is shown in Table 6.

The results of the conducted calculations show that changing the bedding angle of the rigid conduit 2α from 120° (high quality of bedding execution) to 60° (low quality of bedding execution) results in a reduction of the safety coefficient of about 30%.

5 CONCLUSIONS

The results of the carried out computational analysis proved that weak soils that can locally occur along the bedding zone might cause a failure of a conduit. The risk of failure is especially significant in the case of rigid conduits for which the presence of a weak soil lens resulted in the reduction of the safety coefficient by 100%. Therefore, any factors that indicate the possibility of the occurrence of heterogeneous soils along a particular section of a projected conduit route should result in a reduction of the distance between the test holes that are used for soil-water condition investigations.

From the conducted analysis, it can also be concluded that the quality of soil compaction in the bedding zone of a conduit has a significant impact on the safety of the conduit. The safety coefficient can be reduced by approximately 10% in the case of flexible conduits and by approximately 30% in the case of rigid conduits. This is due to a negligent execution of the bedding zone. The formation of a conduit's correct bedding angle requires the use of well-compacted soils in the bedding layer and backfill, and these soils should have an optimum humidity. If it is necessary to transport such soils over long distances, it may be more advantageous to replace the pipe with a pipe with greater circumferential stiffness. In the case of rigid pipes, pipes with a higher load-bearing capacity or different conduit foundations e.g. the use of a concrete bench, can be used for such cases.

REFERENCES

[1] January 2000, German guidelines ATV-DVWK—A 127P *Static Calculations of Drains and Sewers.*

[2] May 2011, Standard PKN-CEN/TS 15223:2011 *Plastics piping systems—Validated design parameters of buried thermoplastic piping systems.*

Underground Infrastructure of Urban Areas 4 – Madryas et al. (Eds)
ISBN 978-1-138-55953-0

New trenchless technology for small diameters and long drives: Jet pump in HDD, E-Power and direct pipe

Michael Lubberger & Dymitr Petrow-Ganew
Herrenknecht AG, Schwanau-Allmannsweier, Baden-Württemberg, Germany

ABSTRACT: It is well-known that HDD uses the borehole as a discharge line to remove the cuttings out of the borehole. A slow velocity of the mud is not productive when large cuttings are transported as they tend to settle along the borehole. Remaining cuttings left in the borehole lead to large overcut dimensions to avoid pipes getting stuck during pullback. Furthermore, excessive pumping of mud can lead to frac-outs when the fluid pressure is higher than the maximum allowable borehole pressure. The complexity of the mud design increases which results in costly mud monitoring and adjustment on jobsites.

With the development of a downhole jet pump system for HDD operations, Herrenknecht has set a milestone in the pipeline industry. In conjunction with a new tooling series, the downhole jet pump system allows drilling and cutting faster at lower cost and risk. The full face hole opener has been developed for hard rock drilling with HDD, where time-consuming reaming passes are reduced to one single step.

The successful concept of the jet pump system for HDD has now been transferred to Pipe Jacking technologies. The long existing demand for increasingly long drives in simultaneously smaller diameters, has now been met with the development of the AVNS technology. The AVNS tunnelling machine (minimum ID 350 mm) with powerpack and slurry jet-pump in the machine is designed for up to 1,000 m drive length, depending on geological conditions. Furthermore, this machine concept can also be used for Direct Pipe® and opens up the lower diameter range from 20" to 30" pipelines.

This paper presents the use of the jet pump systems in HDD, Pipe Jacking and Direct Pipe operations and gives an overview of applications fields and recent case studies.

1 INTRODUCTION

The HDD technology is widely known as a standard method to lay all kinds of utilities by trenchless methods underground without harming the environment. The technology covers borehole diameters from 100 mm up to 1800 mm (4 to 72-inch). As a fairly young technology the limitations with regards to the drilling length, diameter and application range are still extended every year. Nevertheless, there are limitations of the HDD process (e.g. stability of boreholes in gravel areas) due to the frac-out risk.

Frac-out means in general that the drilling fluid is not taking the proposed route through the borehole to the exit point. Instead, the drilling fluid is flowing through the route with the least resistance, which can be through pre-existing fractures within the formation that lead towards the surface, potentially causing serious damage to the environment. This paper will present the development of the jet pump system for the focus on the artificial fractures which are caused by the drilling process itself. Depending on the size of the fractures the drilling fluid circulation is partially or completely lost. With lost circulation you have no control of the process, including the cutting transport, and therefore the drilling operation would need to stop.

Small diameter Pipe Jacking installations using slurry machines, where the excavated material is transported to the separation plant on the surface by a slurry circuit, have been limited in the past in drive length, due to the available slurry pump systems installed in the tunnel or shaft. The new jet pump concept, which has been successfully used in HDD operations, has now been transferred to slurry machines. The jet pump due to its size can now be installed in slurry machines from 457 mm (18") ID of the Jacking Pipes and makes drive lengths up to 1,000 meter possible. For Direct Pipe®, where a slurry machine and a Pipe Thruster is used to install a pipeline in one step, the jet pump opens up the applicable diameter range, starting from 20" instead of 36" pipeline diameter.

2 JET PUMP SYSTEM FOR HORIZONTAL DIRECTIONAL DRILLING

2.1 *Settlement of cuttings during reaming step in HDD*

Frac-outs normally occur when the pressure inside the borehole is higher than the pressure which is generated by the overburden formation and the surrounding ground water pressure. As the borehole is used as a discharge line for the cuttings created by the reamers or hole openers, there are several mechanisms which lead to a clogged borehole which then further leads to an over pressurized borehole leading to frac-out into the formations. Settlement of cuttings can occur due to gravity. This depends on the mass of the cuttings, the viscosity of the bentonite suspension and the flow velocity. The sedimentation of the cuttings leads to a reduction in the borehole cross-sectional area, increasing pressure in the area and increasing the probability of a frac-out. Furthermore, the mass of the sedimentations can be so large in the inclined zones that it starts to slip in the axial direction. The risk of blocking the drill string is very high in these zones.

If the rate of progress during an HDD process is very high, the bentonite suspension needs to carry a much greater volume of cuttings out of the borehole than usual. In this case, some of the cuttings cannot be maintained in suspension because the bentonite suspension is overloaded. Due to the high volume of cuttings in the bentonite suspension, several cuttings will settle at the bottom of the hole which can lead to a clogging of the borehole.

2.2 *New solution to prevent frac-outs and settlement of cuttings*

The new solution is based on the fact, that the above mentioned problems using current technology are using the borehole as the discharge line to the surface. Another reason for the existing problems is the fact that there is only a surface pump which is feeding mud up to the reamer and from there on the pump is also pressurizing the discharge line in order to transport the cuttings from the reamer to the surface. Based on the above findings the following decisions have been made for the design of a new solution:

1. The borehole is NOT a discharge line (except during the pilot bore)
2. The surface pump is NOT used as a discharge pump (except during the pilot bore)
3. Complexity and tasks of the mud are reduced to a minimum.

The solution for the reaming period consists of the findings below. The system will have a steel discharge line and a downhole pump right behind the reamer/hole opener which will pick up the cuttings right after they are produced and transported through the tail string to the surface. This means that the borehole will no longer be used as a discharge line.

As there is normally a tail string behind the reamer, the decision to utilize this part of the string is not difficult to understand, especially when considering the size of the drill pipes which are used nowadays. A standard in large diameter directional drilling is 6 5/8" drill pipes; a new trend especially in rock drills or extreme long drills are 7 5/8" drill pipes. The dimension on these drill pipes offers the ideal discharge line dimensions. Pressure-wise a drill pipe can easily handle the required pressure rating.

The types of pump which can be used downhole are rare. It must be a simple and robust pump which also can transfer torque together with push or pull forces. Furthermore, the pump

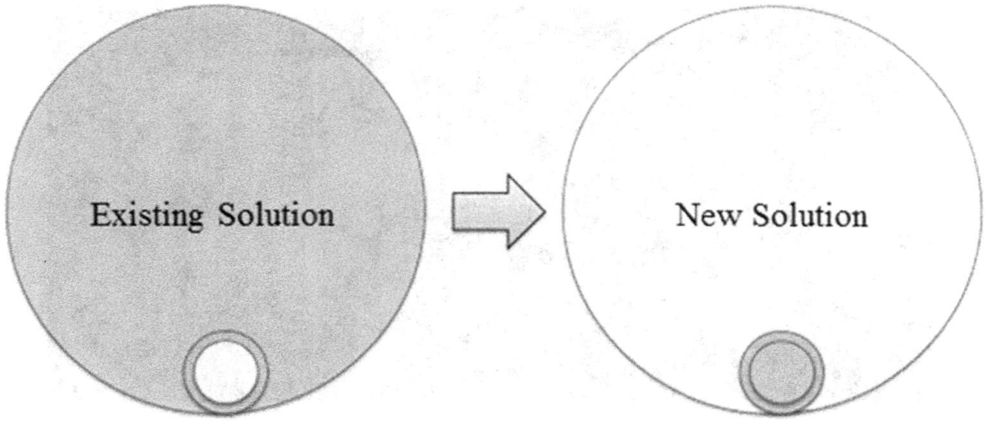

Figure 1. Existing and new solution for the discharge line in the borehole by using the drill string.

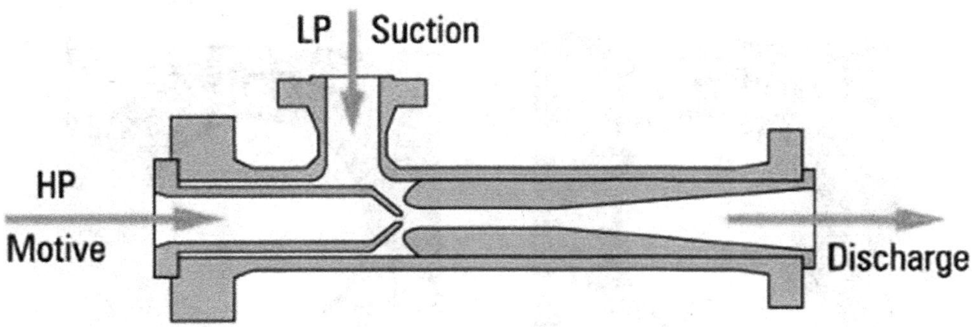

Figure 2. Principle of a jet pump.

must be capable of producing a sufficient pumping head. The power supply must be trouble free and run in wet submergible conditions. The pump must fit at least in a 508 mm/20 inch borehole. There are not many pump types which can handle those requirements. The decision was made in favor of a jet pump which is one of the simplest pumps with no moving parts, but is also one of the most complex pumps in terms of its physical and dynamic behavior.

2.3 *The Herrenknecht Down Hole Jet Pump (DHJP)*

Herrenknecht has developed a new downhole tool which absorbs cuttings from the bottom of the borehole and pumps them via the drill string to the exit side to surface. The power supply for the jet pump is generated by a high-pressure motive flow of a conventional HDD high-pressure mud pump on the entry side. For the mud handling it is necessary to install a mud return line between the pipe site and rig site during the drilling operation. This could be possible with a normal return line on the surface or by using a second pilot hole.

The newly developed Down Hole Jet Pump (DHJP) will cover sizes from 20 up to 72 inch. The main component is the jet pump which is integrated in the collar of the tool. During operation shovels on the DHJP pick up the cuttings from the bottom of the borehole and put them inside the DHJP during rotation. The system can handle a maximum cutting size of 1½". Cuttings which are too big for the DHJP system are guided through a bypass system along the outside of the DHJP to the area behind the downhole tool assembly into an optional cutting catcher.

Figure 3. Herrenknecht Downhole Jet Pump (DHJP) used with full face hole opener for one reaming step.

Figure 4. Settlement of cuttings in the borehole with/without the use of the Down Hole Jet Pump.

The major goal for the tool during the development was to create a maximum clean borehole. During operation with the DHJP tool it is possible for the operator to monitor the cuttings on the separation plant in real time. It is possible to create a more than 98% clean borehole with the new DHJP.

The main advantage of the system is that the cutting transport is inside the drill string. Inside the borehole is only a minimum of mud movement around the downhole tool assembly. The pressure inside the borehole is the same as the surrounding ground water level and therefore there is no possibility of a frac-out risk due to over pressurizing the borehole. Furthermore a "reduced" mud program can be used with fewer tasks for the mud. The remaining task will be the stabilization of the borehole by building up a filter cake.

With this highly innovative HDD tool it is possible to operate in different modes, described below:

Cleaning run mode

This mode is used after the successful reaming operation. Instead of using a conventional barrel reamer for the cleaning run, the DHJP could be assembled into the drill string to

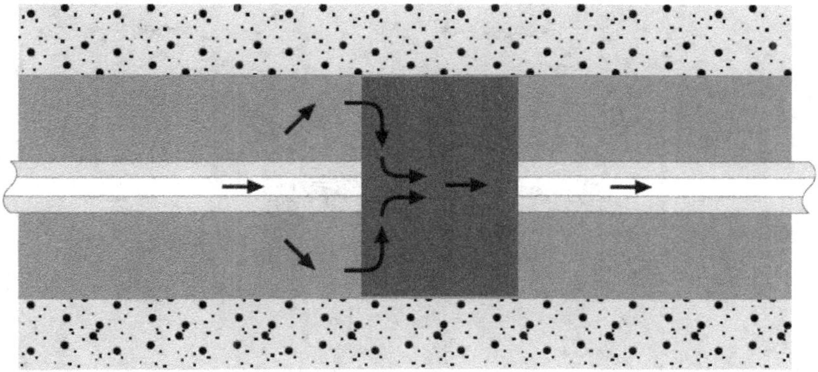

Figure 5. Cleaning run mode.

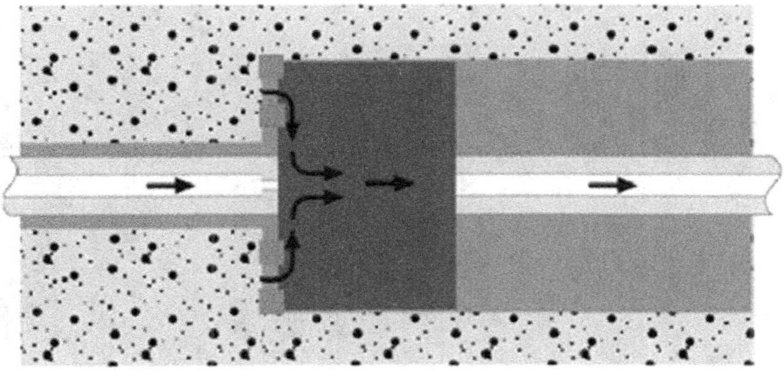

Figure 6. Soft ground tool mode.

directly remove the settled cuttings. The main advantage of this method is that the driller directly sees the success of the cleaning run and identifies areas where more or less cuttings are settled. It is therefore possible to keep the risk of a stuck product pipe caused by cuttings in the borehole to a minimum.

Soft ground tool mode
The DHJP can also be used as a soft ground reamer, and is equipped with normal soft ground tools. In this mode it is not necessary to run another cleaning run after the reaming procedure as the borehole is ready for the product pipe after the reaming procedure. The pump enables so-called full-face reaming which means reaming in one pass to the final diameter.

Tandem mode
For hard rock jobs, the DHJP could be directly coupled behind the new developed Herrenknecht Full-face Hole Opener. These hole openers are designed to drill from pilot hole directly to the final diameter. The high volume of cuttings which are generated by the full-face hole opener are picked up with the DHJP and transferred to the exit side.

Pipe pull mode
It is also possible to connect the DHJP to the product pipe during the pullback operation. Therefore it is necessary to put a "pump-through" swivel in front of the product pipe which links to the DHJP. A discharge pipe is necessary inside the product pipe towards the exit side.

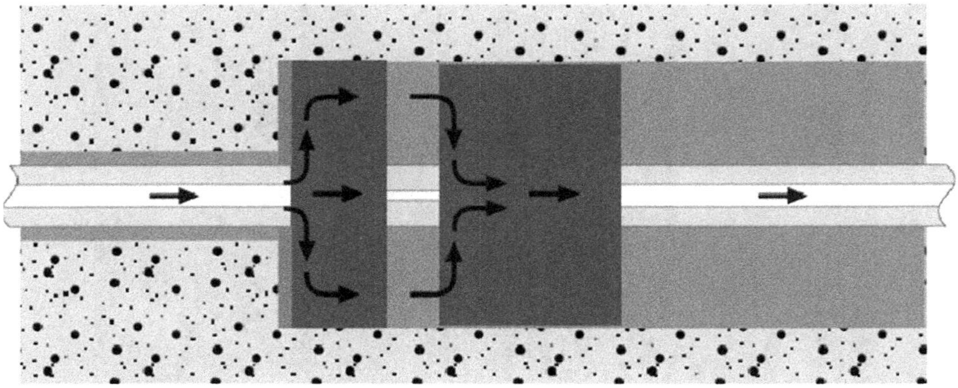

Figure 7. Tandem mode.

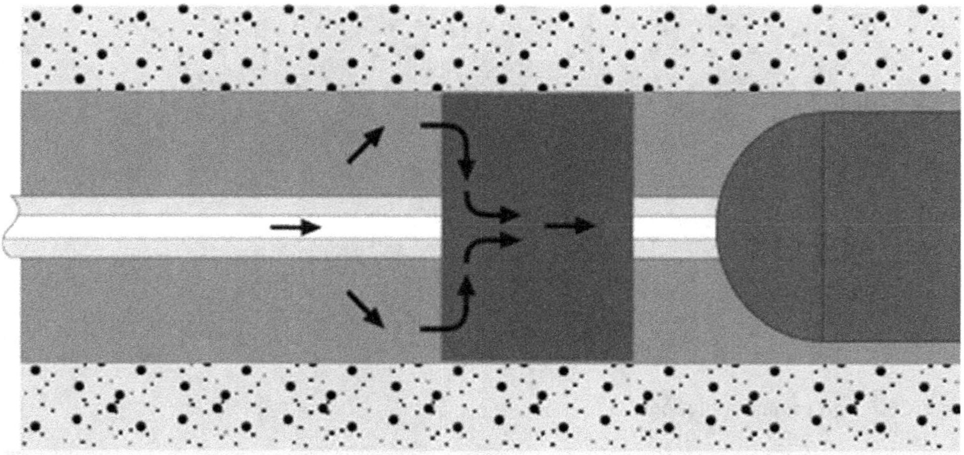

Figure 8. Pipe pull mode.

3 JET PUMP SYSTEM FOR PIPE JACKING

3.1 *First use of jet pump in slurry machine*

After successful tests of the jet pump as a slurry pump for an AVN machine in 2015 on the Herrenknecht yard, the system had its first productive use in a Microtunnelling project in Hannover, Germany, in summer 2016. For the construction of a surface water tunnel, an AVN 700 with jet pump was used to install one drive of 120 meter length in a depth of 4.5 meter. To avoid clogging of the cutter head due to the clayey and silty ground conditions the machine has been operated with the high-pressure nozzle system. The jet pump has been integrated in the third machine can. Tunnel face support has been assured through the feed pump. With its good performance, the jet pump technology had its successful jobsite premiere.

3.2 *AVNS technology—slurry tunnelling machine with jet pump*

The AVNS tunnelling machine is a new development using a jet pump, integrated in the machine, as a slurry pump. This proprietary development from Herrenknecht AG facilitates a conveying capacity of 1000l/min while simultaneously requiring very little space. The hydraulic unit integrated in the tunnelling machine is a further unique selling point.

This eliminates the coupling of hydraulic lines otherwise typical at this construction size, along with the associated losses of performance.

A measurement system ensures that the bore alignment is precisely maintained. This continuously determines the position, direction, and tilt of the tunnelling machine. The measurement system aims towards a magnetic field generated on the surface level, and is thus also suitable for very long drives. The machine can be used on a spectrum of up to 30 MPa during excavation of loose soils and soft rock.

3.3 *Use of AVNS350 in E-Power Pipe® technology*

With the new legislation for grid extension in Germany, underground cable installation has come more into focus since end of 2015. As an economical alternative to open-trench cable installation, trenchless methods are considered to reduce the impact on surroundings to a minimum. While established pipeline and tunnel installation methods are intended to be used predominantly in cases involving sensitive crossing, it was necessary to develop a new technology for individual installation of underground cables. The main challenge: inserting the cables into the ground at low depth and at a defined distance to extend across the longest drifts possible. As existing methods could not cover these requirements, Herrenknecht has developed the new E-Power Pipe® technology.

E-Power Pipe® allows underground cables to be installed close to the surface at a minimum overburden of only 2 m across long drifts. The AVNS350XB tunnelling machine with an excavation diameter of 505 mm is used, dimensioned for drive lengths above 1.000 m. The AVNS machine technology is a new development by Herrenknecht AG. The machine can achieve considerably greater advance lengths in a small diameter range due to its use of a new power supply and the introduction of jet pump conveyance technology.

Figure 9. Assembly of the AVN slurry machine and jet pump in machine can 3.

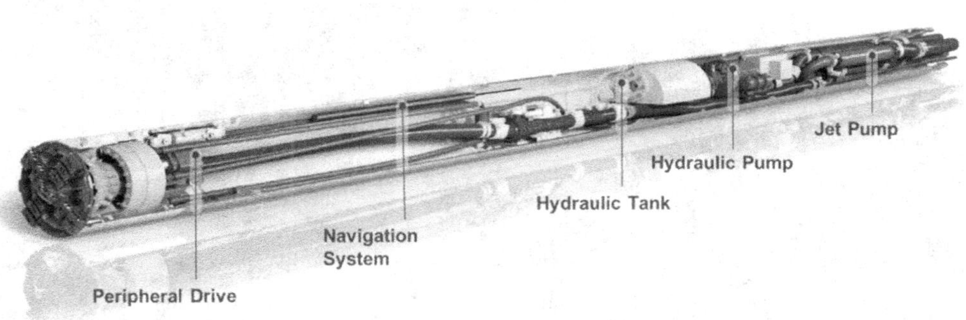

Figure 10. AVNS 350 XB with its components.

The E-Power pipe® method

The jacking frame is installed from the starting point and connected with the tunnelling machine via several thrust pipes. The tunnelling machine is pushed with thrust force in the direction of the destination point along the specified alignment. This is disassembled following the tunnelling machine's cutting action. Subsequently, the plastic guard tube is connected with the thrust pipe in the borehole and withdrawn by retracting the thrust pipes. The guard tube is mechanically and thermally connected to the floor during the insertion process while adding backfill material. This creates a connection between the cable guard tube and the ground that is free of cavities. The underground cable is inserted after the guard tubes are completely in place, and this is not a part of the process.

The E-Power pipe® components

The requirements for a new type of installation process for underground cables with a closed construction method have led to the development of several separate elements, which combine to provide an innovative and environmentally-friendly method for carefully installing underground cables. The AVNS 350XB tunnelling machine is a new development that satisfies the requirements related to mandatory installation depth, drive length, and precision of the procedure. A jacking frame with 10 m stroke and 350 tonne thrust force has been developed specifically for this thrust pipe. The jacking force is transferred with a pressure ring which enters the thrust pipe via a locking mechanism and creates a positive connection. This connection also safeguards retraction of the thrust pipe. The quantity of pipe exchanges and consequent time needed must be kept a low as possible for high tunnelling performance to be achieved. Furthermore, the thrust pipes must be joined with high tensile strength in order to safeguard insertion of the cable guard tube. These characteristics have led to the development of a new thrust pipe, which facilitates almost continuous tunnelling due to its length and significantly expedites the coupling of the water circuit with a sleeve system integrated into the pipe.

3.4 *E-Power pipe® reference project borken, Germany*

After testing and presenting the new technology to the public in December 2016, the E-Power Pipe® equipment arrived on its first jobsite in Borken, Germany in January 2017. The pilot bore of the first of three 300 meters long drive could start on Feb 2nd, 2017. With a performance of up to 126 meters per day on the pilot bore and a maximum daily performance of 266 meters for the pull-in of the casing pipes, it proved out its high performance capacity.

Figure 11. The E-Power pipe® method for trenchless installation of underground cables.

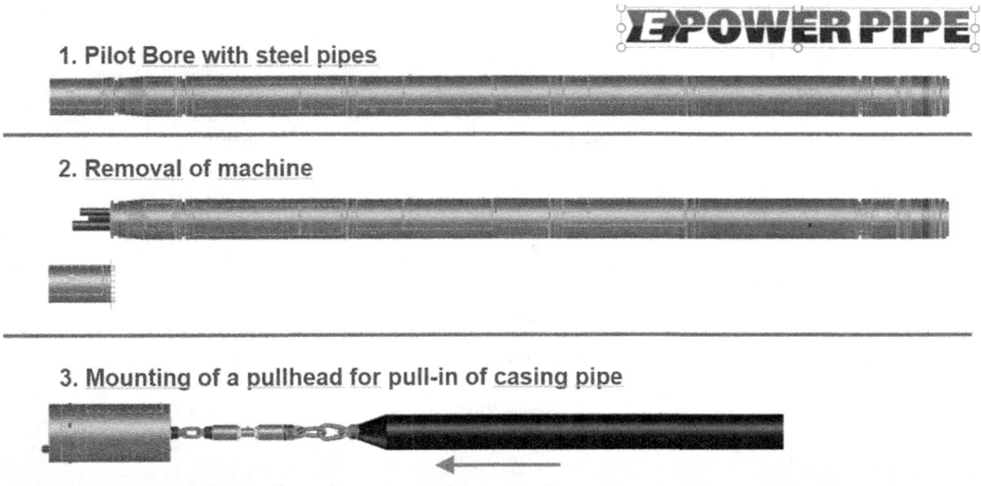

Figure 12. Overview of E-Power pipe® installation steps.

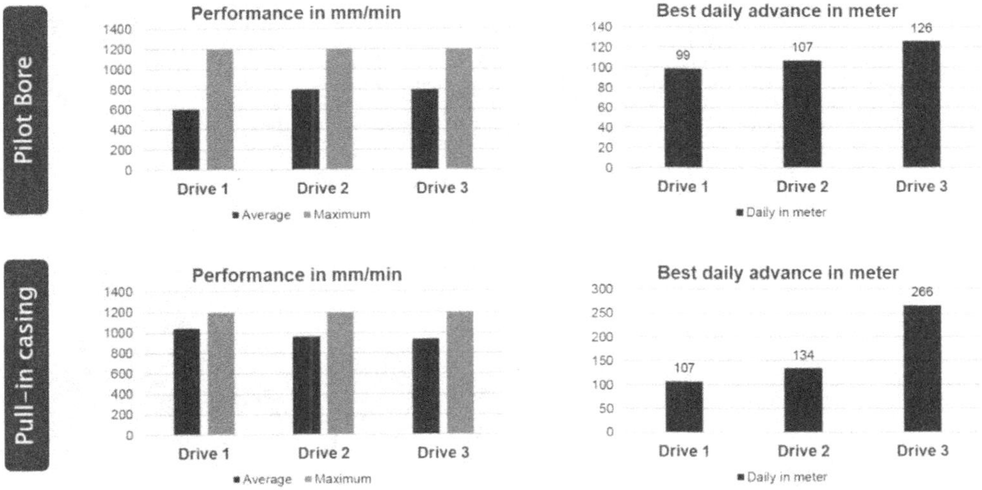

Figure 13. Performance overview of E-Power pipe, 3 × 300 m drives in Borken, Germany.

4 OUTLOOK

With the integration of the Jet Pump in two well-known Technologies (Pipe Jacking & HDD) the limits of both technologies have been stretched again in order to go longer with smaller diameter pipe(lines) (Pipe Jacking, Direct Pipe®) and to lower the frac-out risk significantly, having cleaner holes and to have the ability to work also in dry boreholes (HDD).

This new development will have an effect in several industries not only for the Pipeline industry where there is now a possibility to lay small diameter pipelines trenchless in one step without building first a larger casing tunnel. One of the main advantages are also seen on the environmental side where improvements on the CO^2-Footprint are obvious and the probability of inadvertent returns of drilling fluids will be reduced.

The presented development also influences the commercial side of the drilling business where saving can be seen by risk reduction and the use of smaller equipment.

Underground Infrastructure of Urban Areas 4 – Madryas et al. (Eds)
© 2018 Taylor & Francis Group, London, ISBN 978-1-138-55953-0

Large tunnel boring machine diameters for today's infrastructure systems

M. Herrenknecht
Herrenknecht AG, Schwanau-Allmannsweier, Baden-Württemberg, Germany

Karin Bäppler
Head of Business Development and Geotechnics and Consulting, Germany

Dymitr Petrow-Ganew
Herrenknecht AG, Schwanau-Allmannsweier, Baden-Württemberg, Germany

ABSTRACT: Numerous large diameter tunnelling projects with TBM diameters exceeding the 12 meter mark were completed successfully in the past with acceptable and also outstanding progress in challenging conditions. This has had the effect of encouraging the design of more and more large diameter tunnel projects that are accepted and trusted by the public not only based on outstanding progress but also on the fact that these large scale projects were realized reliably in time, within budget and also fulfilling the demand for high quality standards.

This paper provides an overview on the state of the art mechanized tunnelling technology that is demanded and perspectives on tunnelling projects with large diameters. Highlighted are some projects that provided significant improvement in delivering underground structures which contribute to resilient and sustainable urban solutions.

1 INTRODUCTION

The ongoing trend of designing larger diameter infrastructure tunnels is supported by the successful completion of large and very large diameter tunnelling projects. In the 1990s a 10–12 meter diameter TBM was considered to be the size limit for tunnelling machines whereas today TBMs with diameters in the range of 15 to 17 meters are based on previous experience acquired in proving the functionality of very large diameter TBMs. Examples of very large bores are the 15.6 meter diameter EPB Shield that was used between August 2011 and July 1013 in line with the project schedule for the Galleria Sparvo highway tunnel in Italy and the 17.6 meter diameter Mixshield that finished a 650 m long section of a large-scale subsea road tunnel project in Hong Kong in November 2015. Since the 1990s numerous very large diameter bores were successfully completed for underground infrastructure systems. Mainly these applications have been for road tunnel constructions but also quite a number of very large diameter multi-purpose tunnels and metro tunnels were built since then. All projects have needed to deal with the complexity of geology, site constraints and tight time schedules. The paper highlights three large to very large diameter projects that differ not only in diameter, machine type, geology but also on the specific project conditions. These projects are a large diameter metro tunnel project that was executed in Spain with a 12.06 meter diameter EPB Shield and two of today's largest soft ground TBMs, the 15.6 meter diameter EPB Shield that excavated a 4.8 km long tunnel section in Italy and the 17.6 meter diameter Mixshield that was used in Hong Kong.

2 EXAMPLES OF LARGE TO VERY LARGE DIAMETER TBMS FOR METRO AND ROAD TUNNELS

Mechanized tunnelling is continually opening up new dimensions in tunnel profile, tunnelling length and complexity of subsurface conditions. Larger diameters are accompanied by large face areas that have to be dealt with in respect of geotechnical aspects but also in respect of technical aspects of large diameter TBM design developments such as for example cutterhead design, cutterhead intervention, excavation process and supply logistics. [1] The cited large to very large diameter TBM projects address the machine design that was adapted to the specific project demands.

2.1 *12.06 meter diameter EBP shield for Barcelona Metro Line 9, Spain*

With a population of about 1.6 Million, Barcelona is after Madrid the second largest city in Spain. The purpose of the Metro Line 9 in Barcelona is the construction of a peripheral line that surrounds the city of Barcelona. This line includes 52 stations of which 32 are deep shafts and 19 are transfer stations. Of the total of 52 km of the Metro Line 9 a section of 28 km was bored using TBMs. Part of this underground section was excavated and lined with a 12.06 meter diameter Herrenknecht EPB Shield that started tunnelling at the end of August 2013. At that time it was the largest EPB Shield ever applied. The machine was operated by the UTE GORG Joint Venture (Dragados, Nesco, ACS, Comsa and Sorigué) along a 10 km section between the stations Gorg and Sagreda and between Zona Franca to Zona Universitaria. A special feature of Line 9 is the single tube metro system where the two tracks run side by side in one large diameter tunnel tube along the metro corridor. In the station area the tracks run one upon the other and a specific feature is the double deck configuration in the station area that was first implemented for the Metro Line 9 in Barcelona.

With most of the metro projects that are built with mechanized tunnelling technology, the stations are built first using open cut construction or cut and cover method. After the tunnelling machines reach the future stations they are pushed or pulled through the stations to continue their next tunnelling sections.

For the Metro Line 9 the procedure was different. The large diameter of the EPB Shield was chosen due to the high urban density along the metro corridor. This required the construction of the stations underground with the integration of the passenger platforms within the tunnel profile in the station area. That's why first the EPB Shield excavated and lined the tunnel and then later, the underground stations were constructed.

The subsurface conditions along that metro corridor are characterized by variable geological conditions with gravel and sand in a clayey matrix, clay, boulders and a 200 m long section with the presence of granite banks. Based on the predicted geological conditions the cutting

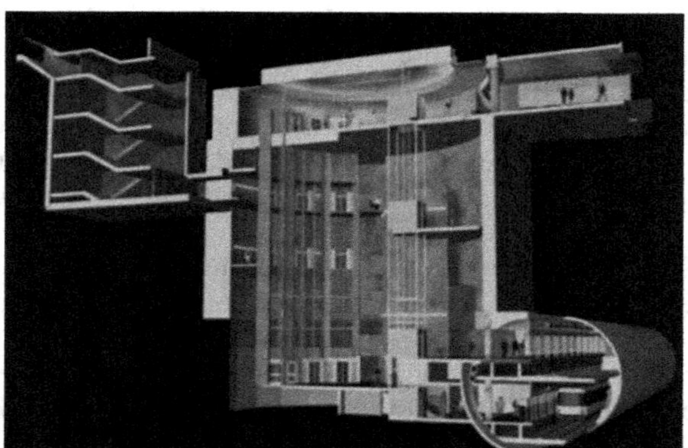

Figure 1. Stations of Metro Line 9, layout of the platforms and the vertical shafts.

Figure 2. Metro Line 9 Barcelona: 12.06 meter diameter EPB Shield for variable geological conditions.

wheel was of closed design with an opening ratio of 35%. It was equipped with soft ground tools and also disc cutters to handle the rocky parts and boulders.

Reliable information on the condition of the soft ground tools and buckets, especially those working hardest on the periphery of the cutting wheel, is vital for an efficient excavation and for avoiding possible damage to the tools and steel structure of the cutting wheel, and this information was obtained by equipping the cutting tools with a wear detection system. Herrenknecht engineers developed an electronic tool monitoring system capable of providing data on the soft ground tools and buckets lips online and thus alerting the TBM operators in real time. This system was integrated in two buckets and four soft ground tools of the large diameter EPB Shield that was used in 2003. The heart of the system was a new tool support with an integrated sender electronic that was permanently connected to the soft ground tool through induction loops. Thus it can be detected whether the wear limit is reached. The sensor/ sending unit was electrically connected to a power supply. If the probe is intact a certain current is maintained, but if the probe is damaged due to the wear of the cutter this is sensed by the sending unit which is inductively connected to the receiver and the cutter probe through a small gap. This leads to a significantly higher fault current that illuminates a LED to inform the machine driver about the possible wear.

A further requirement that resulted from the specific routing of the metro line was that the machine had to deal with tight curve radii of 300 meters. So therefore the EPB Shield was designed with a shield articulation. Design related solutions such as shield articulation, overcut and articulated attached back-ups enabled the excavation of the required tight curve radius of 300 meters.

Due to the large diameter and the predicted geological conditions along the tunnelling section, the EPB Shield was installed with a torque of 45 million Newton meters. It comprised the greatest torque ever used in a TBM until this time. With the large diameter EPB-Shield used for the construction of the Metro tunnel in Barcelona, the maximum limit regarding machine diameter and torque for soft ground shields was still not reached.

The increasing dimensions the projects is shown for example with the successful completion of projects such as the M30 highway tunnel in Madrid (EPB Shield ∅15.2 m) Shanghai Changijiang Under River Tunnel Project in China (Mixshield ∅15.43 m), and the Galleria Sparvo Tunnel in Italy (EPB Shield ∅15.6 m).

2.2 *15.62 meter diameter EBP shield for the Galleria Sparvo highway tunnel in Italy*

In 2011 the largest EPB Shield that was designed and built by Herrenknecht started tunnelling for a twin tube road tunnel in Italy. The tunnel profile carries three lanes of traffic. The very large diameter EPB Shield had a diameter of 15.62 m and was designed to suit the predicted

Figure 3. Galleria Sparvo road tunnel – 15.62 m diameter EBP Shield.

specific project conditions along the planned alignment of the Galleria Sparvo road tunnel that is located between Bologna and Florence.

The demand on the design of the EPB Shield was based in general on the predicted geotechnical issues and was related to topics such as the control of tunnel face support, mixed face conditions, cohesive soils, methane gas all along the alignment and possible squeezing conditions.

With focus on such a large diameter it has to be mentioned that the larger the tunnel diameter, the higher the probability of a heterogeneous tunnel face and possible variation of soil or rock constituents that can even vary from ring to ring thus that the geological distribution along the alignment remains an uncertainty. Compared to shields with liquid supported tunnel face, large diameter EPB Shields require a higher torque. The cutting wheel torque is affected by machine and process technical factors such as drive and bearing unit, the design of the cutting wheel and the rotational speed of the cutting wheel. Part of the torque is consumed between the tunnel face and the cutting wheel. The 15.62 meter diameter EPB Shield had a maximum torque of 125 million Newton meters. In cohesive soils the EPB Shields can face clogging affects that must be counteracted by an appropriate soil conditioning. Special focus here in the center area of the cutting wheel because of a more limited muck flow and lower cutting speeds in the center area than in the outer part of the cutting wheel. A special focus in respect of appropriate conditioning is also in the working chamber. Injection points and foam lances were thus also installed in the outer cutting wheel area on the front face as well as additional foam and water injection openings in the center plate of the main drive to ensure an adequate conditioning of the muck in the working chamber.

A special hazard potential was the presence of methane gas so the machine was therefore specially designed to cope with this hazard potential. Gas detectors were coupled to switches that shut down the power to the machine if gas concentrations are measured above the threshold levels and portable measuring/alarm devices were used to measure the concentration of combustible gases. Furthermore a continuous feed of large volumes of fresh air was provided to dilute any gas. During TBM advance the excavation chamber was always completely filled with muck to prevent an inflow of material into the working chamber in case of instable tunnel face conditions were encountered. But for this project the main reason for an overall closed mode EPB operation was to prevent the possible hazard of formation of a combustion chamber due to potential gas presence in the rock mass. In the back-up area technical measures were taken which exclude any concentration of methane. To eliminate the risk of explosion and to counteract the prevailing risk of gases in the shield and back-up area where people are working the machine design included a double-walled enclosure for the back-up conveyor belt from the screw discharge gate to the transverse conveyor belt on back-up number three with permanent ventilation inside and outside of this system. In between

Figure 4. Illustration of the interface of material handling from screw conveyor to encapsulated conveyor belt [2].

this encapsulated double shell, overpressure was present so that any gas would be pushed back in the channel. The area from the transfer belt conveyor and loading chute to the tunnel belt was not covered thus that from this point on the completely built tunnel was equipped with fully explosion proof equipment. The air quality and the tightness of the system were permanently monitored. This together with the constant level monitoring of the working chamber to guarantee that the chamber is always completely filled to avoid the danger of gas pocket formation, allowed a controlled excavation process through the sections with a potential for gas presence.

A further demand from geology that was considered in the TBM design was the possibility of facing converging (squeezing) ground. This was dealt with the TBM design that considered technical measures to reduce this risk by installing sufficient high thrust force, the conical shield design, lubrication of the shield shell and a reliable overcutting.

Projects in general, especially large to very large diameter tunnel projects, demand specific solutions for tailor-made tunnelling equipment where a standardization of TBMs is not feasible. This is in respect of the loads that have to be considered resulting from the subsurface conditions and also in respect of material handling systems and logistics of large-scale machines.

2.3 *Largest TBM with a diameter of 17.6 meter in use for part of the bored tunnels of Tuen Mun-Check Lap Kok link in Hong Kong*

An efficient road link is being created between mainland China and the International Airport. The centerpiece of this road tunnel is a huge traffic feeder that the Dragages-Bouygues Joint Venture approached with a unique design concept. This has led to a previously unseen TBM configuration with a supersize diameter of 17.6 m. The Mixshield manufactured by Herrenknecht started tunnelling in May 2015 and excavated a 650 long tunnel section with prevailing hydrostatic pressures of up to 4bar. The section is part of the Tuen Mun-Chek Lap Kok Link (TM-CLKL) in Hong Kong that comprises twin bored tunnel sections of 4.2 km. The overall demand along the bored tunnels sections for the TBMs is to deal with high hydrostatic pressures exceeding 5bar and the associated highly unstable ground conditions that are mainly characterized by alluvium (mainly sand with alternations of clay and silt), completely to highly decomposed granite, slightly decomposed to fresh granite and marine deposits (sand and clay). In total 3 TBMs were supplied for these twin tube bored tunnels. Tunnelling operation will face about 50% mixed face conditions of rocks and soil and about 50% full face in alluvium. Rock strength of the granite was predicted to be in the range of 70 to 170 MPa. Apart from high support pressures and expected high wear in the granite, clogging potential in the clayey soils and pockets of biogenic gas within the marine and/or alluvial deposits were considered when designing the TBMs. Based on the geological and hydrogeological conditions, three Mixshields were specified; one of supersize diameter of 17.6 m and two with diameters of 13.6 m.

The mega TBM started its operation for the 650 m long section of the feeder tube in May 2015. After the 17.6 m Mixshield has reached its target in an intermediate shaft in November 2015, the planned transformation of the machine in the size began. The main drive and back-up system of the supersize TBM was kept and implemented in the 13.6 meter diameter Mixshield that continued tunnelling operation for the remaining bored sub-sea tunnel section. The parallel bored section of the road tunnel towards the airport is excavated and lined using the 13.6 meter diameter sister machine.

Figure 5. Supersize TBM for TM-CLKL, Mixshield Ø17.6 m diameter.

Figure 6. Breakthrough of the 17.6 m diameter Mixshield.

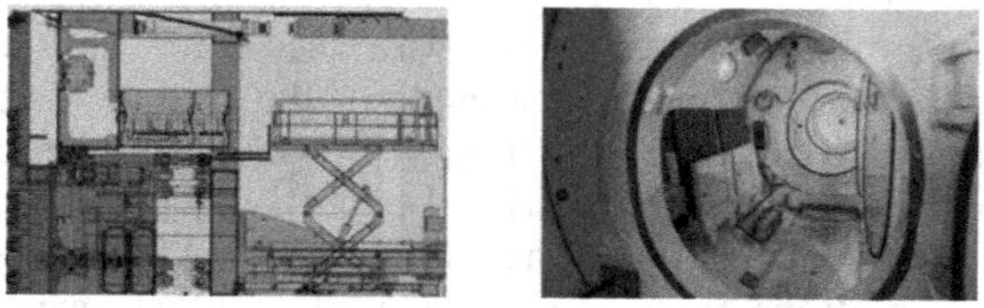

Figure 7. Supersize TBM for TM-CLKL, concept of hyperbaric interventions with permanent pre-chamber.

Due to the demands from geology and specific project conditions, the TBMs were specified and designed to minimize routine maintenance. Thus maximum reliability and good access to all components can be provided. With focus on required cutterhead interventions for regular inspection and maintenance of cutting tools and the cutterhead, redundant systems are needed for inspection and replacement procedures especially with focus on hyperbaric intervention. The machines were prepared for worst case conditions and designed to inspect the cutterhead with prevailing high water pressure and unstable face. Thus the machine is equipped with all the necessary basic installation for chamber access in saturation mode. In addition to the piping and connections that are required for saturation access this included a permanent pre-chamber in the shield to which a transport shuttle can be connected when needed. A clear passage to transport the shuttle through the back-up to the pre-chamber was also foreseen in the design and the transfer under pressure shuttle was available on site.

2.4 *Large TBM of 12.5 m Mixshield for Słowacki Road Tunnel in Gdansk*

To improve the transport infrastructure in the greater Gdansk area the 10 km long Słovacki route connects the airport with the highway to Warsaw and the deep water seaport. The twin tunnels with a length of 1076 m each of the final part cross the Martwa Wisła River and will accommodate a two-lane roadway and an emergency pedestrian verge in each. The tunnel had to be excavated in mixed-face conditions in grounds from the Quaternary Period through sand and gravel with maximum water pressure at the tunnel crown of 2.2 bar. Sedimented boulders were expected along the tunnel route. Notwithstanding the tunnel starts with a steep gradient of 4 percent, the minimum overburden under the river was only 8 m at certain points.

The order was given for the € 221 million tender of the Spanish-Polish joint venture consisting of Obrascón Huarte Lain S.A. (OHL) and PBG Group. Due to the bankruptcy of PBG the job was later taken over by OHL completely. The order for the ∅ 12.56 m Mixshield TBM S-745 was given to HK in November 2011. The scope of supply was not only the TBM itself but the separation plant and the slurry circuit, compressor plant for air supply, rolling stock equipment and steel support for the starting situation were also supplied.

Due to the same TBM diameter being required as that of the Biel Project TBM, and the variable Mixshield design, the existing EPB Shield was refurbished and all required slurry equipment, such as submerged wall and pressure bulkhead, submerged wall gate, ∅ 1.200 mm stone crusher and suction pipe, had to be designed and assembled in the existing shield structure. The existing closed cutting wheel was equipped with 17" single and twin wedge lock discs, scrapers and buckets. The ∅ 6 m cutting wheel drive with fixed centre plate was fitted with 10 × 350 kW electric motors. The backup structure was refurbished and one of the existing but not required gantries abandoned. The last gantry was changed to the required pipe layer. The slurry circuit was designed for 2,100 m³/h output.

Figure 8. Breakthrough of S-745 Slowacki Tunnel Gdansk.

Launching of the TBM for the first drive was on the 15th of June 2013. An average advance rate of 42.6 m/week (1st tube) was achieved and increased up to 76.6 m/week for the 2nd tube. The final breakthrough was made on the 9th of June 2014. The best weekly performance was 132 m. The high ring building quality in combination with the first Mixshield job by a Spanish contractor are the foot print of this large TBM project.

3 CONCLUSION

The selected reference projects for today's infrastructure systems described in this article highlight the demand of the design of large to very large diameter tunnel boring machines with innovative solutions such as the double deck configuration of the underground stations in confined urban area, the construction of a three lane road tunnel or the TM-CLKL in Hong Kong for a huge traffic feeder tunnel section.

Looking back in the history it is shown that numerous very large diameter tunnelling projects with TBM diameters even exceeding the diameter range of 14 meters have been completed successfully generally with acceptable or outstanding performances in variable and challenging conditions. This shows that these supersize diameters are accepted and trusted by the public. Feasibility studies are already underway for future large-scale infrastructure projects that will even exceed the already proven diameter ranges of up to 17.6 m. Supersizing underground infrastructure is a fundamental trend. Beside bold visions, moving from record to record requires proven expertise and trustworthy partnership between all involved experts from the conception of the project start until breakthrough.

REFERENCES

[1] Burger, W. 2015. Super Diameters–Design aspects for very large TBMs, *Rapid Excavation and Tunneling Conference Proceedings 2015.*

[2] Herrenknecht, M. and Baeppler, K. 2013. Tunnelling experiences of the largest EPB Shield to date for the Galleria Sparvo highway tunnel, *World Tunnel Congress 2013 Geneva,* Underground—the way to the future! G. Anagnostou & H. Ehrbar (eds), © 2013 Taylor & Francis Group, London.

[3] Wehrmeyer, G 2014. Mixshield Tunnelling in Urban Areas, Underground Infrastructure of Urban Areas Wroclaw, *Conference Proceedings 2014.*

Underground Infrastructure of Urban Areas 4 – Madryas et al. (Eds)
 ISBN 978-1-138-55953-0

Interaction of buried flexible pipelines with soil

B. Kliszczewicz
Silesian University of Technology, Gliwice, Poland

ABSTRACT: The subject of the paper concerns an analysis of interaction between flexible pipelines with soil. The thermoplastic flexible pipes (PVC, PE) are widely used in water, waste and gas systems. Their proper functioning in complex loading conditions and durability depends on the suitable identification of loads, the strain estimate, good assortment of pipe and the appropriate installing in the soil. The character of the loads acting on the buried pipes is very complicated. They include: the soil load, surfaces loads including live loads, internal pressure and the impact of underground mining exploitation (continuous and discontinuous deformations). The conditions of pipeline laying: in the common trenches, in the areas of low-bearing soils, flood areas or in regions of landslides are important too. The calculation methods of flexible pipelines (ATV method, Scandinavian method), which are available, make it possible to analyse the characteristic cross section of pipe (the strain plane state). In this methods, the interactions between pipelines and soil during pipe deformation under loads, are represented by specific distributions of vertical and horizontal soil pressures. Far better possibilities of the flexible pipelines behaviour in soil analysis are provided by numerical methods (MES). With this methods the pipeline—soil system is analysed. The pipeline—soil system, depending on the rigidity relation of the both components (substructures), reveals different interaction. Some examples of numerical analysis of straight pipe segments (2D and 3D) as well as curve segments (3D) are presented in the paper. Spatial numerical analyses of the pipeline—soil system, which reflect particular situations, related to the functioning of pipelines under acting live loads and strongly variable soil conditions are undertaken with ZSOIL acad. ver. 11.15 software. An elasto-plastic model with isotropic strengthening (Hardening Soil Small) is used for describing soil behaviour. Results of the numerical analysis are presented as maps of displacements in the soil as well as maps of displacements of pipes.

1 INTRODUCTION

The systems of underground infrastructural utilities, being an indispensable part of developed areas, are comprised of complexes of utilities, technical facilities and lines, intended to deliver auxiliary utilities (water, gas, heat) or to dispose utility water and storm water. The correct functioning of such systems is dictating the appropriate living standards of residents, i.e. comfort of life as well as a necessary standard of sanitary and health conditions. For this reason, the correct design, construction and operation of all constituent elements of underground utilities' systems, including pipelines, is an important modern engineering issue.

Pipelines are usually constructed with traditional trench methods or with more technologically advanced trenchless technologies. In the first case, a pipeline is laid within a trench area onto the prepared subgrade and then backfilled with soil layers, which are gradually compacted. A pipeline laying method should be adapted to soil conditions existing in the excavation area. Native soil stored when excavating a trench is usually used for filling the trench completely. Complex systems of internal loads (internal pressure) and external loads (earth load, transport load) and, in specific conditions, loads of the forced kinematic character (impact of underground mining exploitation), are acting on a buried pipeline. Pipeline behaviour under the influence of the acting systems of loads and interactions and the resulting soil reaction is called a pipeline-soil interaction or, more often—interaction of a pipeline-soil system (Kliszczewicz 2014). The last term highlights especially the fact of an interaction between two completely different sub-structures: a pipeline construction and soil. A reaction of the pipeline coating to the acting loads and interactions is complex and very different. The character and range of pipeline-to-soil system interactions depend on multiple factors such as: pipeline material and geometric parameters (type of the pipeline material solution, pipeline diameter and rigidity), soil and water conditions and a pipeline placement method, pipeline laying technology in soil (trench and trenchless methods, placement in embankment area), as well as a system of loads and external impacts (Sandford 2000).

The material solutions employed and the related pipeline rigidity have a special effect on the behaviour of pipelines laid in soil. The three main groups of such solutions include: metals, concrete and plastics, characterised by highly differentiated strength parameters, influencing their geometric forming capabilities (diameters, wall thickness, length of pipe segments), as well as the ability of transmitting effectively loads and potential deformations of the cross-section. Adequate computation methods allowing for a comprehensive analysis of the pipeline—soil system are required to evaluate the effort state of a buried pipeline. Soil conditions and active loads should be analysed prior to selecting such methods, and an adequate computation model should be consequently applied. A rectilinear section of a buried pipeline can be analysed as a bar construction (1D model), as a pipe ring with unit length (2D model) or as a spatial construction (3D model). The available computation methods can be employed if one of the quantified calculation models is applied (Kliszczewicz 2016a). A 2D model, representing the characteristic section of pipeline, is used most often; it can be analysed with analytical methods: classical method (Kuliczkowski 2004, Madryas et al. 2002), Scandinavian method (Janson 2010) or a method given in directives ATV (ATV-DVWK-A127P 2000). The main difference between such methods is that different load distributions of a pipeline ring are adopted and that pipeline rigidity is considered with different methods. Numerical methods, used more and more often, are an alternative for analytical methods. A special computer system enabling to create a virtual pipeline—soil model has to be used in the analysis to apply such methods. Both, 1D (Mokrosz & Paszkiewicz 2013) or 2D as well as 3D (Bildik et al. 2006, Gerscovich 2008, Kliszczewicz, 2013) models are utilised in numerical analyses. Numerical methods are mainly distinct for: comprehensive analysis of a pipeline interworking with the surrounding soil, extensive variants of pipeline rigidity, ability to use soil conditions and loads; moreover, pipeline deformation and effort state can be analysed and variations in stresses and deformations in soil. A pipeline construction is modelled in numerical analyses most often in terms of elasticity, and soil—using constitutive models considering the plastic characteristics of soil. Numerical analyses are of an incremental—iterative character.

This work presents the possibilities of applying numerical analyses to examine the behaviour of buried pipelines. The analysis was undertaken with ZSOIL acad. ver. 11.15 (Truty et al. 2011) software. A flexible pipeline—soil system is the subject to analysis. The system, depending on the adopted soil conditions and loads, was considered in the flat state of deformation (2D model) as a space structure (3D model). The analyses presented relate to straight pipeline sections and to the pipeline route curve. An elastic-plastic model with Hardening Soil Small isotropic strengthening was used for describing soil behaviour (Benz 2006, Schanz 1998, Truty 2008, Kliszczewicz 2016b). The soil parameters applied in the numerical analyses performed were assumed by using active tables available in ZOIL software, enabling to determine specific HSS model parameters based on the declared subsoil characteristics. A behaviour of a PVC pipeline laid in homogeneous subgrade and layered subsoil is examined in the analyses.

2 NUMERICAL ANALYSIS 2D OF PIPELINE-SOIL SYSTEM MODEL

A 2D model of a pipeline—soil system represents a pipeline ring with unit length (flat state of deformation), separated from a long, rectilinear section of a line with geometric parameters, and with load and soil conditions unchangeable along its length. A half of the system is analysed due to system symmetry (Fig. 1). The model consists of a rectangular area of soil with dimensions of 1.8 m × 3,5 m. In this part of the area, adjoining the axis of symmetry, a trench zone 0.6 m wide and 2.5 m deep was separated. A subcrust layer (0,2 m thick) and bedding layer (0,6 m thick) were distinguished in this zone. Two layers, 0,2 m thick each, representing a road surface and road substructure, were introduced below the top edge of the model. An evenly distributed load with intensity of 30 kN/m^2 was introduced on the top edge of the model along a 0,4 m long section. A PVC pipeline with a diameter of 400 mm and wall thickness of 15,3 mm is laid 2,3 m deep.

The model is made up of 1,364 nodes and 1,232 elements, including 1,218 four-node *quad* elements in the *Continuum* (soil) zone, 14 *Beam* (pipeline) elements and 74 *interface* elements. Articulated and sliding supports enabling vertical displacements of the model nodes were introduced in the nodes located on the vertical edges of the model; the nodes on the bottom edge are unable to be displaced vertically and horizontally (immobile supports). The model grid was thickened in the trench zone, in the direct adjacency of the pipeline. The model was used in analyses examining the impact of subcrust compacting on pipeline deformation and, hence, its interaction with soil. The selected material parameters of soil and pipeline used in two variants of the analysis, differing in subcrust compaction, are compiled in Table 1. It was assumed that, following pipeline installation and provision of bedding, the trench space will be charged with native soil. The subcrust layer was modelled with an elastic-perfectly plastic Mohr-Coulomb module (Young modulus $E = 160$ MPa, internal friction angle $\phi = 30°$, cohesion $c = 5$ kPa), and the surface layer was modelled within elastic range (Young modulus $E = 1600$ MPa). An elastic model was used for the pipeline (long-term Young modulus $E = 2700$ MPa).

A pipeline construction process was simulated in the both variants of the analysis in the following stages: 1) undisturbed native soil, 2) execution of excavation, 3) execution of subcrust layers, 4) pipeline laying, 5) pipeline bedding, 6) filling the remaining excavation space, 7) provision of subcrust layer, 8) execution of surface layer, 9) loading the part of surface (a load growing linearly from the value of $q = 0$ to $q = 30$ kN/m^2, and time $t_1 = 0$ to time $t_5 = 2$). The analysis is of an incremental and iterative character, with increment $\Delta t = 0,1$.

The results of both variants of the analysis are shown in Fig. 2 as diagrams of vertical displacements of main points of the pipeline cross-section (top, side and bottom points) caused by a simulated technological process (steps 0–10) and the acting load (steps 10–12). The maximum values of vertical displacements are as follows:

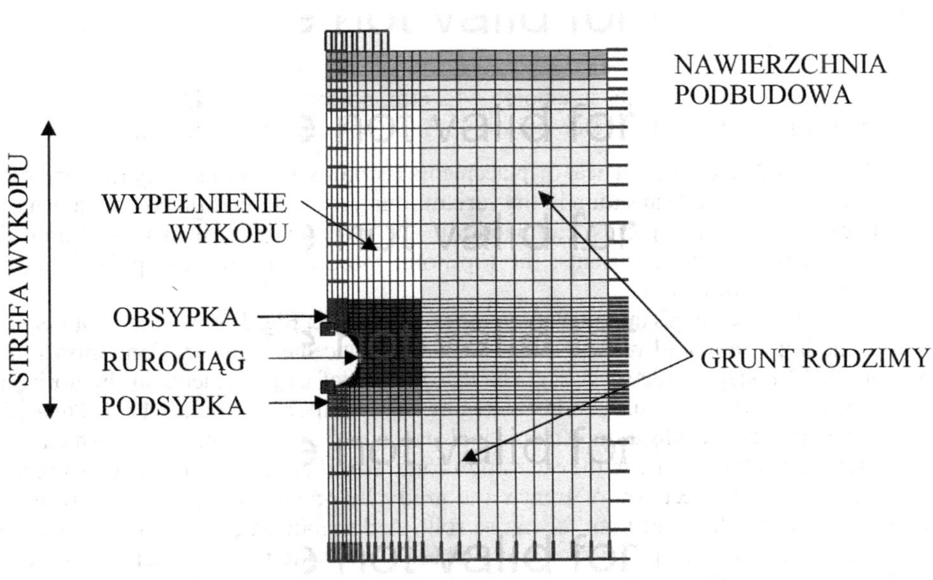

Figure 1. View of 2D numerical model of pipeline-soil system.

Table 1. Materials parameters of 2D model pipeline-soil system.

Zone	Soil type	Young modulus unload./reload. E_{ur}^{ref} [MPa]	Friction angle ϕ [°]	Dilatancy angle ψ [°]	Cohesion c [kPa]
Native soil	Clay (very stiff, low plasticity)	21,0	28,5	9,5	20
Subcrust (Variant I)	Sand (medium density, well-graded)	120,0	34,5	4,88	10*
Subcrust (Variant II)	Sand (very loose, well-graded)	55,0	25,8	0	5*
Cover	Sand (medium density, well-graded)	120,0	34,5	4,88	10

*Values taken in view of the computing.

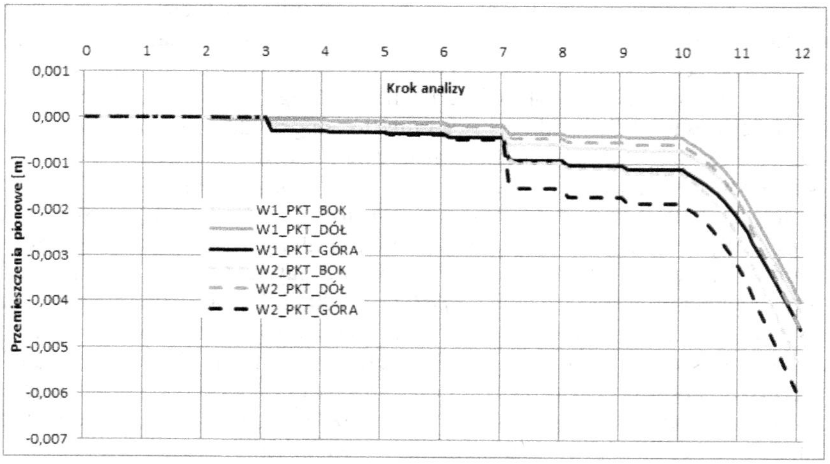

Figure 2. Vertical displacement diagram in main point of pipelines in sequential analysis steps (variants 1 i 2).

- variant 1: top point – 4,6 mm, side point – 4,2 mm, bottom point – 4,0 mm,
- variant 2: top point – 6,1 mm, side point – 5,4 mm, bottom point – 4,6 mm.

The determined values of vertical displacements (variants 1 and 2) signify the pipe cross-section ovalisation. Such ovalisation in the variants analysed is differentiated and depends on support conditions (sand subcrust compaction). The same tendency in cross-section deformation is observed in both variants: the top point more than a bottom point is moving towards the vertical direction.

In order to depict changes taking place in the ground space, Fig. 3 shows a map of displacements in the horizontal and vertical direction. The displacement distributions shown relate to variant 2 (12th step of the analysis). The distribution of displacements in the horizontal direction indicates their cumulation in the region of the pipe's side point. This is related to the occurring ring pipe ovalisation. An area is also seen in the native soil zone, in which native soil presses against the ground in the excavation zone. This is a result of conducting a pipeline construction process (execution of trench and gradually providing the bedding layers with material parameters different than for native soil). A distribution of vertical displacements indicates the deflection of the surface and subbase layers (due to a load acting on the surface), propagating to the ground layers situated lower and reaches the direct surrounding of the pipe.

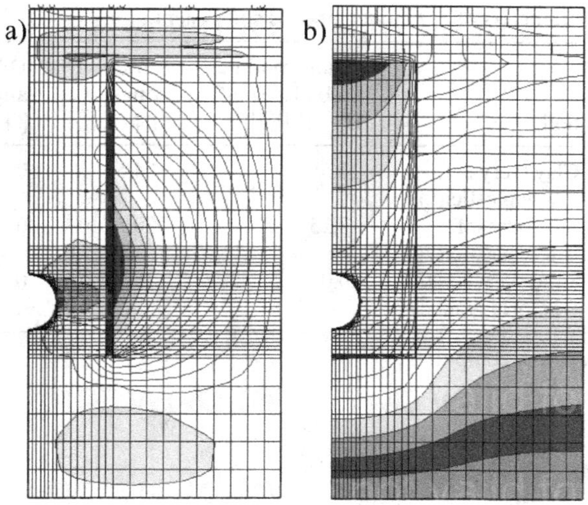

Figure 3. Maps of displacements (variant 2, 12th step of analysis: 12): a) horizontal displacements, b) vertical displacements.

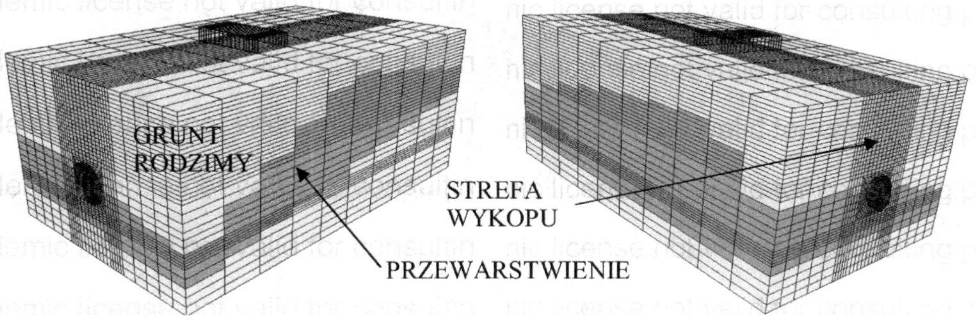

Figure 4. View of 3D numerical model of the pipeline straight segment-soil system.

3 NUMERICAL ANALYSIS 3D OF PIPELINE-SOIL SYSTEM MODEL

3.1 *3D numerical model of the pipeline straight segment-soil system*

Due to the existence of non-uniform subsoil stratification in a 3D model of a straight pipe segment—layered subgrade system, it is not possible to reduce the aspect to the flat state of strain. The model is shown in two views in Fig. 4. Soil mass dimensions: 4,0 m × 3,0 m × 10,0 m. Trench dimensions: 1,1 m × 2,7 m × 10,0 m. A PVC pipe with a diameter of 500 mm and wall thickness of s = 15,3 mm is laid in a trench on a 0,2 m thick soil layer. Three material zones were introduced in the ground area in the model, i.e. native soil, soil with a top layer with variable thickness and soil filling the excavation. Table 2 lists material parameters of the zones (HS Small model). The material parameters of the pipe are identical as in a 2D model (item 2).

The model is made up of 13 575 nodes and 11 840 elements, including 11 560 eight-node elements in the *Continuum* (soil) zone and 280 *Shell* (pipeline) elements. Articulated and sliding supports were introduced on vertical planes of the model, and immobile supports on the bottom plane.

The analysis programme provides for a uniform load of part of the top surface of the model. The load value is growing linearly from the value of $q = 0$ in time $t_1 = 0$ to the value of $q = 50$ kN/m^2 in time $t_2 = 2$. The analysis is of an incremental and iterative character, with increment $\Delta t = 0,1$.

Table 2. Materials parameters of 3D model pipeline-soil system.

Zone	Soil type	Young modulus unload./reload. E_{ur}^{ref} [MPa]	Friction angle ϕ [°]	Dilatancy angle ψ [°]	Cohesion c [kPa]
Native soil	Clay (very soft, low plasticity)	17,5	24,5	0	20
Weak interbeddings	Silt (very soft, low plasticity)	3,25	30,5	0	20*
Soil filling in the excavation	Sand (very loose, poorly graded)	20,0	25,8	0	5*

*Values taken in view of the computing.

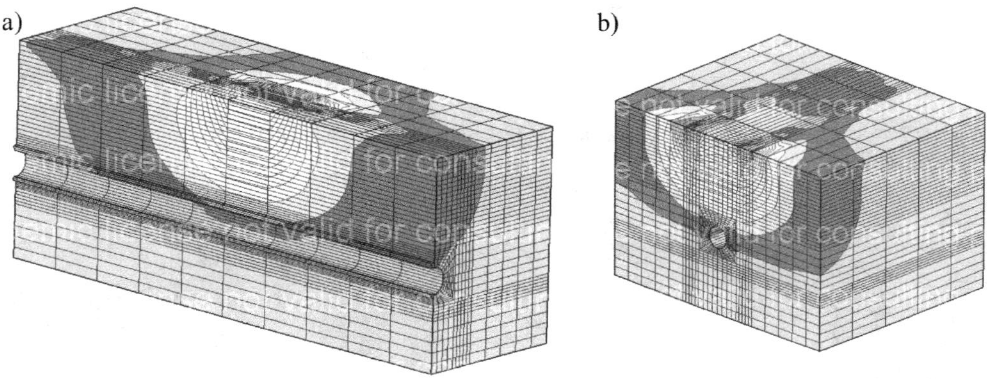

Figure 5. Map of the vertical displacement of soil: a) longitudinal section of model, b) cross section of model.

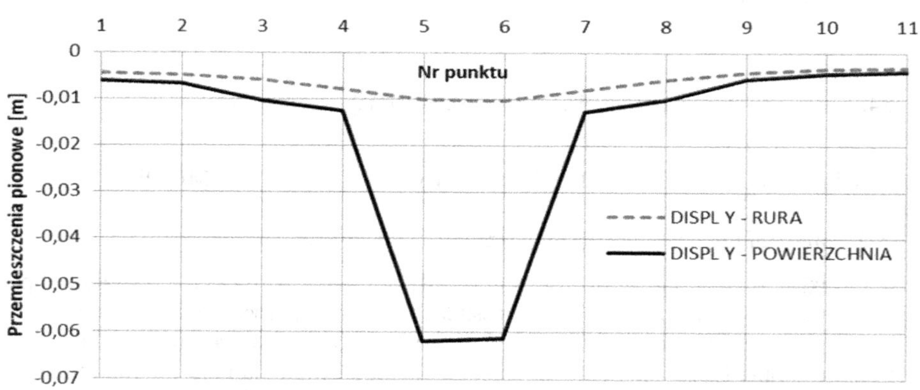

Figure 6. Vertical displacement map of the upper surface of model and pipe crown line (load value $q = 50$ kN/m^2).

A deformation of the model surface and of a certain area of the entire subsoil mass is the effect of acting loads. The zone of vertical displacements, due to the existing interbeddings, is very irregular (Fig. 5) and also includes, along a certain part, a direct surrounding of the pipeline.

A reaction of the soil and pipeline to the acting loads is very different. This is presented in a diagram (Fig. 6) showing land surface deformation along the line lying over the top generating line of the pipeline and the deformation of this generating line. The values shown in the diagram pertain to time $t_2 = 2$ (load value of $q = 50$ kN/m^2). The maximum value of vertical displacements of the top surface of the model in the load activity area is 0,062 m.

The maximum value of vertical displacements of the top generating line of the pipeline is much lower and is 0,010 m.

The deformation of the pipeline coating is of spatial character. This is shown in Fig. 7, with a map of vertical pipe coating displacements and one of its deformed cross-sections.

3.2 *3D numerical model of the curve pipeline-soil system*

A model of a curve section pipe-soil system presents a rectangular body of subsoil with dimensions of 9,0 m × 3,25 m × 9,0 m and of PVC pipeline with a diameter of 500 mm, wall thickness of 14,6 mm and curve angle of 90° (Fig. 8). The pipeline is laid at a depth of 2,25 m.

The model is made up of 13 575 nodes and 5 628 elements, including 5 148 eight-node elements in the *Continuum* (soil) zone and 480 *Shell* (pipeline) elements. The dimensions and shape of the elements are different, grid densification was used in the pipeline placement zone. There are 6,212 nodes in the model. Boundary conditions of the model (articulated and sliding supports and immobile supports) are arranged along the nodes on the vertical and bottom horizontal surface identically as in the model described in section 3.1. An evenly distributed load was introduced on the part of the top model surface, growing linearly from the value of $q = 0$ in time $t_1 = 0$ to the value of $q = 22.5$ kN/m^2 in time $t_2 = 2$. The analysis is of an incremental and iterative character, with increment $\Delta t = 0.1$. The following material parameters of homogeneous soil were adopted (clay, low rigidity, high plasticity): Young modulus unload./reload. $E_{ur}^{ref} = 1175$ kPa, internal friction angle $\phi = 20.3°$, dilatancy angle $\psi = 0°$, cohesion c = 30 kPa. Deformations of the model surface, shown in Fig. 9, occur as a result of the load activity.

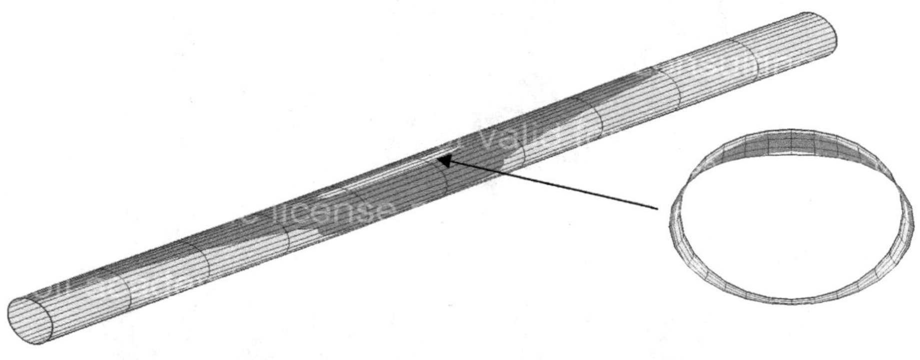

Figure 7. Vertical displacement map of the pipe model surface (load value $q = 50$ kN/m^2).

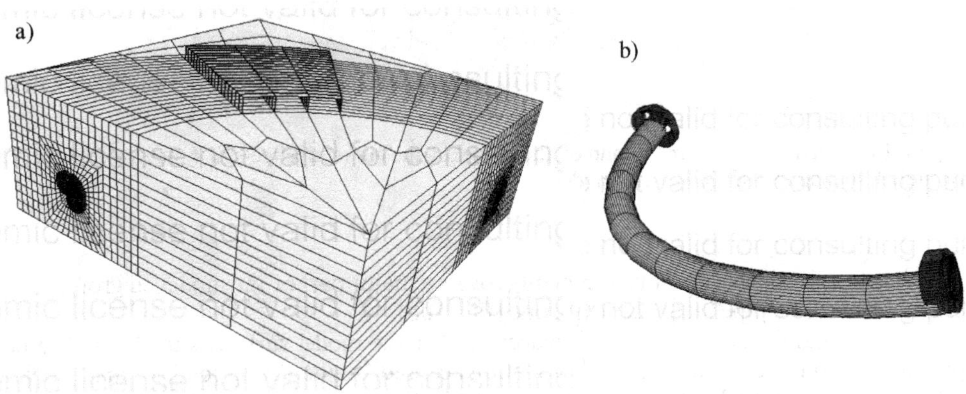

Figure 8. 3D numerical model of the curve pipeline-soil system: a) general view of model, b) view of curve pipeline model.

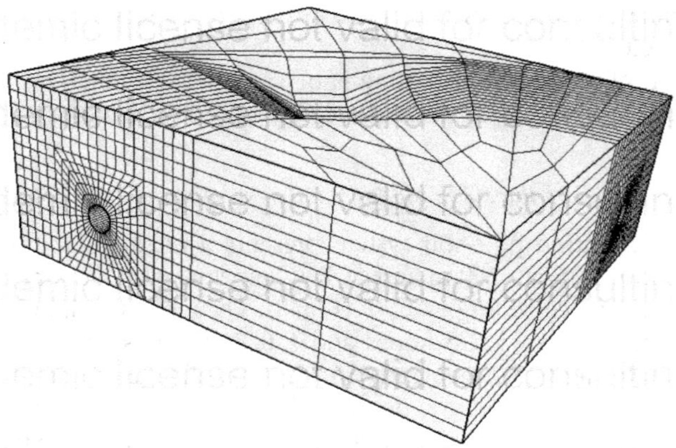

Figure 9. Deformed 3D model of the curve pipeline-soil system.

a)

b)

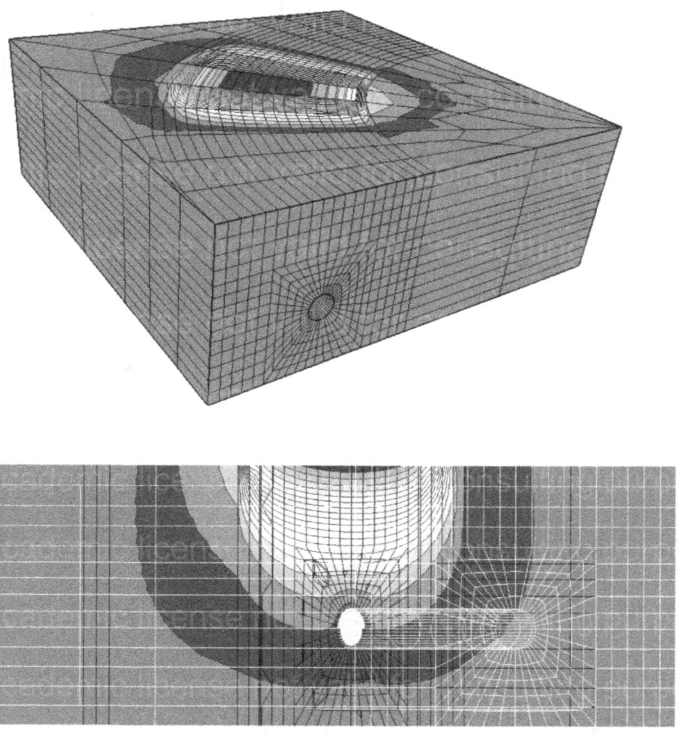

Figure 10. Vertical displacement map of the model of the curve pipeline-soil system: a) general view of model, b) diagonal section of the model.

The zone of model nodes' displacements covers a major part of the modelled ground space (Fig. 10a) and the pipeline placement zone (Fig. 10b).

The maximum value of vertical displacements of the model surface is 0,172 m. The bent pipeline section is also subject to deformations caused by the acting load. A map of vertical components of such displacements is shown in Fig. 11. Displacements with varied intensity exist in the central part of the pipeline. The maximum value of vertical displacements is 0,0138 m.

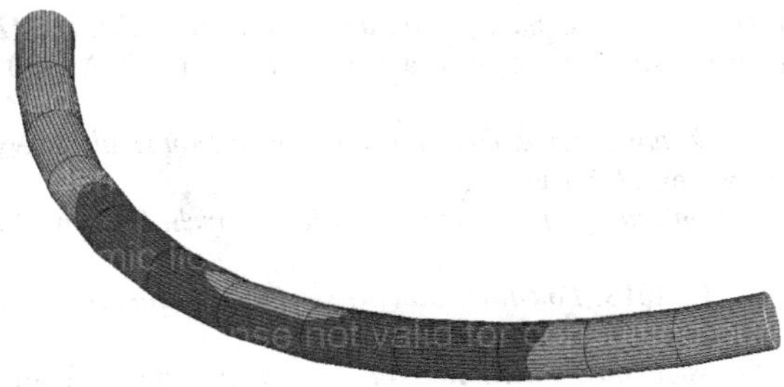

Figure 11. Vertical displacement map of the curve pipeline model.

4 SUMMARY

Numerical methods are an effective tool for investigating a behaviour of buried pipelines in a load activity zone. Pipelines' behaviour, hence their deformation and strain, are different for inflexible and flexible pipelines. In case of flexible pipelines, a character of pipeline behaviour is dictated by a relationship between pipeline stiffness and ground. This claim necessitates to conduct comprehensive analyses of the pipeline—ground system, not analyses of the loaded pipeline only. Numerical analyses, the examples of which are shown in this work, relate to a flexible pipeline laid in various soil and load conditions. Variants were also examined of a straight and bent pipeline section (pipeline route curve), interworking with the surrounding (homogeneous or layered) soil. The created pipeline—soil system models were adapted to a possibility of considering the problem in a flat state of deformation (2D model) or in a space state (3D models). An important characteristic of the numerical analyses performed is their incremental and iterative nature and the application of an advanced constitutive soil model (elastic-plastic model with Hardening Soil Small isotropic strengthening).

A programme of the numerical analyses performed comprises the examinations of influence of such factors as: a technological process (pipeline construction in a trench with vertical walls), subsoil layers (existence of soil layers with varied thickness), pipeline shape (straight and curved sections in plan) and the type and material parameters of subsoil. Although the results obtained are up-to-date for the introduced material parameters of soil and pipeline only, they reveal that a pipe model is sensitive to the changing external conditions (growing loads, subsoil character).

REFERENCES

ATV-DVWK - A127P, 2000. *Obliczenia statyczno-wytrzymałościowe kanałów i przewodów kanalizacyjnych*. Wyd. Seidel-Przywecki.

Benz, T., 2006. *Small-strain stiffness of soil and its numerical consequences*. Phd, University Stuttgart.

Bildik, S. et al., 2012. *Parametric studies of buried pipe using finite element analysis.* The 3rd International Conference on New Developments in Soil Mechanics Engineering. Near East University, Nicosia, North Cyprus.

Gerscovich, D.M.S. et al., 2008. *Numerical simulation of the mechanical behaviour of buried pipes in trench.* The 12th International Conference of International Association for Computer Methods and Advances in Geomechanics (IACMAG), India, Goa.

Janson, L.E., 2010. *Rury z tworzyw sztucznych do zaopatrzenia w wodę i odprowadzania ścieków*. Borealis. Wyd. IV. Polskie Towarzystwo Producentów Rur i Kształtek z Tworzyw Sztucznych, Toruń.

Kliszczewicz, B., 2013. *Numerical 3D analysis of buried flexible pipeline*. European Scientific Journal, vol. 9, No. 36.

Kliszczewicz, B., 2014. *Interakcja podziemnych rurociągów o różnych sztywnościach z gruntem*. Monografia nr 534, Wyd. Politechniki Śląskiej, Gliwice.

Kliszczewicz, B., 2016a. *Modelowanie numeryczne współczesnym narzędziem analizy statyczno-wytrzymałościowej podziemnych rurociągów*. Nowe technologie w sieciach i instalacjach wodociągowych i kanalizacyjnych. Instytut Inżynierii Wody i Ścieków. Politechnika Śląska, Gliwice.

Kliszczewicz, B., 2016b. *Konstytutywne modele gruntu stosowane w analizach MES interakcji rurociągów z gruntem.* Nowoczesne miasta. Infrastruktura i środowisko. INFRAEKO 2016. Oficyna Wydaw. Politechniki Rzeszowskiej.

Kuliczkowski, A., 2004. *Rury kanalizacyjne. Tom II. Projektowanie konstrukcji.* Wyd. Pol. Świętokrzyskiej. Mon., Studia, Rozprawy – nr 42. Kielce.

Madryas, C. et al., 2002. *Konstrukcje przewodów kanalizacyjnych.* Oficyna Wyd. Pol. Wrocławskiej. Wrocław.

Mokrosz, R., Paszkiewicz, T., 2013: *Podstawy statyki sieci ciepłowniczych z rur preizolowanych.* Wyd. Politechniki Śląskiej, Gliwice.

Sandford, T.C., 2000. *Soil-structure interaction of buried structures.* Transportation in the New Millenium, University of Maine.

Schanz, T., 1998. *Zur Modellierung des mechanischen Verhaltens von Reinbungsmaterialien.* Mat. Inst. für Geotechnik 44. Universitat Stuttgart.

Truty A. et al., 2011. *ZSoil.PC 2011. User Manual.* © Zace Services Ltd, Switzerland.

Truty A., 2008. *Sztywność gruntów w zakresie małych odkształceń. Aspekty modelowania numerycznego.* Czasopismo techniczne, z. 3.

Underground Infrastructure of Urban Areas 4 – Madryas et al. (Eds)
 ISBN 978-1-138-55953-0

Subway line optimization through risk management

D. Kolic
Geodata Tunel doo, Zagreb, Croatia

ABSTRACT: Subway lines have major parts locates underground and they are complex projects that include different structural parts, need a lot of financial resources for development and require a long time for construction. Such projects are therefore very much sensitive to many unpredictable moments that may cause additional costs. Risk management analyses and measures used for the control of the project development do not consider safety of labour only but investigate possibly discrepancies from the planned project schedule and respective planned costs.

1 INTRODUCTION

Subway lines carrying passenger traffic consists of a set of large structures as tunnels and stations but also bridges, ramps and their combination (Kolic 1997). During the project development it is important to define the optimal structural shape and combination of structures that will enable regular traffic conditions and minimize required construction costs.

During last decade different types of risk analyses have been applied and methods developed that enable better understanding of procedures that could cause development of additional costs above planned project budgets.

The method FAUST is one of methods that have been developed, applied and checked on different mega transit projects like subway lines that are complex in development and possible source of causing additional costs. It has been developed considering and optimizing structural capacity of structures within the transit lines. It analyses different types of structures along the transit line and evaluates the overall final construction costs of each combination of structures.

Optimization of structures is based on variation of geometrical and structural characteristics of structures and enables consideration of different construction technologies as well.

The evaluation of different structures within one transit line is based on comparison of their construction prices using criteria of economically most feasible solution within prescribed or approved technologies.

The examples of application are showing good results that are corresponding with results of constructed transit lines and experiences on cost-overruns that were known at the end of the construction.

2 RISK MANAGEMENT

2.1 *Purpose of risk management*

At the beginning of application of methods of risk management in construction industry most of application were directed and concentrated on safety of labour during the construction (Kolic 2002). During the last decade it became clear that methods of risk analysis could bring far more results and cover wider spectre of problems that may appear during complex construction processes. This way methods of risk analysis started to be used for the evaluation of: feasibility of construction process in general (Kolic 2001), for the check of construction procedures (AGS 2005), as an optimization tool on complex projects (Kolic 2009), as a cost control tool evaluating pre-calculated construction costs (AGS 2005) and as a numerical tool to discover paths that could lead toward cost over-runs (Tsamboulas 2000).

2.2 *Methods of risk analysis*

Risk management uses different risk analyses methods that have own restrictions and possible areas of application. Very often we can find that some of methods is applied even not appropriate for the sector or problem where it could be used. Othersides the practice develops some procedures that are applied in the name of risk analysis but they are still just simple procedures without boundaries of application and prescribed conditions of use and sectors of the use where they can be applied with usable and applicable results. Approved risk analysis methods that are known and used in the sector of civil engineering are grouped in 2 main groups for analyses as qualitative and quantitative methods (FTA 2004). In the group of qualitative methods are:

- PPA – Preliminary Risk Analysis or Potential Problem Analysis (also called Opportunity Analysis)
- HAZOP (Hazard and Operability studies)
- FMEA (Failure Mode and Effect Analysis)/FMECA (Failure Mode and Effects Criticality Analysis)
- Tree based analysis
- Fault tree analysis
- Event tree analysis
- Cause Consequence Analysis
- Management Oversight Risk tree
- Safety Management Organization Review Technique and as well different methodologies for analysis of dynamic system like: GO method,
- Diagraph/Fault Graph, Markov modelling
- Dynamic Event Logic Analytical Methodology or Dynamic Event Tree Analysis Method.

Quantitative Methods are mostly using principles of action of some qualitative methods like described in tree-analyses or are following principles of methods with probabilistic approach. Application of different risk analysis methods is very often followed by severe mistakes that are giving wrong final results. Very often identification of hazards and description of risks in risk register are the source of severe mistakes:

1. Copying of last risk register for a new project (it is fast and useful way to start new analysis but new project can be very much different from the last one)
2. Defining risks for RA that are already included in project design or BoQ (very often happens that the one who is performing risk analysis is not informed about solutions already implemented in the design or foreseen in the BoQ and is repeating coverage by measures and costs in RA)
3. Using wrong method for RA (sometimes wrong methods are used for areas where they are not giving proper results or a "method" is self-developed and is no method but just a helping procedure or a technique).

3 METHOD

The risk analysis method used recently for projects of subway lines is using methodology presented in the practice after name of optimization module **FAUST** (Feasibility And Utilization of Underground STructures). The method is based on the application of qualitative and quantitative risk analysis methods and can be used in different phases of the project development. The detailed description of the method is given in some published works in the past (Kolic, 2009) and the algorithm with major milestones is presented in the flow chart below (Fig. 1).

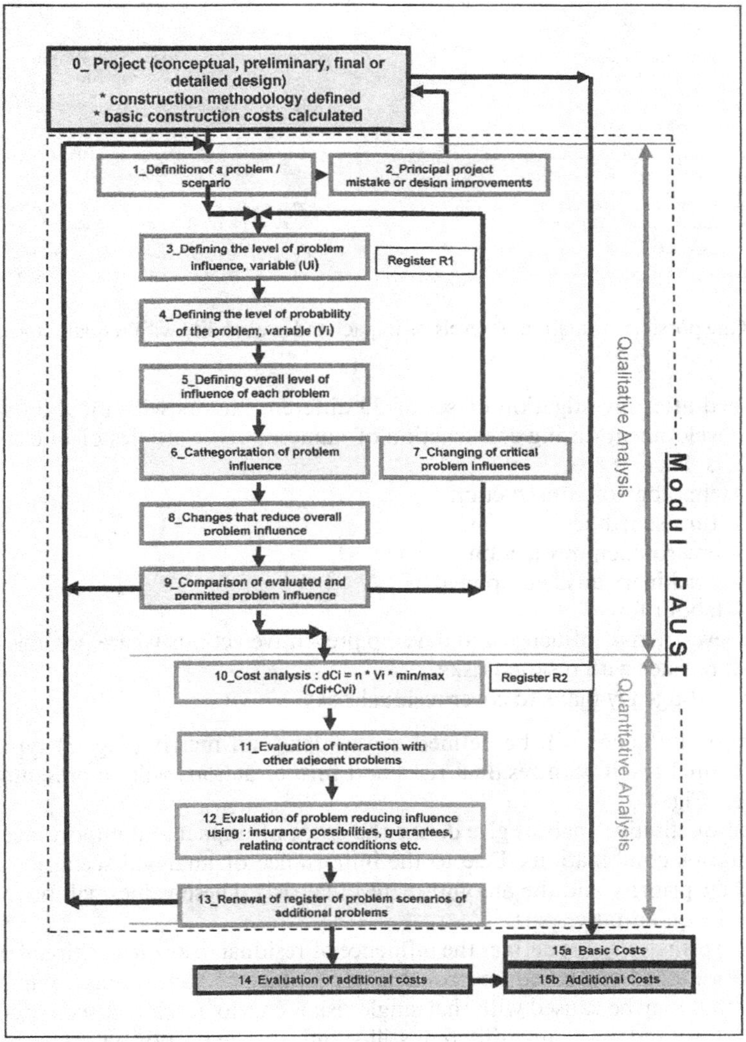

Figure 1. Optimization module FAUST in detailed flow chart.

3.1 *The way of application*

The method uses qualitative and quantitative methods to analyse the problem in analysis. Both parts are based on evaluation of risk registers: R1 – for qualitative analysis and R2 – for quantitative analysis (see flow chart, Fig. 1).

Qualitative analysis part uses method of "Preliminary" or "Potential-Problem-Analysis" (PPA). The severity of risks is evaluated over their "HS-hazard severity" and "Pi index—probability of appearance". Procedure of the analysis can be described over below mentioned steps:

1. Define the key requirements of risk analysis
2. List and explore all potential problems and make a risk register R1 (list of scenarios = risk register)
3. List possible causes for each potential problem
4. Assessment and evaluation of problems (over evaluation of "hazard severity" and "probability index")* evaluation of UI-index and Vi–index is performed over impact of for the project most important factors on the construction process (Fig. 2). These factors

Level of impact „Ui"	grade	Level of probability „Vi"	grade
Small impact on construction works, no repairs, no time delays	1	Very rare case to appear on the length of the structure n/L=0.01-0.1 / L (improbable)	1
Small impact on construction works, smaller repairs, time delays up to 1 week	2	Rare case to appear on the length of thestructure n/L=0.1-1 / L (remote)	2
Higher impact on construction works, causing serious repairs, time delays up to 1 month	3	Case to appear once on the length of the structure n/ L = 1 / L (infrequent)	3
Higher impact on construction works, causing serious repairs, time delays up to 12 months	4	Case to appear several times on the length of the structure, n/L=1-3 puta / L (probable)	4
Fatal impact on construction works, causing break of works/contract	5	Case to appear several times on the length of the structure, n/L=3-10 or more/L (real/very probable)	5

Figure 2. One possible evaluation of levels of impact and probability within qualitative analysis part.

are defined after investigation of set of 30 different factors with most influence on the project development during construction of numerous case studies of international similar projects. Factors are:

"TI" – tehnology of construction
"VG" – time schedule
"ZO" – environment protection
"SO" – machinery ansd equipment
"RS" – labour force

5. Evaluate level of risk influence and develop preventive actions where possible
6. Make risk register with residual risks
7. Develop contingency plans to cover residual risks

The level of influence will be defined in the risk level matrix (Fig. 3) typical for PPA method and final result with residual risks and further actions will be presented in the R1 Risk register (Fig. 4).

Results of qualitative analysis give define the grade of impact and importance of separate hazard scenarios combinations. Due to the importance of analysed scenarios risk register will show their priority and the amount of residual risk. These values will be further evaluated within the quantitative part.

Quantitative analysis part defines the influence of residual risks on additional project costs. Cost estimation has been based on probable appearance of some scenario and amount of direct costs that may be caused with that single risk scenario. Each risk scenario will happen within some time and may cause different still-stands on the part or on entire project. These costs have been evaluated as time-dependent costs also. Beside the probability of appearance of some scenario or a set of scenarios, time-dependent costs are defined in some range, based on their maximal possible and minimal possible appearance. Additional costs are evaluated for each negative risk scenario by the equation:

$$dC_i = n * V_i * min/max\,(Cd_i + Cv_i) \tag{1}$$

... whereas separate values are equal to:

DC_i	additional cost for each negative scenario
n	number or repeating one scenario along the project length/duration
V_i	probability that scenario will take place
Min/max	min and max value of the calculated amount in brackets
Cd_i	part of direct cost of one scenario
Cv_i	part of time-dependent costs of one scenario
Ac	total additional costs for all negative risk scenarios

The summation of all influences coming from all negative scenarios will give the total amount of additional construction costs:

$$Ac = Summa\ dC_i \tag{2}$$

HS vs. PI	Pi = 1	Pi = 2	Pi = 3	Pi = 4	Pi = 5
0 –2.5	Neglig.	Neglig.	Neglig.	Neglig.	Accept.
2.5 – 5.0	Neglig.	Neglig.	Accept.	Accept.	Accept.
5.0 – 7.5	Neglig.	Accept.	Accept.	High	High
7.5 - 10.0	Neglig.	Accept.	Accept.	High	High
10.0 –12.5	Neglig.	Accept.	Accept.	High	Severe
12.5– 15.0	Accept.	Accept.	High	High	Severe
15.0-17.5	Accept.	Accept.	High	Severe	Severe
17.5-20.0	Accept.	High	High	Severe	Severe
20.0-22.5	Accept.	High	Severe	Severe	Severe
22.5-25.0	Accept.	High	Severe	Severe	Severe

Figure 3. Risk level matrix of PPA method.

Western part
L = 3014 m

Qualitative Analysis
Risk register 1

Oresund crossing
total L = 7845 m

nr.	HAZARD sector	HAZARD description	Negative scenario difficulties that may appear	Impacts "Ui" TI	VG	ZO	SO	RS	Summ impact "UI"	Prob appear "VI"	sever.	Possible risk mitigation and actions	Action to mitigate risk	
A	Location conditions	a.4 sea depth on the location of work of floating crane "Svanen"	deepening and widening of a traffic profile for vessels during construction	3	3	3	2	2	13	3	accep.	measurement of seabed section, deepening of the seabed at location where floating crane works	accep.	
		a.9 earthquake	earthquake influence during construction	1	1	2	1	1	6	2	accep.	seizmoligacl measuremtns of last 30 years showing small impact, area of low impact	accep.	
		a.10 wind	wind influence during construction	3	3	1	3	3	13	4	sev	in design wind plans investigated with intensities of (> 61 m/sec) and directions of influence	corr.	
		a.11 waves and sea streaming	wave and streaming impact during construction	2	2	2	2	2	10	3	accep.	in design defined height of wave defined with 2.5 m for analysis	accep.	
		a.12 ice impact	ice impact on construction	2	2	2	2	2	10	3	accep.	in design defined ice thickness with 0.8 m for analysis	accep.	
A2	Geological conditions	a.2 waeak soil layers for foundation foreseen on limestone	non bearable layers of soil pn the location of foundations of columns and pylons	2	2	1	2	2	9	3	accep.	cleaning sea bed down t othe level of the wearing layer of limestone	corr.	*
A3	Geotehncial conditions	a.4 sea bed preapration on foundation location	difficulties at foundation locations for column caissons	2	2	2	1	1	8	4	sev	provide parallel machinery and working power groups for excavation wit hpossibility to use underwater blasting	corr.	
B	Traffic conditions	b.6 connection on existing road network	roads do no t exists but are part of another contract	0	0	0	0	0	0	0		included in another part of the structer out of this part of crossing, on thepart "coast-to-coast"		
		b.7 traffic beyoinfd the bridge on traffic channels	providing traffing channels beyond the bridge deepening of a sea bed if required	2	2	2	1	1	8	3	accep.	design to foresee the batrimetrical measurement and to adopt the section for traffic of vessels	dop.	
		b.8 alignment change	alignment layout : in sens e of the approach to the coast, monotony in traffic	2	2	2	1	1	8	3	accep.	design to foresee the alignement layout with "S" or "C" curve	dop	
C	Structural requirements analysis and design	c.2 earthquake zones	project details against earthquake	1	1	1	1	1	5	2	accep.	check structural details against possible earthquake	corr.	*
		c.3 wind zones	detalils against wind impact	2	3	1	3	2	11	3	accep.	to check the impact of the wind on the structure and improve the design on this impact	dop.	
		c.4 wave stroke	deatils against waves	2	3	1	2	1	9	3	accep.	to check the impact of waves and streaming by analysis and design the structure on this impact	dop.	
		c.5 vessel stroke to columns stupove	structural elements and protection concept against collision with vessels	2	2	2	2	1	9	3	accep.	to analyze the impact of a vessel stroke and design structure on this impact	dop.	
		c.8 erosion protection of ("scour") columnsoko stupova	erosion protection of columns	1	2	2	1	1	7	3	accep.	to define the construction of protective dams for columns and pylons and erosin protection	dop.	
		c.9 protection islands around columns (2x3) and pylons(2 as vessel stroke protection	protection island in fornt of columns and pylons against vessel collision Imact	1	2	2	1	1	7	3	accep.	to define the construction of protective islands against vessel stroke to columns and pylons	dop.	

Figure 4. Risk register R1 with residual risks and further actions planned.

3.2 *Overall project construction costs including additional costs*

Direct and time-dependent costs together show the overall amount and the range of construction costs for each scenario that remained unsolved after analysis together with some probability of its appearance. This way they can be placed on the list of priorities and pertinent back-up solutions and contingency plans may be developed in order to limit their influence on entire project.

Western part
L = 3014 m

Qualitative Analysis
Risk register 2

Oresund crossing
total L = 7845 m

nr	HAZARD sector		Hazard description	Nr of appear on L	Direct costs min (-) / max (+)	Time delay zastoj		RS+SO Delay 1 wd	Time-dependent costs		Probability of appear.	Total additional costs	
				(-)	€	from-(wd).	.-to (wd)	(€ / wd)	from-(€)	.-to(€)		from-(€)	.-to(€)
1	2	3	4	5	8	7	8	9	10=9*7	11=9*8	12	13=5*12*(8+10)	14=5*12*(8+11)
A	Location conditions	a.4	sea depth on the location of work of floating crane "Svanen"	1	50.000,00	30,00	90.00	3.500.00	105.000	315.000	1,00	155.000,00	365.000,00
		a.9	earthquake	0	0,00	0,00	0,00	3.500,00	0	0	0,10	0,00	0,00
		a.10	wind	1	50.000,00	30,00	90,00	3.500,00	105.000	315.000	1,00	155.000,00	365.000,00
		a.11	waves and sea streaming	1	50.000,00	30,00	90,00	3.500,00	105.000	315.000	1,00	155.000,00	365.000,00
		a.12	ice impact	1	50.000,00	30,00	90,00	3.500,00	105.000	315.000	1,00	155.000,00	365.000,00
A2	Geological conditions	a.2.2	waesk soil layers for foundation foreseen on limestone	1	50.000,00	30,00	90,00	3.500,00	105.000	315.000	1,00	155.000,00	365.000,00
A3	Geotehncial conditions	a.3.4	sea bed preapration on foundation location	3	50.000,00	30,00	90,00	3.500,00	105.000	315.000	1,00	465.000,00	1.095.000,00
B	Traffic conditions	b.6	connection on existing road network	0	0,00	0,00	0,00	0.00	0	0	0,00	0,00	0,00
		b.7	traffic beyoinfd the bridge on traffic channels	1	25.000,00	7,00	30,00	1.125,00	7.875	33.750	1,00	32.875,00	58.750,00
		b.8	alignment change	1	25.000,00	7,00	30,00	1.125,00	7.875	33.750	1,00	32.875,00	58.750,00
C	Structural requirements	c.2	earthquake zones	0	0,00	0,00	0,00	0,00	0	0	0,00	0,00	0,00
	analysis and design	c.3	wind zones	1	25.000,00	7,00	30,00	1.125,00	7.875	33.750	1,00	32.875,00	58.750,00
	Probability improbable	c.4	wave stroke	1	25.000,00	7,00	30,00	1.125,00	7.875	33.750	1,00	32.875,00	58.750,00
	0.01-0.1 remote 0.1-1	c.5	vessel stroke to columns stupove	1	25.000,00	7,00	30,00	1.125,00	7.875	33.750	1,00	32.875,00	58.750,00
	infrequent 1.00 probable	c.8	erosion protection of ("scour") columnsoko stupova	1	25.000,00	7,00	30,00	1.125,00	7.875	33.750	1,00	32.875,00	58.750,00
	od 1-3 real 3-10, >10	c.9	protection islands around columns (2x3) and pylons(2) as vessel stroke protection	1	25.000,00	7,00	30,00	1.125,00	7.875	33.750	1,00	32.875,00	58.750,00

Figure 5. Risk register R2 with final range of costs for evaluated risks.

4 CASE STUDY FOR FEASIBILITY, COST CONTROL AND OPTIMISATION

Application of risk management is now mostly in use as a tool for checking of feasibility of projects, as a cost control device and for the project optimization and rationalization. In all cases on the projects presented in the article the analysis consists of qualitative and quantitative part. Qualitative part defines risk registers and possible negative scenarios that may appear. Qualitative evaluation is performed using a PPA method with the evaluation of severity of risks and their probability of appearance and with defining measures for the risk mitigation. Remaining part of risks is further evaluated in the quantitative part of the analysis with the definition of the fixed and time dependent part of the costs giving finally overall range of costs that may appear above existing BoQs for basic costs. Example of Metro 4 Line in Budapest will show one application.

4.1 *Metro 4 line Budapest: Undercrossing Danube river*

The section has a total length of 7,3 km and consists of 10 Metro Stations and running tunnels. The station length is based on the platform length of 80 m; the length of running tunnels varies between 300 m and 1.400 m. The excavation diameter of tunnels is 6 m.

The project of the Metro 4 Line in Budapest had one section that had to under-cross the river Danube and connect to underground stations on both Danube banks, station Gellert on Buda side and the station Fövam on the Pest side (Kolic 2001). Figure 6 shows the 3D simulation of the Metro 4 Line undercrossing the Danube River and underground stations Gellert (Buda side, down) and Fövam (Pest side, up).

The solution with undercrossing Danube River was chosen as it does not impact the views to the bridge from XIX century that is under heritage protection and located near to the location of the crossing. Under the river bed there were several mineral and hot water wells that had to be remained protected and in full operation as the water were regularly used for the well-known hotel and spa centre on the Buda side.

Figure 6. 3D simulation of the Metro 4 Line undercrossing the Danube river and underground stations Gellert (Buda side, down) and Fövam (Pest side, up).

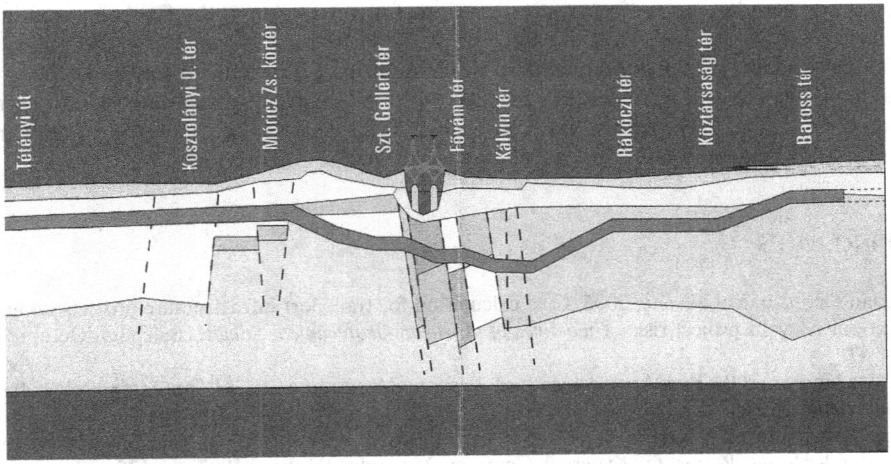

Figure 7. Longitudinal geological profile of the Metro 4 line with faults in the area of under-crossing Danube river.

The risk analysis had to investigate technological obstacles and dangers when underpassing the river Danube with the TBM and to prove the feasibility of the mechanized solution against water ingresses and pertinent accidents that may happen. Results were giving input toward definition of technology procedures that are to keep the tunnel drive safe. Figure 7 shows the longitudinal geological profile of the Metro 4 line with faults in the area of under-crossing the Danube River.

Results considering construction technology and technical difficulties with pertinent solutions for overcoming them have input into description of the TBM machine later proposed and ordered for the project and to other safety measures during construction that have been successfully applied and have secured safe undercrossing with limited influence on wells, river bed, settlements and adjacent station boxes on both river banks. Further, outcome results about technology have been evaluated in order to define possible costs deviations and causes for additional costs that may appear during construction and that have not been considered in a regular BoQs. Cost analysis have therefore been additional second analysis considering risk with respect to influences of tunnel drives and construction on cost control and defining volumes of financial reserves for contingency measures secured for the project.

5 CONCLUSIONS

Herewith presented capacity of the risk management using risk analysis methods and modules like "FAUST" (Kolic 2009) shows the ability to predict the total construction project costs of entire subway lines or some critical parts of lines including different types of structures and different technologies. The method used in "FAUST" is based on the evaluation of the negative risk scenarios based on the character of the structural solution and on the information about the conditions on the location of the crossing.

Negative risk scenarios have been developed for the specific structural options but are based on the experience of similar conditions or limitations on other known and already constructed similar structures. The quality of estimation and prediction is based on the range and quality of available existing and proved project information.

The analysis can seriously change relations among different structures that constitute the subway line and could be a decisive factor in the definition of the most feasible shape and type of structures included defining better part of unknown, unpredicted or unexpected projects costs and make project cost estimations far more near to the final required budget size level.

The method has shown good result on the estimation of different subway and other transit lines showing that the optimization module "FAUST" covers estimation for the complex structures included in the transit lines (Kolic 2016) and shows to be usable for different structures and different methods of technology applied for the construction.

REFERENCES

Austrian Geotechnical Society, 2005. Cost calculation for transport infrastructure projects taking into account relevant project risks. *Guidelines of Austrian Geotechnical Society*, Salzburg, October 2005, pp. 47.

FTA Washington, 2004. Risk Analysis Methodologies and Procedures. *FTA White paper on Risk Analysis*, June 2004: pp. 128.

Kolic, D., Paterson, J. & Knight, J. 1997. Subway Singapore, Contract C710: Risk Analysis Report for the Tender Stage. Report for Kvaerner-Gammon-Mayreder JV, April 1997, pp. 125.

Kolic, D. & Gulyas, L. 2001. Risk Analysis for the Metro 4 Line of City of Budapest; *Int. Conference "Safety, Risk and Reliability—Trends in Engineering"*, Malta, March 21–23, 2001, pp. 665–670.

Kolic, D. 2002. Risk Analysis Methodology for Underground Mass-Transit Projects., *Transportation Research Board,* 81th Annual Meeting, Washington D.C., January 13–17, 2002, pp. 11.

Kolic, D. 2016. Tunnelling on New Railway Line Divaca—Koper, 2nd Track, *East European Tunnelling Conference 2016, Prague May 23–25, 2016*: pp. 28.

Kolić, D. 2009. The FAUST Method for Optimization of Strait Crossings, 5th Conference on Strait Crossings, June 21–24, 2009, Trondheim, Norway, pp. 6.

Tsamboulas, D., Panou, K. & Abacoumkin, C. 2000. Method to Assess the Attractiveness of Transportation Infrastructure Projects to Private Financing, *TRB 2000*, Washington D.C., January 13–16, 2000, pp. 17.

Underground Infrastructure of Urban Areas 4 – Madryas et al. (Eds)
© 2018 Taylor & Francis Group, London, ISBN 978-1-138-55953-0

The development of CIPP sleeves used in the renovation of sewage conduits

A. Kolonko
Wrocław University of Science and Technology, Wrocław, Poland

ABSTRACT: The renovation of damaged sewage conduits is a real problem. The demand for their renovation has led to the establishment of many specialized companies and also to the development of new and more effective methods. Currently, the dominant method on the market is a method that uses sleeves that harden on site, the so-called CIPP (Cured-in-Place Pipe), which has been known since 1971. In recent years, new variations of this method have been developed that aim to shorten the renovation time and reduce costs. The evolution of this method and the stages of its development are the subject of the paper.

1 INTRODUCTION

The technical condition of sewage conduits in Poland and many European countries, despite significant expenditure on their renovation, is still unsatisfactory. It can be expected that in the coming years this state will systematically improve due to EU environmental requirements and financing projects from this area. Due to the fact that the renovation of sewage conduits with the use of CIPP sleeves, which are defined in the current standard (ISO 11296-1, 2011) as cured on site sleeves, has become the dominant technology covering more than 50% of the market according to (Schaff, Lenz, 2009), it is worth analysing the development of this technology. On the Polish sewage conduit renovation market there are more new companies and new types of sleeves, and therefore the quality of cured sleeves should be especially controlled as it determines their durability. The occurrence of the Polish version of the above-mentioned standard PN-EN ISO 11296-4: 2011 – Plastic piping systems for the renovation of underground non-pressure rain and sanitary sewage networks, Part 4: Implementation of cured on site sleeves (PN-EN ISO 11296-4: 2011) – has became an additional impulse for such analysis.

The particular usefulness of CIPP sleeves in the renovation of sewage conduits is related to the lack of limitations of this method in terms of cross-sectional shape. This is due to the lack of stiffness of the sleeve before curing and its low thickness, which enables the reduction of the cross-section of the sewer that is under renovation to be minimalized.

However, it is important to realize that there are situations when the renovation of a damaged sewage conduit using CIPP sleeves is not the optimal solution. The decision on the choice of renovation method should be based on the analysis of the results of the technical condition of a sewer. This analysis should be carried out on a sewer that has been previously cleaned. More details on the concept of renovating sewage conduits and their diagnostics can be found in studies (Madryas, Przybyła, Wysocki, 2010). It should be noted that sleeves are a structural element and should be designed while taking all operating loads and the technical condition of a damaged sewer into account.

2 STATISTICAL DATA

According to (Central Statistical Office, 2015), in 2015 the length of the sewage network in Poland reached almost 150,000 km, with the number of connections to buildings amounting to about 3.1 million pcs. When compared to 2014, the length of the built or reconstructed sewage network increased by about 6,800 km – i.e. by 4.8%, while the number of connections increased by more than 160,000 – i.e. by 5.5%. The percentage of users of the sewage network in the period between 2005 and 2015 increased from 59.2% to 69.7% (an increase of 10.5 pp). In cities, 89.8% of the population were using the sewage network (an increase of 5.0 pp), and in rural areas this figure was equal to 39.2% (an increase of 20.4 pp). As can be seen, rural areas are where new sewage systems are being constructed, whereas in urbanized areas, financial outlays are mainly spent on the renovation of existing sewage conduits.

Interesting statistical data concerns Germany, where the assessment of the technical condition of networks and their inventories has been systematically carried out for many years. As it turns out, Germany has about 575,000 km of sewage pipelines and at least 1,000,000 km of connections and other conduits for sewage disposal in private areas. These numbers may in practice only be subjected to minimal changes, because the sewage system availability rate in Germany stands at 96% [Winkler, 2010]. Previous studies of the technical condition of sewers (Berger, Falk, 2009) showed that, despite expenditure, it is still in many cases unsatisfactory. It is estimated that about 20% of the total length of sewage networks require renovation. Therefore, the demand for renovations in Germany alone is for over 100,000 km of conduits. On this basis, it is estimated that in the whole of Europe about 400,000 km of sewers require renovation. This shows how big the market in the area of sewer renovation is. In Poland, it is assumed that the technical condition of the sewage network is similar to that of Germany and that about 30,000 km of sewage pipelines may require renovation. This does not take into account the hundreds of thousands of connections in which the technical condition is rarely checked.

3 THE ESTABLISHMENT OF A METHOD TO RENOVATE SEWAGE CONDUITS WITH THE USE OF CIPP SLEEVES

The widely used method of renovating sewage conduits with the use of liners that are curing on site (the so-called CIPP sleeve) is already over forty-five years old. The method is known in English language literature as CIPP (Cured In Place Pipe). Technical issues related to this method are regulated in detail by standards, among others by (PN-EN ISO 11296-4: 2011).

The CIPP sleeve was used for the first time to renovate a sewage conduit in 1971 in Hackney, East London. A 70-meter section of an egg-shaped masonry sewer (1175 × 610 mm) was subjected to renovation. The work was led by the inventor of the sleeve, Eric Wood. Eric Wood, Doug Chick and Brian Handler founded Insituform Pipes and Structures Ltd. Company and introduced the patented technology on the market. Moreover, they have been working to improve this technology and also the materials that are used.

It is worth mentioning that the first implementation of this technology involved pulling the CIPP sleeve into a sewer and then blowing it. The method of sleeve installation by inversion only became possible in 1973 when the felt sleeve was covered with impermeable foil. Eric Wood submitted his technical solution to the patent office in the United Kingdom on August 21, 1970. The US patent application took place on February 22, 1977. Insituform Pipes and Structures Ltd. Co. gave many licences to executive companies, but initially only in UK. In 1976, licenses were granted to companies from Australia, USA and many countries of Western Europe. In 1994 the patent expired, which became an impulse for the development of trenchless sewage conduit renovation methods. An important moment in the development of the new revolutionary renovation technology was the patenting of the resin vacuum-impregnated CIPP sleeve (Sterling R et al., 2012) on December 28, 1982. Nowadays, the basic Insituform system is designed to renovate conduits of up to 3000 mm in diameter (Insituform Insights).

4 THE CONSTRUCTION OF CIPP SLEEVES

In the case of CIPP sleeves, the matrix that is a support for resin may be made of fibres of various materials. It initially consisted of a nonwoven fabric made of needle-punched polyester fibres with a well-resinous felt structure and was covered with polyurethane (PU), polyethylene (PE) or polypropylene (PP) foil.

In the case of such CIPP sleeves, the matrix alone does not significantly affect the strength characteristics of the cured sleeve. They mainly depend on the type of used resin. In the 1990s, a new generation of sleeves was developed and their matrixes were first based on glass fibres, and later on carbon fibre or aramid fibres with a very high tensile strength. This allowed the thickness of the sleeves, and hence their weight, to be significantly reduced, which greatly facilitated and accelerated their installation. This has a particular importance in the case of the renovation of large cross-sectional conduits when the resin-coated felt sleeves have a mass of many tons.

Sleeves with a felt structure have a thickness that ranges from 3 to 50 mm, which is obtained by the increase of the number of nonwoven layers. In practice, the number of layers varies from 1 to 7. Depending on the type of matrix fibres and the type of resin for a particular case, the thickness of the sleeve may vary considerably. Fiberglass and carbon fibre sleeves are much thinner than felt and are therefore lighter, which enables long lengths to be installed. It should be emphasized that CIPP sleeves are a structural element and should be designed with a consideration of all operating loads and the technical condition of the damaged conduit. This should be carried out by a qualified designer and on the basis of relevant guidelines.

The photographs in Figures 1a and 1b show the process of impregnating a CIPP sleeve with resin and the venting of it with a vacuum system. Figure 2 shows samples of various CIPP sleeves after curing (Muenchmeyer, P.G., Gemora, 2007).

a) b)

Figure 1. a) The process of CIPP sleeve impregnation with resin; b) Venting the CIPP resin impregnated sleeve with a vacuum system.

Figure 2. Samples of various CIPP sleeves after curing.

5 RESINS USED IN CIPP SLEEVES

5.1 *Resin types used*

In practice, three types of resins are used for CIPP sleeve prefabrication: polyester, epoxy and vinyl ester. Other types of synthetic resins such as alkyd resins or bisphenol resins, which among others are described in study (Kuliczkowski et al., 2010), play a marginal role in the CIPP sleeve market. All basic synthetic resins are thermosetting and are suitable to be used as a component of CIPP sleeves. Studies have shown that polyester resins are the most commonly used in practice. Their market share ranges from 80 ÷ 90% (Dilg, 2008).

Polymeric resins are obtained, as are linear polymers, by the polymerization of the corresponding input chemical compounds that are called monomers. Polymeric resins are usually a mixture of linear, branched and cyclic polymers and oligomers with a relatively low degree of polymerisation. Due to this, most resins have a liquid or semi-liquid consistency (Pęczek, Kłosowska-Wołkowicz, Królikowski, 2011).

5.2 *Polyester resins (UP)*

Polyester resins are a group of synthetic resins in which the main constituents are polyesters of all kinds. The most commonly encountered are two-component unsaturated polyester resins in which the crosslinking process is based on a radically initiated reaction between the carbon-carbon multiple bonds that are present in the structure of these substances. The cross-linking scheme of unsaturated polyester resins is shown in Figure 3 (Pęczek, Kłosowska-Wołkowicz, Królikowski, 2011).

Cross-linking is most common in the context of polymers. When long linear polymer molecules contain reactive side groups, they can react with each other. This results in a denser or less dense three-dimensional network, in which all or almost all of the original polymer chains are chemically related.

Polymers with a relatively low molecular starting weight, which are cross-linked to a small extent, form gels. Not very dense cross-linked polymers of an appropriate structure and molecular weight often obtain rubber properties. The cross-linking process that leads to the formation of rubber is called vulcanization. Dense cross-linking leads to the achievement of brittle and hard duromers. Cross-linked polymers become insoluble. They can only absorb a solvent, and as a result of this they swell.

Polyester resins used in the prefabrication of CIPP sleeves are usually isofosphate resins. They are moderately reactive, rigid after hardening and corrosion resistant. They are characterized by high viscosity when compared to other resins. They contain the addition of colloidal silica to prevent them from draining down from the top part of a sleeve during the curing process. They show good utility in the renovation of sewage networks and are eagerly

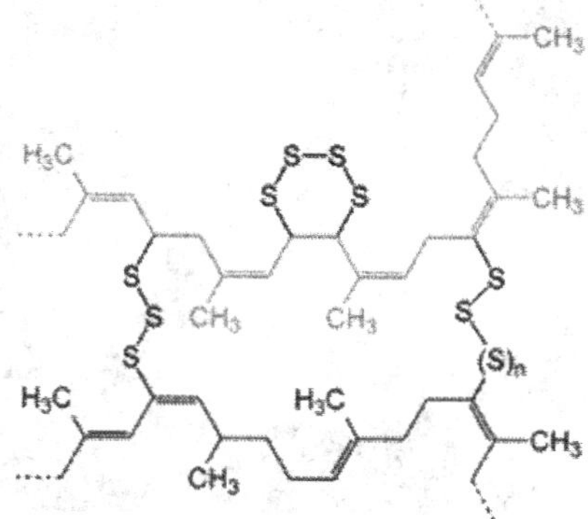

Figure 3. The cross-linking scheme of unsaturated polyester resins.

used due to their considerably lower price when compared to epoxy and vinyl ester resins (Sterling R et al., 2012).

Polyester resins are characterized by linear and volumetric shrinkage. The size of shrinkage can reach up to 6% according to (Kuliczkowski A. et al., 2010). This is undoubtedly an unfavourable feature from a point of view of the behaviour of the cured sleeve as a structural element. Negative effects of the shrinkage can be counteracted, however they cannot be completely eliminated. The effects of longitudinal shrinkage can be relatively well minimized by the development of various anchorages in the areas of socket joints, manholes, side drains etc. located along the length of the hardened section.

Volumetric shrinkage is largely neutralized during the process of curing and cooling of a sleeve by maintaining the internal pressure. However, avoiding the formation of a circumferential ring gap is practically impossible and its existence must be taken into account in static-strength calculations. The German design guidelines (ATV-DVWK-M127- 2, 2000) of CIPP sleeves take such a gap into consideration as a possible imperfection. It is therefore wrong to assume that the sleeve is stacked to the inner surface of the conduit that is being renovated. This is not only due to the shrinkage of the resin, but also to the lack of appropriate conditions for the sleeve to be able to stick to the usually damp and greasy wall of the sewage conduit. The disadvantage of polyester resins, in addition to the above-mentioned shrinkage, is that they contain styrene – a harmful organic solvent with a very unpleasant odour. The disadvantages also include a relatively low resistance to operate at high temperatures. Such a situation may occur in the case of industrial sewage discharge.

Modified polyester resins that contain special initiators can be hardened by UV radiation. The process involves the dragging of a set of UV-emitting lamps through the sleeve filled with pressurised air. The advantages of polyester resins, apart from being the cheapest, include their longer reaction time and the resulting possibility of an industrial prefabrication of the CIPP sleeves, better chemical resistance and easier treatment than is the case with epoxy resins (Dilg, 2008).

The styrene contained in polyester resins, as a reactive organic solvent that is used for their curing, constitutes 30 ÷ 45% of the total resin weight. In theory, the entire styrene should react during curing of the resin. The results of laboratory tests (Akzo Nobel information materials) indicated the importance of the optimal selection of polyester resin components and also the importance of achieving an appropriate heating temperature for residual styrene that is not bonded with a resin and is likely to pose risks to people and the environment. It turned out that depending on the chemical composition of the polyester resin, the amount of residual styrene could vary several times.

The dependence between the curing time and temperature is also very clear. The research of Akzo Nobel company showed that for two specific polyester resin compositions, the curing time at 60°C was equal to 12.6 hours or 24 hours, at 100°C it was equal to 13.5 minutes or 9.0 minutes, while at 140°C it was equal to 3.5 minutes or 2.1 minutes, as shown in Table 1 (Akzo Nobel Information Materials).

In practice, ensuring that the required temperature is achieved should be controlled by special sensors located between the conduit and the sleeve. In the case when a sleeve is cooled from the outside with infiltrating ground water, the heating process must be continued until an appropriate temperature is reached. The resin will otherwise not be sufficiently hardened and the sleeve will not achieve its design strength parameters and unbound styrene will be emitted.

Table 1. The dependence between the time of the occurrence of the maximum temperature of the exothermic reaction and resin composition and also the dependence between the time of the occurrence of the maximum temperature of the exothermic reaction and curing temperature.

	Time to Peak		
	at 60°C (hours)	at 100°C (min.)	at 140°C (min.)
1 phr Trigonox C 1 phr Trigonox C + 0.5 phr	24	13.5	3.5
Perkadox CH-50X	12.6	9.0	2.1

5.3 *Non-styrene resins*

There are two main possibilities of replacing polyester resins that contain styrene. The first of these is the use of other reactive monomers in vinyl ester or unsaturated polyester resins. The second option is the use of epoxy resins. There are already polyester resins without styrene or other volatile organic compounds available on the market. They were used in the CIPP sleeves that were used to renovate conduits that drain rainwater to the Toronto River (Sterling et al., 2012). Until now, such resins have been too expensive and their practical use on a broader scale is a matter of the future.

5.4 *Epoxy resins*

The real alternative to polyester resins without styrene are epoxy resins. The undoubted advantages of epoxy resins are their low shrinkage and the lack of styrene in their composition. Moreover, these resins are characterized by very good strength properties and are also less sensitive to not reaching an appropriate curing temperature in the hardening process. This is very important in the case of the most commonly used polyester resins. At lower temperatures, the epoxy resin hardening process is prolonged, however, the designed strength parameters are achieved.

A disadvantage of epoxy resins, apart from their high price, is the relatively short time of the starting of the curing reaction after the CIPP sleeve is soaked in resin. This, however, can be prevented by designing a composition of epoxy resin that is adapted to existing conditions. In such a case, a workforce with an appropriate competence is required. In practice, the impregnation of the sleeve with epoxy resin is carried out on site in specialized vehicles. Very modern vehicles, which are currently available on the market, allow a CIPP sleeve with a high quality to be achieved (Krasowski Information Materials). A limitation during their prefabrication in vehicles is the size of the sleeves, which does not exceed DN1000. This means that the CIPP sleeves that are impregnated with epoxy resin are not used to renovate man-entry sewers.

6 INSERTION OF CIPP SLEEVES INTO A DAMAGED SEWAGE CONDUIT

6.1 *Insertion using the inversion method*

After 1973, when the CIPP felt sleeve was coated with an impermeable foil, its installation inside a damaged conduit was carried out using inversion. The sleeve was inserted through an existing manhole or a vertically mounted tubular section with a special ring that allowed the sleeve to be inverted. The sleeve that is mounted to the ring, during the inversion process, is inverted under the hydrostatic pressure of water, introduced into a conduit and then attaches to the conduit's wall with its side that is impregnated with resin. After inserting the entire length of the sleeve (inliner), the water that fills it is heated to about 80°C in order to thermally cure the resin that the sleeve has been soaked in.

Once the curing process and cooling of the water is finished, the water is pumped out and the end of the sleeve is cut off. The introduction of a CIPP sleeve using the inversion method can be conducted with the use of compressed air. In this solution, hot steam is used to cure the resin. In hardened CIPP sleeves, in the places where pre-existing drains were closed, appropriate openings are cut using remote controlled robots that are supervised by a TV camera. Renovation of a conduit and its connections may be carried out using short sleeves or hat CIPP profiles. After inspection with the use of a TV camera and appropriate inspection tests that were specified in the standard (PN-EN ISO 11296-4: 2011, PN-EN 1610: 2002), the conduit can be put into operation.

6.2 *Insertion by pulling*

The CIPP sleeves that are currently used are most often cured with UV rays. In such a case, the sleeves are pulled into sewage conduits with the use of a winch after the sewer was previously lined with a strip of foil that protects the pulled sleeve from damage. After pulling, both ends of the sleeve are closed. However, on one of the ends, a lock to allow the introduction of a chain of UV lamps is assembled. The filling of the sleeve with compressed air causes it to be attached to the inner surface of the damaged conduit. Before curing, the geometry of the inflated sleeve can be verified through its entire length with the use of a TV camera. This is important, because any folds that could have resulted from the twisting of the sleeve during its insertion can be eliminated. Such an adjustment is not possible in the case of insertion that uses the inversion method and hydrostatic pressure of a water column.

7 METHODS OF CIPP SLEEVE HARDENING

7.1 *The external thermal method*

The oldest method of hardening CIPP sleeves is the thermal curing method. The heat supply is provided by heating the water that fills a CIPP sleeve or by letting through hot steam. The disadvantage of this method, especially when using polyester resins, is the risk of ineffective hardening. This most often occurs in cases when the sleeve is cooled from the outside by infiltrating ground water, and as a result the resin is not heated enough. The solution provided by standard (PN-EN ISO 11296-4: 2011) is the obligatory usage of properly spaced thermal sensors that are located on the CIPP sleeve and conduit contact, and also the continuation of the heating process until a suitable temperature is reached over the entire length of the installed sleeve. The resin will otherwise not be sufficiently hardened and the sleeve will not achieve the designed strength parameters. An additional negative effect is the above-mentioned emission of unbounded styrene. It is important to note that currently only inversion technology that uses hydrostatic pressure allows CIPP sleeves with very large sizes of up to DN3000 to be installed. Other curing technologies that are described below, except for UV curing, are mainly intended for non-man entry conduits.

7.2 *The internal thermal method*

Internal thermal hardening is provided by the IHC™ electro-curing system. This system was developed for the renovation of sewage connections. The basis of this system is a patented (patent no. US 6146576 A) composite liner, which is also called a hybrid composite that has highly conductive carbon fibres with high strength parameters in its structure. A detailed description of the patent can be found on the website (www.google.com.mx/patents/US6146576).

The patent holder is INTRALAMINAR HEAT CURE, INC. The sleeve, which is used in the IHC™ system, is mainly made of polyester fibres and glass fibres with a felt structure and also integrated fiberglass fabric. These materials are also used in other CIPP sleeves. An innovative solution is the introduction of the above-mentioned carbon fibres into the structure of the sleeve. The current passing through these fibres causes the generation of heat, accelerated polymerization of the resin, and hardening of the composite. This sleeve is pre-impregnated with specially designed synthetic resin and its hardening is initiated by raising the temperature. In practice, epoxy, vinyl ester and polyester resins can be used. The resin-impregnated sleeve can be stored for long periods of time when stored at low temperature. The high quality of the IHC sleeve is ensured by its prefabrication. The view of the IHC sleeve during its installation is shown in Figure 4 (Muenchmeyer).

The IHC™ system allows for the curing process of the sleeve to be precisely controlled and it enables the duration of this process to be significantly shortened when compared to conventional CIPP sleeves cured by heat supplied from hot water or steam. Curing the sleeve can be shortened by up to 60 minutes (Muenchmeyer). In traditional systems, the heat flow is much slower as it has to pass through the entire thickness of the sleeve, which is simultaneously cooled by the cold surface of the conduit to which it is attached. A particularly significant slowdown of the curing process takes place when the conduit that is being renovated is located below the ground water level. The water infiltrates through both cracks and other damage and cools the sleeve. As a result, the resin may not be properly cured, and therefore the traditional sleeve will not reach the expected strength parameters.

The great advantage of the described IHC™ system is the uniformity of curing the entire length of the installed sleeve. In traditional systems that use hot water or steam there is a significant temperature difference in the inlet and outlet of the heating medium, and therefore the curing time should be prolonged accordingly.

The chemical composition of the resin used in the IHC™ system increases its viscosity when introduced into the composite. Due to this, the gravitational flow of the resin is avoided and the thickness of the installed sleeve in the upper part of its cross-section is not reduced.

7.3 *The UV curing method*

Modified polyester resins, which contain special initiators in their composition, can be cured using UV radiation. The process takes place while pulling a set of vacuum lamps that emit UV rays through a sleeve filled with compressed air. An exemplary element of the set is shown in Figure 5 (Muenchmeyer, Gemora, 2007).

In this curing system, considerable progress can be noticed and it results from the use of the set of lamps with a much higher power than originally. This allows for a significant acceleration of the curing process.

Another advantage of the UV method of curing the resin is the continuous measurement of the temperature that is reached in the sleeve structure during the duration of the process. The measurement is conducted with the use of special sensors. This allows a fully controlled hardening of the sleeve and the achievement, even in difficult conditions when the sleeve is cooled from outside by the infiltrating ground water, of the designed strength parameters. In such cases, chilled sections are exposed to UV radiation for longer. It is worth noting that the range of diameters of the sleeves that are cured using UV rays has grown significantly over the last few years and is already reaching DN 1800 (Relineeurope Information Materials).

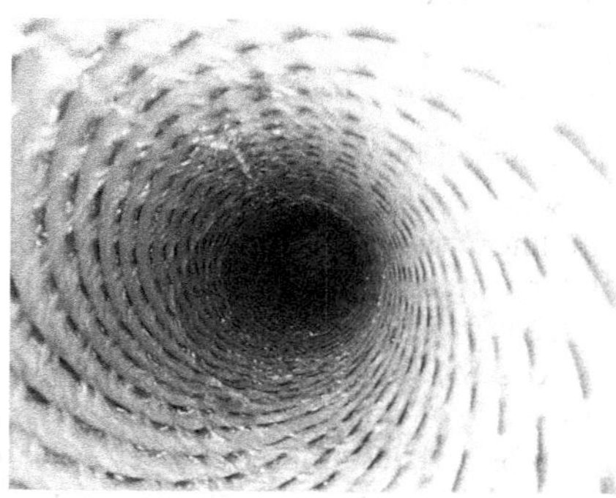

Figure 4. View of the IHCTM sleeve during the installation process.

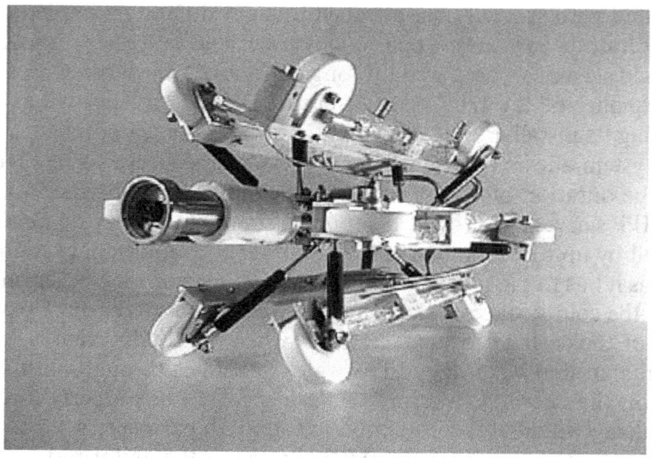

Figure 5. An element of the set of lamps that emit UV rays.

Apart from the advantages of curing CIPP sleeves using UV rays that involve the minimizing of energy consumption when compared to the conventional curing method that uses hot water or steam, it also has a disadvantage related to the harmful UV radiation for operators handling the equipment of this technology. Moreover, the well-known UV emitting devices are based on the fairly out-dated technology of vacuum electron lamps and are therefore unstable. The energy and power generated by the UV emitting device is subjected to changes during the short lifetime of this device, which constitutes a great disadvantage of such devices.

7.4 *The ultrasonic curing method*

The most recently proposed solution for CIPP sleeves includes the ultrasound-initiated resin curing system. In this case, the special resin contains microcapsules in its composition, inside of which there is a catalyst component. When renovating a sewage conduit filled with compressed air, the curing process takes place while a device equipped with ultrasonic transducers moves through the conduit. The ultrasonic energy ruptures the capsules and the catalyst initiates the resin polymerization (curing) process. This system is protected by US Patent No.: US 8,048,360 B2, Date of Patent: Nov. 1, 2011. Until now, there has been no information on the practical application of this CIPP sleeve curing method.

7.5 *The microwave curing method*

Another innovative solution that was described and published in 2012 is CIPP Microwave Curing. This method was patented in China as patent No. CN102649317A. In the case of this method, there is also no information on any attempts of its practical use on an industrial scale.

7.6 *The LED light curing method*

Research on new resin curing methods is still ongoing. In the case of the newest innovations in the area of CIPP sleeves, the most promising seems to be the method of curing resins with the use of visible LED light. This has been confirmed by new international patents. The starting point was to develop a special resin. The DSM Company from the Netherlands developed and presented a photocurable composition of resin, in which the resin is cured by visible light with a wavelength of about 450 nm, i.e. blue light. The commercial photocurable resin was described in the published Dutch patent 1007205, which is also referred to in the published international patent application WO2005/103121 of the above company.

The advantage of such a LED hardening method is its high stability and constant energy efficiency throughout the life of the LEDs. In addition, these diodes have significantly higher energy efficiency than lamps emitting UV radiation. This enables a much bigger number of LED diodes to be used due to the relatively high ration of energy/power to the surface area (information materials of SewerLight). Currently, various research and development studies are being carried out in various countries to develop specialized devices for curing CIPP sleeves with LED light. The Danish company Per Aarsleff A/S developed two versions of the devices for curing CIPP sleeves, which are based on the use of LED diodes. A particularly interesting element of such a device is integrated liquid cooling, which improves its efficiency.

The company patented its solution and received the patent No. 8561662 B2 in 2013. The Polish company Sewerlight started developing such a solution and prepared its own prototype to be patented. The advantage of the invention is that the construction developed according to SewerLight can be built in the form of a compact device that can be used for photocuring CIPP sleeves that are used for renovating conduits with diameters in the range of DN70-300. This range may be increased in future if there is such a demand. Another advantage of this invention is the possibility to integrate the basic device with other devices. This will allow for the visual monitoring of the curing process and the control of temperature, which is the key indicator of correct resin curing. It is worth mentioning that this is a safe method from the point of view of the health of employees who operate the LED curing system.

8 CONCLUSIONS

There is an ongoing development of methods of repairing damaged sewage conduits in order to improve the quality and durability of carried out works. An important criterion is also the reduction of time and costs for particular methods of renovation. Due to the relatively low cost and efficiency, the method of renovating damaged sewage networks using CIPP sleeves will constantly develop. The interest in this method is proved by the progress in the field of used resins and the innovative possibilities of their curing. Work related to the curing of resins with the use of visible LED light is very promising.

This is confirmed by new international patents. Progress in this case is based on better efficiency, reduced energy consumption and the ability to monitor the course of the curing process. It is also important to improve the safety of people carrying out installation works. It is important to remember that the efficiency and durability of renovating damaged sewers with the use of CIPP sleeves is based on the reliable assessment of the technical condition of these conduits, correct load summary, realistic assumptions for static and strength calculations, correct sleeve installation and also appropriate testing within the framework of technical acceptance. More information regarding the quality of hardened CIPP sleeves can be found in article (Kolonko, Madryas, 2011).

REFERENCES

Berger Ch., Falk Ch.: *Zustand der Kanalisation in Deutschland*; Ergebnisse der DWA-Umfrage 2009, Deutsche Vereinigung für Wasserwirtschaft, Abwasser und Abfall e.V. 2009.

Central Statistical Office: Communal Infrastructure in 2015.

Dilg R.: 35 Jahre Schlauchlining Teil 1. Tiefbau, Ingenieurbau, Strassenbau Nr 1-2/2008.

Dilg R.: 35 Jahre Schlauchlining Teil 2. Tiefbau, Ingenieurbau, Strassenbau Nr 4/2008.

Information materials of Akzo Nobel Polimer Chemicals bv.

Information materials of Insituform.

Information materials of Krasowski® Technische Geräte und Systeme GmbH.

Information materials of Reline America Inc.: Styrene and CIPP.

Information materials of Relineeurope.

Information materials of SewerLight.

Information materials of TecLiner.

Kolonko A, Madryas C.: *Notes on CIPP liners*, Gaz, Woda i Technika Sanitarna. No.2/ 2011.

Kuliczkowski A. et al: Trenchless Technologies in Environmental Engineering. Publishing house of Seidel—Przywecki Sp. z o.o., 2010.

Madryas C., Przybyła L., Wysocki L.: *Tests and the assessment of the technical condition of sewage conduits.* Dolnośląskie Wydawnictwo Edukacyjne, Wroclaw 2010.

Muenchmeyer G.P. *Lateral Sewer Rehabilitation Overview. Current technologies.* Information materials of K. R. Swerdfeger Construction, Inc.

Muenchmeyer, P.G., Gemora I.: *Mainline Pipe Rehabilitation Using Cured-in-Place Pipe (CIPP) & Folded Pipe Technology*, 2007 Pumper & Cleaner Environmental Expo International.

Pęczek P, Kłosowska-Wołkowicz Z., Królikowski W.: *Unsaturated polyester resins*, Scientific and Technical Publishing House, Warsaw 2011.

PN-EN 1610: 2002 *The construction and testing of sewage conduits.*

PN-EN ISO 11296-1: 2011 *Plastic piping systems for the renovation of underground non-pressure rain and sanitary sewage networks. Part 1: General issues.*

PN-EN ISO 11296-4: 2011, *Plastic pipe systems for the renovation of underground non-pressure rain and sanitary sewage networks. Part 4: The lining of cured in place sleeves.*

Schaff O., Lenz J.: *Schlauchliner—ganzheitliche Lösung vom Haus bis zur Kläranlage. Stadtentwässerungsbetriebe.* bi UmweltBau Kongressausgabe | 2009.

Static-strength calculations for the technical rehabilitation of sewage conduits using the insertion of liners or the assembly method: 2000.

Sterling R et al.: A *Retrospective Evaluation of Cured-in-Place Pipe (CIPP) Used in Municipally Gravity Sewers*, United States Environmental Protection Agency, 2012.

Winkler U.: *Abwasserkanäle der "vergrabene Schatz".* 10. Deutscher Schlauchliner Tag. Hannover 2010.

www.google.com.mx/patents/US6146576.

Underground Infrastructure of Urban Areas 4 – Madryas et al. (Eds)
 ISBN 978-1-138-55953-0

Performance and structural design of liners in non-circular sewage pipelines

J. Kozubal & A. Szot
Wroclaw University of Science and Technology, Wroclaw, Poland

ABSTRACT: Basics of identified behaviour, performance and general rules of structural design of liners for egg-shaped, pear-shaped and other non-circular cross-sections are presented in this paper. The main causes of failure are pointed out based on experience gained so far in this field and feedback from researchers and manufacturers. Along with it, some accompanying conditions of installation and acting loads are listed. The primary approach to the assessment of liner load-bearing capacity with satisfactory safety margin is also included. Some basic and most common cases are distinguished to cover the area of liner application. Finally, the analytical approach is compared with rigorous FEM calculations.

1 INTRODUCTION

As close-fit lining systems for sewers keep gaining in popularity throughout the world the problem of their structural design needs to be promptly identified and expressed in a concise algorithm for usage in everyday engineering practice. Short period of application of this technology, namely since 1971, along with little experience, research and relative complexity of the problem resulted in slow but progressive development. Still the state of the art on that matter leaves a lot to desire to consider it economical and fully valid, which is expressed by relatively high safety factors of 2 or 2.5. Whereas the issue of circular pipe linings was recently improved, due to rapid demands, non-circular cross-sections were widely neglected, except for a few papers touching on that problem.

The term "non-circular" refers to the following shapes, which were formerly widely used for brick or concrete sewage pipelines:

- egg-shaped [see Fig. 1a]
- high egg-shaped [see Fig. 1b]
- pear-shaped [see Fig. 1c]
- bell-shaped [see Fig. 1d]

Although the algorithm worked out so far for circular liners seems matching the experimental results with reasonable accuracy, it cannot be simply transposed to the usage for non-circular cross-sections. Some aspects, ideas and identified facts, however, may be incorporated into the new approach as it was published elsewhere (Doll et al 1996, Falter 1997 & 1999, Thepot 2000 & 2001).

The liner material is known to be affected by several factors. Since liners are made of plastics of different kind, their deflection depends on the pressure value, the time lapse, temperature and chemical constitution of transported medium (Żuchowska 2000). In addition, certain irregularities of the shape and deviations from the assumed (nominal) shape usually occur. Their influence is identified only for circular shapes but for non-circular ones so they must be worked out independently. The main shape imperfection resulting in significant reduction of critical pressure is circumferential gap due to some side effects of polymerisation or inaccurate liner geometry. Another one, of slightly less importance, is local imperfection, which usually takes form of an inward wave.

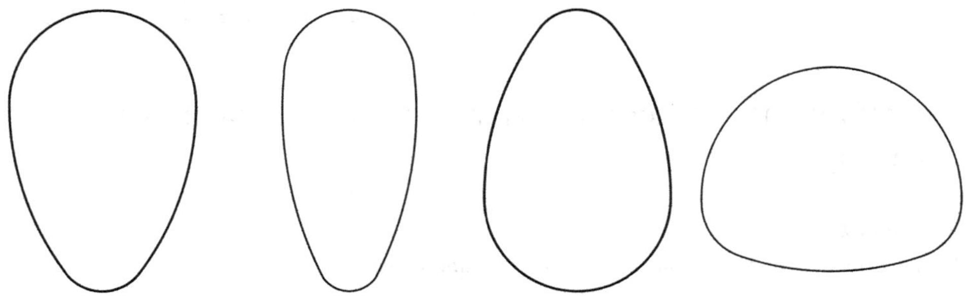

Figure 1. Common non-circular shapes of sewer cross-section.

As it was established the main source of liner failure is, regardless of its shape, snap-through buckling caused by ground water pressure, penetrating the gap between the host-pipe and liner. As the material used for liners shows rheological (visco-elastic) behaviour (Boot et al 1996, Guice et al 1994, Żuchowska 2000) the failure, in the form of complete inward collapse, in practice occurs due to creep deformations of liner under the ground water pressure. Since taking this into account in the rigorous way in an analytical formulation leads to high complexity of the problem and may be solved only by means of iterative methods an equivalent of creep modulus was incorporated (El-Savy et al 1997 & 1998, Falter 1997 & 1999) namely a fraction of the short-term modulus of elasticity. This is called a long-term modulus of elasticity and expresses the strain-stress relation as a secant modulus resulting in transposition into a quasi-elastic behaviour of the model. That approach is widely used now and as it turned out it seems to be sufficient to describe this aspect of liner behaviour (Boot et al 1996 & 1998, Falter 1997 & 1999).

Non-circular liners cause some troubles with the concise formulation of the problem due to rapid increase in number of geometrical parameters comparing to circular ones, where using a few dimensionless parameters reduces the problem significantly. However, an attempt to describe the model regardless of its shape was made resulting with quite good accuracy (Thepot 2000). In this approach circumference of the liner was used for sufficient description of geometry. Particulars of the analytical model are presented in the next chapter.

2 ANALYTICAL APPROACH

The stability of liner under ground water pressure, for both circular and non-circular shapes, is to be thoroughly examined to identify the liner performance. In order to reduce the real model to the analytical one the following assumptions have been made:

- two-dimensional model is considered,
- liner is in plane strain conditions,
- there is no bond between the liner and the host-pipe,
- friction between host-pipe and liner is neglected,
- geometry along the pipe axis is constant in each step i.e. all cross-sections along longitudinal axis behave in the same manner,
- pipeline under consideration is non-pressure one,
- large displacement is valid for both analytical and FEM analysis,
- liner performance is examined within the linear material behaviour and Hook's law is applies here,
- creep modulus is replaced by its quasi-elastic equivalent, namely fraction of the short-term modulus of elasticity (0.25–0.5 of E),
- the ground surrounding the host-pipe does not affect the liner, i.e., the host-pipe is structurally sound but leaking, which allows the ground water to act on the liner.

Since the analytical approach for non-circular shapes involves incorporating both non-deflected (initial) and deflected (post-buckling) shapes the problem appears to be highly complex due to the need to work out several separate formulas describing the critical pressure for each individual case. However, in the process of energy balance formulation some action may be taken to derive the expression for critical pressure in terms of magnitude of the circumference (Thepot 2000 & 2001) so it matches cases of any shape.

There are two presumable post-buckling configurations, namely symmetrical and asymmetrical one [Fig. 2]. Although both of them are equally probable, some authors (El-Savy et al 1997 & 1998, Falter 1997 & 1999) incline to the latter since it complies with both results of the balance of potential energy and intuitive predictions. The asymmetrical mode of post-buckling deflection involves less potential energy so it results in the lower critical pressure. In fact, only the geometry configuration of the real model governs the scheme of failure but in order to cover most practical cases and to keep a safety margin the assumption of asymmetrical mode seems advisable.

The liner installed in a host-pipe is known to be subject to ground water pressure in most real-life cases, which is the main cause of failure. Although, in practice, the load pattern is non-uniform along the circumference of the liner the uniform distribution is accepted in publications presented to date. In this paper, the non-uniform pressure is incorporated into FEM analysis as accurate enough at this stage of problem examination. Both uniform and non-uniform distributions of the liner load are presented in Fig. 3. Moreover, there may be

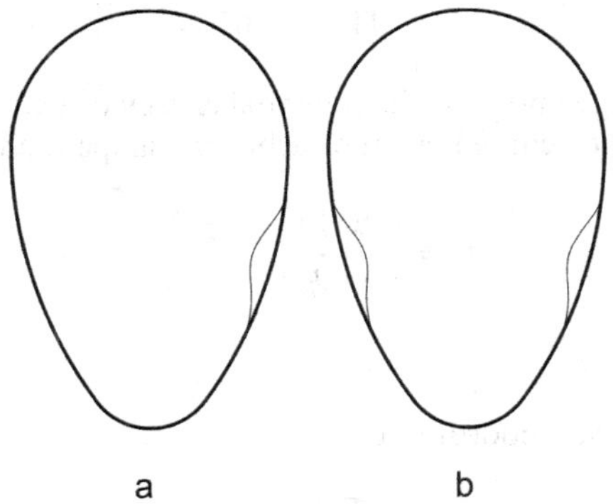

Figure 2. Snap-through form of liner. Asymmetrical (a) and symmetrical (b) deformation.

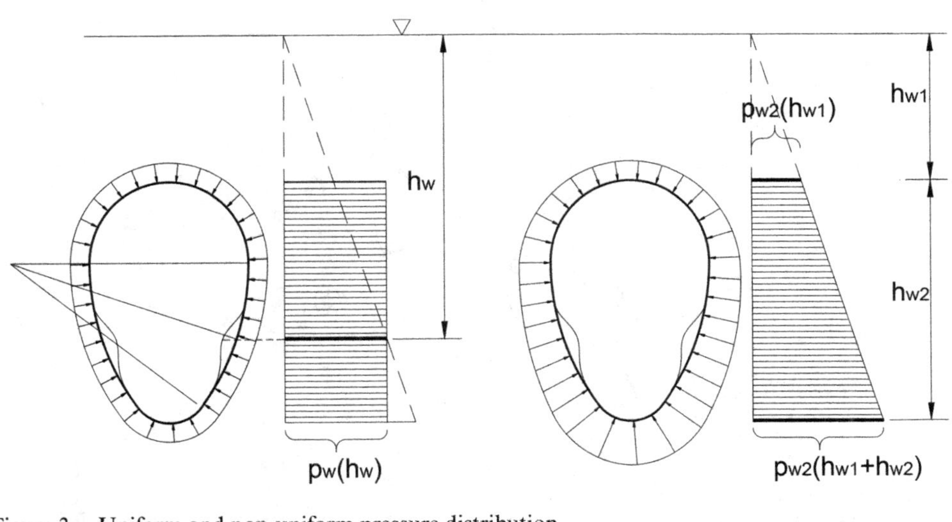

Figure 3. Uniform and non-uniform pressure distribution.

certain circumstances where liner is also subject to ground and overburden load. This occurs where the host-pipe is severely structurally deteriorated which results in transferring some part of loading on the liner by the local contact interaction. Such highly complicated models have not been fully examined so far even for circular liners and require separate approach to the problem to be undertaken in the future.

3 POTENTIAL ENERGY BALANCE

There is a need to assume the parametrical expression for post-buckling deflection in terms of its geometry so that the balance of potential energy could be calculated. After the Glock's approach (Glock 1977), the following formula was accepted:

$$w(\phi, w_0) = w_0 \cdot \cos^2\left(\frac{\pi \cdot \phi}{2 \cdot \phi_0}\right) \tag{1}$$

The parameter φ_0 denotes the half angle of the deflected zone and φ is a parametrical coordinate referring to the buckled zone [Fig. 4]. In further formulation both one-lobe and two-lobe mode of deflection are considered.

The general expression for calculation of the potential energy of deflected model is as follows:

$$\Pi = U - W \tag{2}$$

where U is the strain energy and W is the potential energy of external forces.

For large deformations of curved beam of arbitrary shape these two values are given by:

$$U = \frac{1}{2} \cdot \int_0^C \left(\frac{N^2}{EA} + \frac{M^2}{EI}\right) ds \tag{3}$$

where:

s – curvilinear parametrical coordinate

$$N = \varepsilon \cdot E \cdot A \text{ axial force} \tag{4}$$

$$M = \kappa \cdot E \cdot I \text{ bending moment} \tag{5}$$

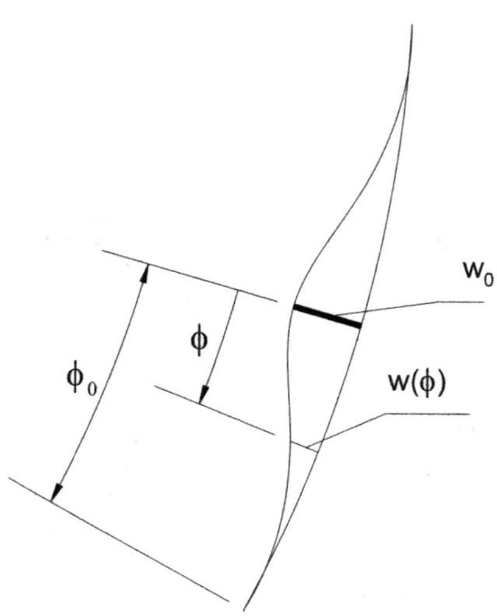

Figure 4. Description of the buckled zone.

The bending strain and axial strain for a curved beam are given by [1,2]:

$$\varepsilon = \frac{w}{R} + \frac{du}{ds} + \frac{1}{2} \cdot \left(\frac{dw}{ds}\right)^2 \tag{6}$$

$$\kappa = \frac{d^2 w}{ds^2} \tag{7}$$

Then, all the parts are to be substituted into (2) and integrated with respect to curvilinear coordinate s accordingly to the shape of the liner cross-section with consideration being given to the assumed geometry of the deflected region (1). The assumption of the uniform averaged axial force along the circumference is accepted after Glock [10] and extended to the non-uniform cross-section shapes, which was examined [5–7] and found valid and accurate enough. The liner remains in the stable position as long as the principle of the equilibrium is maintained. This is valid for the local minimum of the total potential energy, which is expressed as follows:

$$d\Pi = \frac{\partial \Pi}{\partial w_0} \cdot dw_0 + \frac{\partial \Pi}{\partial \varphi_0} \cdot d\varphi_0 = 0 \tag{8}$$

$$\frac{\partial \Pi}{\partial w_0} = 0 \quad \text{and} \quad \frac{\partial \Pi}{\partial \varphi_0} = 0 \tag{9}$$

4 FORMULATION OF RESULTS

Solving the system of two equations finally gives [13]:

$$\varphi_{0crit} = 2.54 \cdot \Lambda \cdot \frac{1}{n^{0.2}} \cdot \left(\frac{I \cdot C}{A \cdot R^3}\right)^{0.333} \tag{10}$$

$$p_{crit} = 2.02 \cdot n^{0.4} \cdot \Psi \cdot E_L \cdot \frac{I^{0.6} \cdot A^{0.4}}{C^{0.4} \cdot R^{1.8}} \tag{11}$$

$$w_{0crit} = 1.35 \cdot \Gamma \cdot \frac{R}{n^{0.4}} \cdot \left(\frac{I \cdot C}{A \cdot R^3}\right)^{0.4} \tag{12}$$

and internal forces at the snap-through point:

$$M_{crit} = 1.2 \cdot \Phi \cdot \frac{E_L \cdot I}{R} \tag{13}$$

$$N_{crit} = 1.26 \cdot p_{crit} \cdot R \tag{14}$$

where:

$$\Lambda = 1 + 0.17 \cdot \delta - 0.007 \cdot \delta^2 \quad \text{for } \delta \leq 10 \tag{15}$$

$$\Psi = \frac{1}{1 + 0.41 \cdot \delta - 0.006 \cdot \delta^2} \quad \delta \leq 15 \tag{16}$$

$$\delta = \frac{1}{n^{0.2}} \cdot \frac{2 \cdot G}{R} \cdot \left(\frac{\pi \cdot A \cdot R^3}{2 \cdot I \cdot C}\right)^{0.6} \tag{17}$$

$$\Gamma = 1 + 0.53 \cdot \delta - 0.012 \cdot \delta^2 \quad \text{for } \delta \leq 10 \tag{18}$$

$$\Phi = \frac{\Gamma}{\Lambda^2} \tag{19}$$

g – gap as a fraction of the radius [mm]
C – circumference of the medium line of the liner cross-section
R – nominal (initial) radius of the buckled zone of the liner
I – moment of inertia of the longitudinal cross-section of liner
A – area of the longitudinal cross-section of liner

Originally, also the influence of ovality on the performance of nominally circular liners has been incorporated in the analytical approach by Thepot [13], but according to the current state of the art, the expression used there is judged to be inaccurate and conservative. Therefore another one is proposed here, based on the recent examination carried out by the author of the present paper:

$$\eta_o = \left[\frac{1}{1+3.54 \cdot \left(\frac{\Delta_o}{R}\right)^{1.23} \cdot \left(\frac{t}{R}\right)^{-0.54}} \right] \tag{20}$$

where:

Δ_0 – eccentricity of the host-pipe quarter centre due to ovality [mm].

This formula makes the reduction due to ovality dependent on liner hoop stiffness represented by thickness thickness, and not only the diameter.

The expressions presented above (excluding 18) cover the cases of both homogenous and non-homogenous material of the liner so they hold also for all the liners of profiled (corrugated) longitudinal cross-section. The expressions may be reduced to the form which is valid only for the homogeonous material and then they may be respectively given by [13]:

$$\varphi_0 = 1.55 \cdot \Lambda \cdot \frac{1}{n^{0.2}} \cdot \left(\frac{t^{0.4} \cdot C^{0.2}}{R^{0.6}} \right) \tag{21}$$

$$p_{crit} = 0.455 \cdot n^{0.4} \cdot \Psi \cdot E_L \cdot \frac{t^{2.2}}{C^{0.4} \cdot R^{1.8}} \tag{22}$$

$$w_{0crit} = 0.5 \cdot \Gamma \cdot \frac{R}{n^{0.4}} \cdot \left(\frac{t^2 \cdot C}{R^3} \right)^{0.4} \tag{23}$$

$$\delta = 11.65 \cdot \frac{G}{n^{0.2}} \cdot \frac{R^{0.8}}{t^{1.2} \cdot C^{0.6}} \tag{24}$$

$$M_{crit} = 0.1 \cdot \Phi \cdot \frac{E_L \cdot t^3}{R} \tag{25}$$

$$N_{crit} = 1.26 \cdot p_{crit} \cdot R \tag{26}$$

where:

t – thickness of the homogenous liner.

The buckling process starts due to even small geometrical imperfection and then spreads as a blister. As the pressure increases both the internal forces and displacements grow up to the crtitical state. The relations of bending moment, axial force and displacement at the top of the wave against the ground water pressure are derived as follows:

$$M(p_w) = \frac{\omega \cdot \frac{p_w}{p_{crit}} \cdot M_{crit}}{1-(1-\omega) \cdot \left(\frac{p_w}{p_{crit}} \right)^2} \tag{27}$$

$$N(p_w) = N_{crit} \cdot \frac{p_w}{p_{crit}} \tag{28}$$

$$w_0(p_w) = w_{0ctit} \cdot \left[1 - \left(1 - \frac{p_w}{p_{crit}} \right)^{0.5} \right] \tag{29}$$

where:

$$\omega = \frac{2.65 \cdot \left(\frac{\alpha}{\varphi_0} \right)^2}{1 + 4.35 \cdot \left(\frac{\alpha}{\varphi_0} \right)^5} \tag{30}$$

All the expressions above refer to any liner regardless of its shape which is convenient for engineers in everyday practice of structural design. Further on the author compares the analytical approach with the FEM model. Two most common non-circular shapes are examined in the next chapter in terms of critical parameters. Unlike for the analytical model, non-uniform distribution of the ground water pressure is assumed in the FEM analysis, which complies with the model in practice.

5 FEM ANALYSIS OF THE EGG-SHAPE AND THE PEAR-SHAPE LINERS

Two liner models have been calculated in order to evaluate the accuracy of the analytical approach presented above. Geometry of these models is presented in Fig. 5.

In order to perform FEM calculations semi-models have been assumed with consideration of appropriate boundary conditions expressing symmetry. This results in reducing the range of comparison between the FEM model and the analytical approach, since only magnitude n = 2 is valid for the former one in this case and it has been assumed that the buckling deformation is symmetrical with respect to the vertical axis of the liner. The assumed, predicted post-buckling shape is presented in Fig. 6.

For the computational process with the use of Finite Element Method the COSMOS/M system was applied. The model is assumed to be in plane strain conditions and elements applied are determined to be linear in terms of material behaviour. There were following types and numbers of finite elements used in the model:

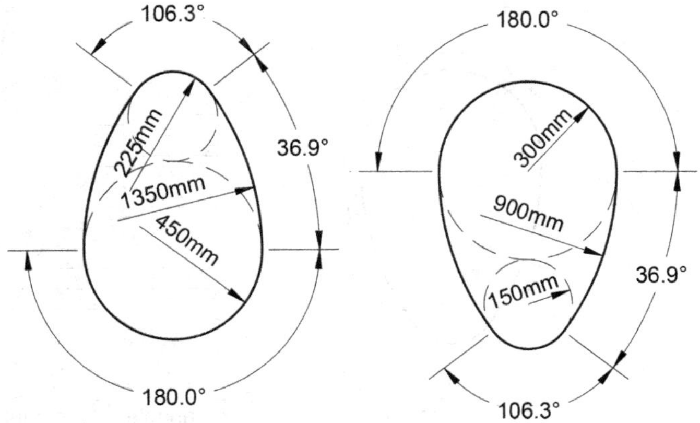

Figure 5. Geometry of two liner cases under examination.

- curved beam elements—for liner description,
- quadrangle plane elements—for host-pipe description,
- node-to-line contact elements—to describe boundary conditions of the liner (contact between the liner and the host-pipe, only inward deflection possible, friction neglected)

The load has been assumed to be non-uniformly distributed along the circumference, unlike in the analytical model. The pressure applied was a variable expressed in terms of the ground water table so its height was the parameter of the FEM analysis. Configuration of the load for calculation purposes is shown in Fig. 7.

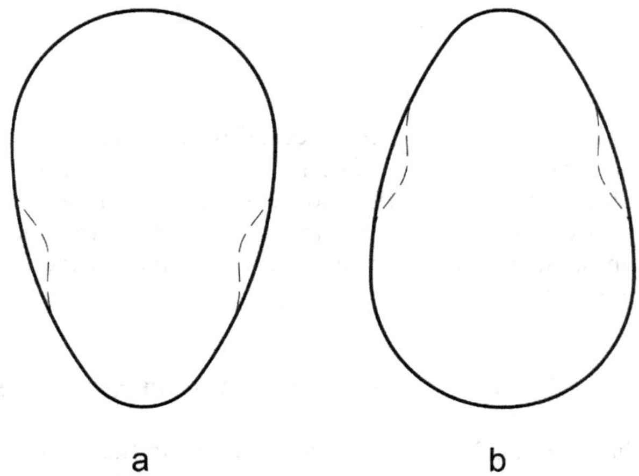

Figure 6. Predicted snap-through forms for egg-shaped liner (a) and pear-shaped liner (b).

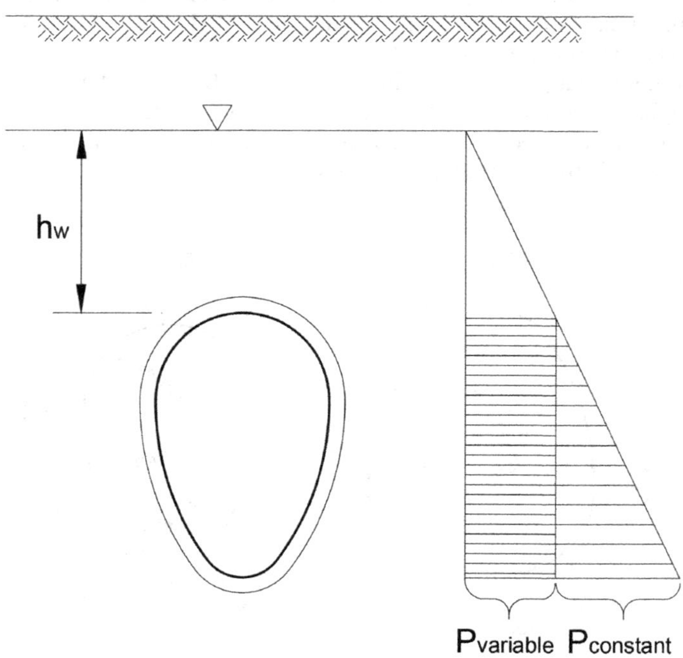

Figure 7. Distribution of the water pressure along the circumference for the FEM analysis.

6 THE COMPARISON OF THE FEM ANALYSIS AND THE ANALYTICAL APPROACH RESULTS

Both cases have been compared in terms of major parameters of the structural design, namely critical height of water table and relations of displacement, bending moments and axial forces against the water table height in pre-buckling state.

The critical ground water table was set up to be the intermediate parameter of the comparison. Since the analytical approach uses the point at the top of the buckling wave as valid for calculating the ground water pressure, certain modification of the analytical model has been introduced. At the point of departure of the analysis the ground water table was set up at the crown of the liner, so the magnitude obtained from expression (11) was converted to water height and the difference in heights between the top of the liner and the blister summit was taken into account.

The characteristic parameters and results of comparison between the analytical and FEM calculation are compiled in Table 1.

Table 1. The comparison of analytical and FEM example results.

	Egg-shaped cross-section						Pear-shaped cross-section					
Parameters												
E_L [MPa]	1500						1000					
ν [–]	0.35						0.35					
R [mm]	900						1350					
t [mm]	12						20					
Results												
	Analytical			FEM			Analytical			FEM		
gap [mm]	0	6	15	0	6	15	0	9	22	0	9	22
P [mm]	2379	2363	2339	2379	2363	2339	3568	3545	3510	3568	3545	3510
h_{crit} [m]	3.99	0.85	0.22	3.86	1.09	0.3	3.37	0.83	0.27	2.88	0.83	0.57
M_{crit} [N·mm]	288.0	345.0	449.7	284.4	607.5	2451	592.6	708.0	833.8	616.8	1206	1472
N_{crit} [N]	51.9	16.3	9.1	50.5	17.2	12.9	65.4	22.2	12.6	62.0	19.7	14.9
d_{crit} [mm]	15.9	58.6	98.3	13.6	93.3	465.2	26.0	88.5	149.5	23.0	121.5	201.1

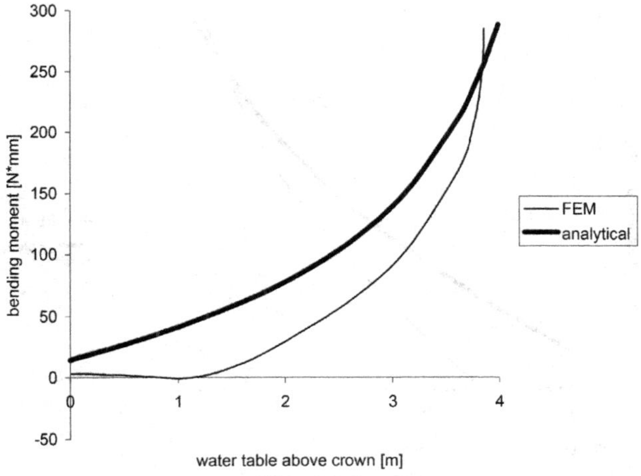

Figure 8. Maximum bending moment in egg-shaped liner at the top of the wave versus water table level above the crown.

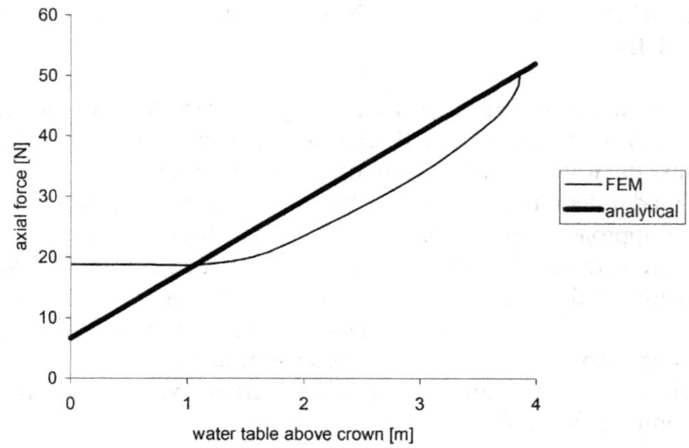

Figure 9. Maximum axial force in egg-shaped liner at the top of the wave versus water table level above the crown.

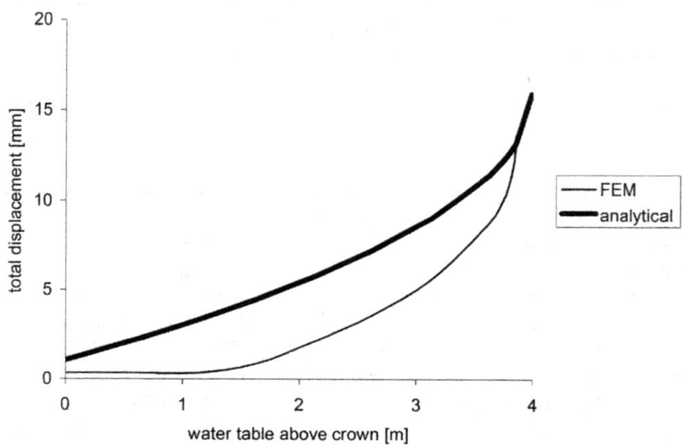

Figure 10. Displacement at the top of the buckling wave of the egg-shaped liner versus water table level above the crown.

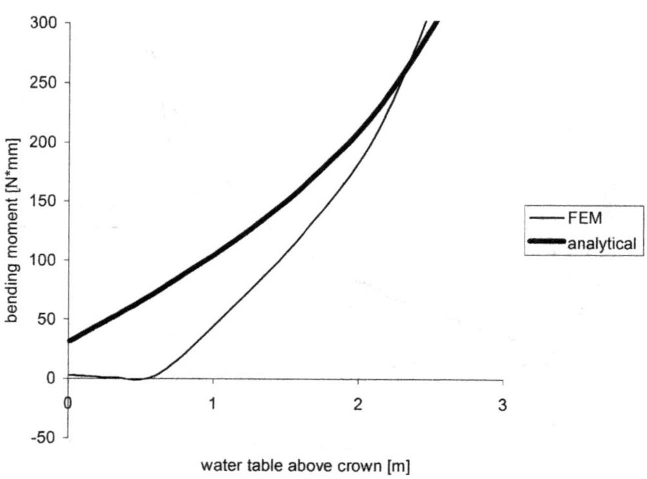

Figure 11. Maximum bending moment in pear-shaped liner at the top of the wave versus water table level above the crown.

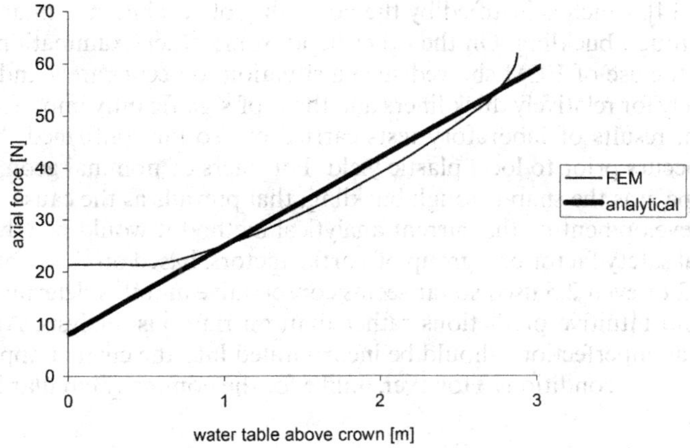

Figure 12. Maximum axial force in pear-shaped liner at the top of the wave versus water table level above the crown.

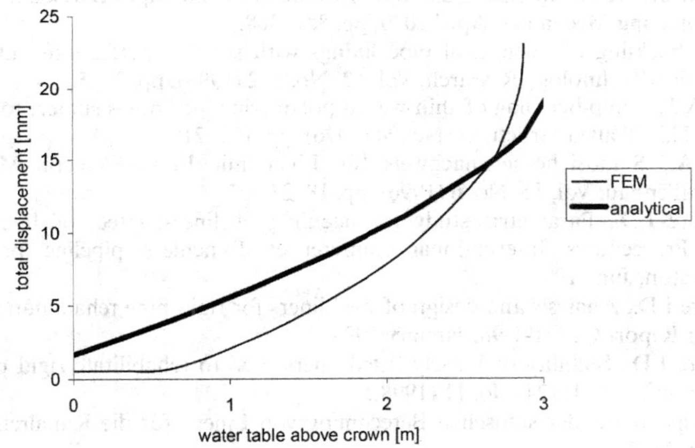

Figure 13. Displacement at the top of the snap-through wave of the pear-shaped liner versus water table level above the crown.

Also the relation of maximal internal forces against water level has been examined for both analytical and FEM approach. The corresponding plots of results are coupled for egg-shaped liner in Figs. 8–10 and for pear-shaped one in Figs. 11–13.

7 SUMMARY

The problem of stability of the polymer non-circular sewer liner subject to the action of ground water pressure has been considered in this paper. In comparison with computational results the first approximate analytical solution turned out to be fairly accurate and useful for practical purposes. The magnitudes of the snap-through pressure calculated in this way were in good agreement with those obtained from FEM analysis. On the other hand, the curves of displacement and internal forces against pressure revealed some discrepancies compared to FEM results. Nevertheless, they seem negligible at this point of analysis advancement since most of the results obtained analytically were on the safe side.

Discrepancies in the course of some parameters against water table height may be neglected since only the final critical results seem valid for the structural analysis. However, the importance of intermediate results like stress or displacement is highlighted by some authors [8–9,13–14], which is justified by the possibility of reaching the material yield point prior to snap-through buckling. On the other hand, some other examination performed by the author with the use of FEM showed such a situation to occur rarely and under specific conditions, namely for relatively thick liners and those of significantly imperfect initial geometry. In addition, results of laboratory tests carried out so far confirmed that in practical cases buckling occurs prior to local plastic yield. For liners of nominal geometry well fitted into the lined pipe, it is the snap-through buckling, that prevails as the cause of failure.

For further development of the current analytical method it would be advisable to work out a valid global safety factor or a group of partial factors, based on the safety analysis. The safety factor of 2 or even 2.5 used so far seems conservative and it is deterministically based on experience and intuitive predictions rather than on rigorous analysis. Also some other initial geometrical imperfections should be incorporated into the current approach to adapt it better to the real-life conditions. However, unlike for the nominally circular liners, this does not seem so easy.

REFERENCES

Akl F., Guice L.K., Omara A.M., Straughan T.W., Instability of thin pipes encased in oval rigid cavity, Journal of Engineering Mechanics, April 2000, pp. 381–388.

Boot J.C., Elastic buckling of cylindrical pipe linings with small imperfections subject to external pressure, Trenchless Technology Research, Vol. 12, No. 1–2 (1998), pp. 3–15.

Boot J.C., Welch A.J., Creep buckling of thin-walled polymeric pipe linings subject to external ground water pressure, Thin-Walled Structures, No. 24 (1996), pp. 191–210.

Doll H., Hoch A., Standsicherheitsnachweis für Liner mit Finite-Elemente-Methode Tiefbau Ingenierbau Straßenbau, Vol. 15, No. 6 (1996), pp. 18–23.

El-Sawy K., Moore I.D., Parametric study for buckling of liners: effect of liner geometry and imperfections, Proceedings International Conference: Trenchless pipeline projects—practical applications, Boston, June 1997.

El Sawy K., Moore I.D., Analysis and design of oval liners for rigid pipe rehabilitation, Geotechnical Research Center Report GEOT-1-98, January 1998.

El Sawy K., Moore I.D., Stability of loosely fitted liners used to rehabilitate rigid pipes, Journal of Structural Engineering, Vol. 124, No. 11 (1998).

Falter B., Neue Aspekte bei der statischen Berechnung von Linern für die Kanalrenovierung, WAP, No. 1 (1999), pp. 24–30.

Falter B., Structural analysis of sewer linings, Trenchless Technology Research, Vol. 11, No. 2 (1997), pp. 27–41.

Glock D., Überkritisches Verhalten eines starr ummaltelten Kreisrohres bei Wasserdruck von außen und Temperaturdehnung, Der Stahlbau, No. 7 (1977), pp. 212–217.

Guice L.K., Straughan T., Norris C., Bennett R., Long-term structural behaviour of pipeline rehabilitation systems, internal report, Trenchless Technology Center, 1994.

Merkblatt ATV M127 Teil 2 Statische Berechnung zur Sanierung von Abwasserkanälen und -leitungen mit Lining- und Montageverfahren, GFA Verlag, Hennef, 2000.

Thepot O., A new design method for non-circular sewer linings, Trenchless Technology Research, Vol. 15, No. 1 (2000), pp. 25–41.

Thepot O., Structural design of oval-shaped sewer linings, Thin-walled Structures, No. 39 (2001), pp. 499–518.

Żuchowska D., Polimery konstrukcyjne, Wydawnictwo Naukowo-Techniczne, Warszawa, 2000.

Underground Infrastructure of Urban Areas 4 – Madryas et al. (Eds)
© 2018 Taylor & Francis Group, London, ISBN 978-1-138-55953-0

Dents in the walls of PVC-U sewers in the initial phase of their operation

E. Kuliczkowska
Kielce University of Technology, Kielce, Poland

ABSTRACT: The objective of the study was to analyse dents in PVC-U sewer pipes observed in the initial phase of the operation called the "burn in" period. It is a part of the "bath curve" that shows the whole service life of pipes, and which also includes two further subsequent phases, namely normal operation and "wear out" period. CCTV investigations were conducted in eleven cities. The analysis focused on quantitative data concerning size and frequency of dents observed in the walls of PVC-U sewers. Classification of dents was proposed, based on the dent size defined as the dent depth to the pipe diameter ratio. The reasons for dent occurrence were indicated. It was explained why trenchless rehabilitation of PVC-U pipes with dents cannot be performed using close-fit methods.

1 INTRODUCTION

The defects that occur in sewer pipelines are listed in the European Standard EN 13508-2 (2003). They affect the structural safety of sewer pipelines, e.g. cracks, fractures or corrosion are responsible for structural failures. Defects also influence sewer operation in terms of hydraulic capacity, e.g. roots of trees blocking the flow of sewages can cause operational failures. Additionally, defects are likely to pose a threat to the environment, e.g. sewage exfiltration from sanitary sewers is a source of groundwater pollution.

Dents, one of the most common types of structural defects, are discussed in many papers. For example, Allouti et al. (2012) analyzed the transmission pressure pipeline failure in Europe. The authors indicate that mechanical defects, such as dents and gouges, are the major cause (49.6%) of failures. Other possible reasons include material failure, corrosion, ground movement and hot-tapping errors. Dents occur not only in pressure steel pipes (Allouti et al. 2014), but also, more frequently, in gravity-flow PVC sewers.

In PVC-U sewers, a number of other defects are observed. The cause of pipe failure in PVC-U sewers were discussed by Scott et al. (2013), who analyzed a premature failure. From the material assessment, it is evident that root intrusion is a cause of failure in fractured pipes. Other causes include loads exceeding the structural capacity, exposure to aggressive effluents and excessive temperatures. The graphs provided by Fachverband der Kunstoffrohr-Industrie (2000) illustrate the effects of temperature on the permissible stress in the pipes as the function of time. It can be concluded that the higher is the temperature of sewage, the shorter pipe service life can be expected. Even chlorinated PVC, which are designed to withstand higher operational pressures and temperatures, are not resistant to weather factors such as UV, moisture and high ambient temperature (Merach, 2007).

PVC-U sewers that have been in operation for a longer period of time, show some defects that are the same as those observed in, e.g. in vitrified clay or concrete sewers, including root intrusion into sewers (Ostberg et al., 2012), cracks and many other damages described in Kuliczkowska (2015).

The typical defects of PVC-U pipes, not found, e.g. in vitrified clay or concrete pipes include the following:

- dents in the pipe wall, analyzed in this paper
- loss of stability of the pipe wall (buckling), discussed, e.g. in Kuliczkowska and Gierczak (2013),
- excessive deflection of the pipes, examined, e.g. in Kuliczkowska and Zwierzchowska (2016).

This paper aims to analyze the size of dents observed in newly laid PVC-U sewage pipelines, which are formed during and immediately after pipeline installation and commissioning. Also, the paper examines the frequency of dent occurrence in different Polish towns, in which dents were surveyed with the CCTV method.

2 METHOD AND MATERIALS

Investigations into PVC-U pipes with the use of the CCTV method, conducted by researchers from Kielce University of Technology, began in different cities of Poland in 1991. The first CCTV system, used at the very beginning, was portable and consisted of a camera, monitor, camera control panel, removable rod cassette with epoxy push rod, videocassette recorder, video printer, electric generator and rubber pipe plugs. Those had a large range of dimensions and were used for blocking the wastewater flow during inspections. The black and white forward view camera placed on skids was pushed along the sewers ranging from 100 mm to 1,000 mm in diameter by means of a flexible cable to the distance of about 100 m.

Starting from 1998, the second CCTV system was employed. Portable, and much more modern, the system allowed inspections of sewers that ranged from 150 mm to 1,000 mm in diameter. The trolley, with a camera mounted on it, could be driven to the distance of up to 100 m; the rotary color camera with the rotation range 360° vertically and 270° horizontally ensured a thorough visual inspection of the sewer inside.

The third and the latest CCTV system has been used since 2011. The system, mounted inside a van and provided with the specialist equipment for monitoring pipe tightness, is designed to inspect sewers 135-2,000 mm in diameter, measure longitudinal slopes, width of cracks and fractures, size of deformations, missing parts of sewer wall surface and other.

The pipes surveyed with the CCTV method were made from polyvinylchloride (PVC-U) which belongs to the group of thermoplastic materials. It is polymer of vinyl-chloride, which is made of ethylene and chlorine, and to which thermal stabilizers and lubricants are added for pipe production. In addition to hardness and form stability, PVC-U shows a high degree of resistance to acids, alkalis, alcohols, oils, fats, aliphatic hydrocarbons and benzene (Janson 1999; Stein, 2005; Kunstoffrohrverband e V., 1999), but it swells or dissolves in esters, ketones, chlorinated hydrocarbons and some other solvents. Despite the latter characteristic PVC-U can be used for sewer pipes.

PVC-U sewers, examined using the CCTV method, were manufactured in Poland. Depending on the needs, those were L class pipes that have SDR = 51 and SN = 2 kN/m^2, N class pipes with SDR = 41 and SN = 4 kN/m^2 or S class pipes with SDR = 34 and SN = 8 kN/m^2. The pipes were available in different lengths ranging from 500 mm to 12,000 mm. For pipes 250 mm to 500 mm in diameter, the construction lengths were 2000 mm, 3000 mm and 6000 mm.

PVC-U pipes are designed for the service life of 50 years, on the assumption that sewage temperature is 20°C. The graph provided in Kunstoffrohrverband e V. (1999) indicates that at 20°C, in the initial period PVC-U 250 pipes withstand stress of approx. 46 N/mm^2, but after 50 years, this value is only approx. 25 N/mm^2. However, at the temperature of 60°C the initial values of stress are approx. 20 N/mm^2 and after 50 years only approx. 5.9 N/mm^2.

The crucial parameters describing material properties of PVC-U include the following:

- density γ_R (N/mm^2) – 14.00,
- short term modulus of elasticity E_R (N/mm^2) – 3000,
- long term modulus of elasticity E_R (N/mm^2) – 1500,
- short term bending tensile strength σ_R (N/mm^2) – 3000,

- short term modulus of elasticity σ_R (N/mm^2) – 3000,
- tensile strength σ_Z (N/mm^2) – 45–55,
- yield limit σ_S (N/mm^2) – 55,
- impact strength (kJ/m^2) – 4,
- linear coefficient of expansion ($10^{-4}K^{-1}$) – 0.8,
- thermal conductivity (W/mk) – 0.15.

A review of the standards applicable to PVC-U pipes and the other material properties are found in the Kunstoffrohrverband e. V. (2000).

PVC-U pipes are designed to permissible stress, and permissible deflection. The design must account for the condition of the loss of stability (buckling) when safety factor is assumed to be 2.0 (for the safety class B for failure probability of 10^{-3}) or 2.5 (for the safety class A for failure probability of 10^{-5}). The possibility of dent formation is not taken into consideration because of the assumption is that pipes are installed in trenches backfilled with sand. In backfill material of this kind dents are not observed.

3 THEORY

3.1 *The ROCOF function*

Figure 1 presents a diagram of pipe technical condition over time. The ROCOF function describes the rate of failure occurrence (Rostum, 2000). The ROCOF function is the time derivative of the expected cumulative number of failures and is defined as:

$$\text{ROCOF} = d/dt\ E\,[N(t)] \tag{1}$$

where E [N(t)] denotes the mean number of failures in the interval (0, t].

The curve shown in Figure 1 represents repairable systems, where the system can fail several times. The ROCOF function is often high in the initial phase. Failures observed at this stage may result from:

- design errors, e.g. assumption of too low ground or traffic loads, or too small thickness of the pipe wall,
- manufacturing defects, e.g. use of poor quality materials with to low strength parameters, improper storage, or improper quality control of pipes,
- installation faults, e.g. human errors, transit damages, poor workmanship including improper backfilling of pits, improper bedding of pipes, impact damages or improper quality control.

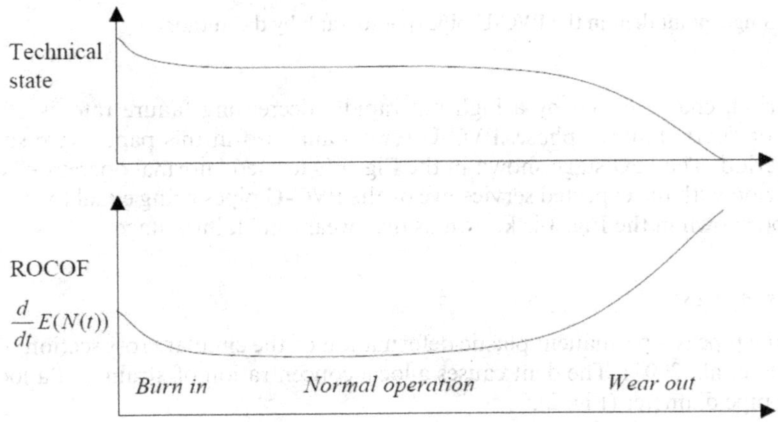

Figure 1. Service live of pipes (Rostum, 2000).

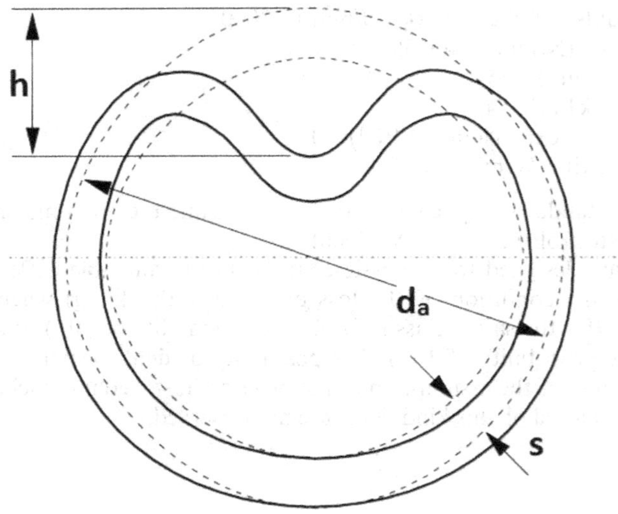

Figure 2. The dimensions of a dent.

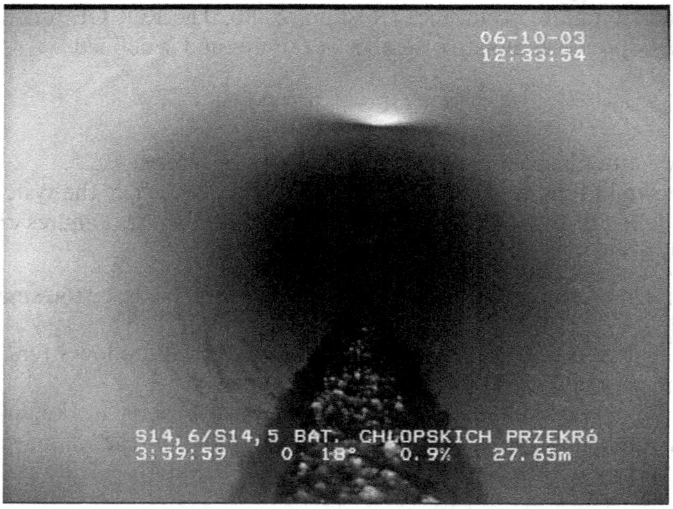

Figure 3. Single point dent in the PVC-U pipe (photograph by the author).

This period, characterized by a high but rapidly decreasing failure rate, is often called "burn in" or "early failure" phase. PVC-U sewers analysed in this paper were surveyed in this first period. The next stage shown in the Fig. 1 is termed "normal operation" or "stable failure" period with the expected service live of the PVC-U pipes being equal to 50 years. The third period shown in the Fig. 1 is known as the "wear out" failure stage.

3.2 *Dents in pipes*

A dent in the pipe is a permanent plastic deformation of the circular cross section of the pipe (Macdonald et al., 2007). The dent causes a local concentration of strain and a local reduction in the pipe diameter (Fig. 2).

Different types of pipeline dents are defined (Allouti et al., 2012), depending on their impact on pipeline wall dimensions and geometry. Dent types include the following:

– smooth dent: is a dent that caused a smooth change in curvature of the pipe wall,
– kinked dent: is a dent that caused an abrupt change in curvature of the pipe wall,
– plain dent: is a smooth dent that does not lead to wall thickness reduction,
– unconstrained dent: is a dent that is free to rebound elastically when the indenter is not removed,
– constrained dent: is a dent that is not free to rebound because the indenter is not removed.

The dent depth is the most significant factor affecting the burst strength and the fatigue life of a plain dent. In CCTV testing of PVC-U sewers single point dents (Fig. 3), multiple point dents on the same pipe (Fig. 4), linear dents (Fig. 5) and surface dents (Fig. 6) were observed.

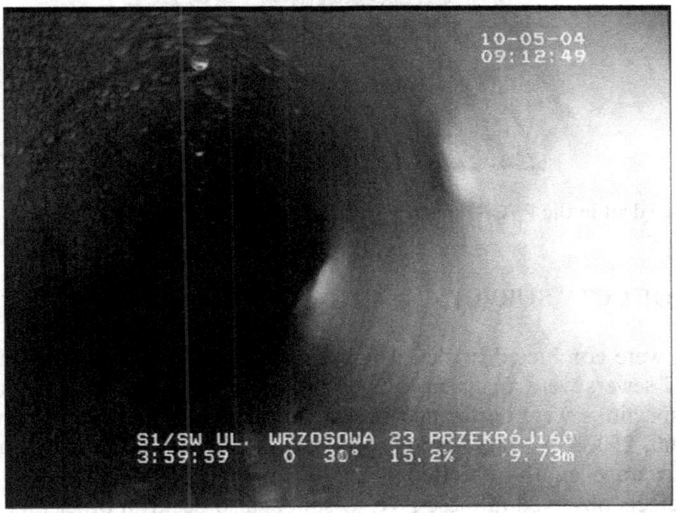

Figure 4. Multiple point dents in the PVC-U pipe (photograph by the author).

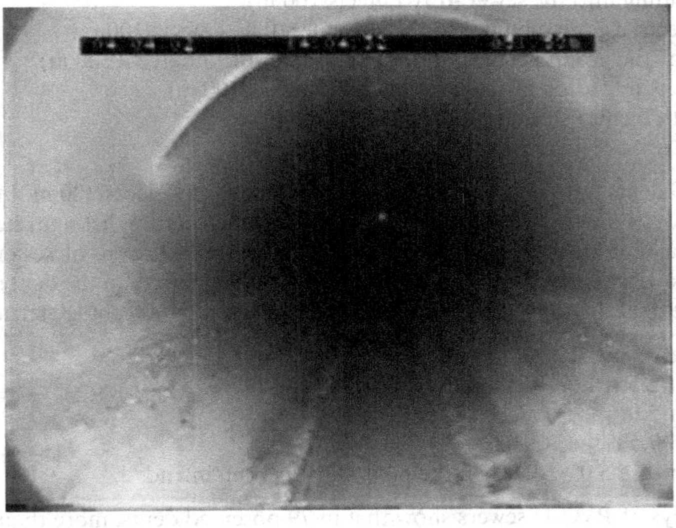

Figure 5. Linear dent in the PVC-U pipe (photograph by the author).

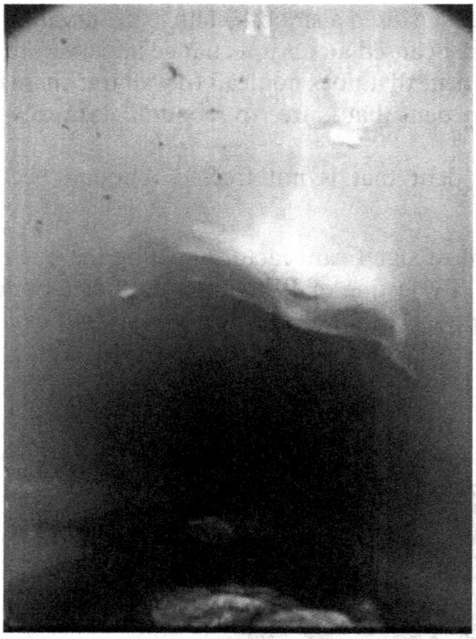

Figure 6. Surface dent in the PVC-U pipe (photograph by the author).

4 RESULTS OF CCTV SURVEYS

CCTV surveys were conducted in eleven Polish cities. A total of 11,388.6 m of randomly selected PVC-U sewers were analyzed to determine dent size and frequency of occurrence. Forty-three inspections were performed in different cities and streets for sewer diameters ranging 160, 200, 250, 315, 400, 500 and 630 mm. A total of 336 sections of sewers were inspected. The average length of section was 33.89 m.

Structural defects observed in these PVC-U sewers also occur in other pipes, e.g. vitrified clay and concrete pipes (Kuliczkowska, 2015; Kuliczkowska, 2016), such as circumferential cracks (0.01 pieces/100 m), diagonal cracks (0.03 pieces/100 m), or missing parts of sewer wall (0.10 pieces/100 m). The operational defects revealed in the surveys included:

- laterals intruding into the sewer (0.10 defects/100 m),
- external objects, e.g. pipe built into the structure (0.01 defects/100 m),
- infiltration of the groundwater into the leaky sewer (0.17 defects/100 m),
- sealing gaskets intruding into the sewer (0.12 defects/100 m),
- post infiltration encrustation (0.06 defects/100 m),
- longitudinal joint displacement (3.10 defects/100 m).

A total of 79 dents were registered, which amounted to 0.69 pieces/100 m. In most classifications, five groups of structural and operational sewer defects are distinguished. It has been proposed that a dent, denoted as d_p, should be included in the groups of sewer defects listed below. The classification of sewer defects into five groups is based on the dent size defined as the ratio of the dent depth to the sewer diameter, expressed in percentage terms:

- group G1 for $d_p < 2.5\%$,
- group G2 for $2.5\% \leq d_p < 5.0\%$,
- group G3 for $5.0\% \leq d_p < 7.5\%$,
- group G4 for $7.5\% \leq d_p < 10.0\%$,
- group G5 for $d_p \geq 5.0\%$, or when the pipe wall has been intruded.

CCTV surveys of PVC-U sewers show that in 79 observed dents, more than half of those were classified as belonging to G1 (14.3%) and G2 (41.4%). The third group G3 contained

26.6% of observed dents. The last two groups comprised the most serious dents, which together constituted 17.7% (16.4% of dents were classified as G4 group and only 1.3% as G5 group).

A dents could also be described by providing its circumferential area, or the ratio of the dent depth to the pipe wall thickness. In the case of linear dents (Fig. 5), the length along which they occur should also be given. However, the most important parameter characterizing the dent is the dent depth, because it most affects the probability of failure occurrence.

The results of CCTV survey indicate that dents occur as early as several days, weeks or months after pipe laying. When some of the surveyed pipes were exhumed, it was clear that the dents were mainly caused by the presence of stones or other hard materials pushing against the wall of the pipes.

In the case of dents occurrence, PVC-U sewers cannot be rehabilitated by means of the most commonly used trenchless technologies, i.e. close-fit lining with CIPP, HDPE, PP, PVC-U or GRP liners, which operate as non-structural or fully-structural components. Because of their visco-elastic properties, the liners will be subjected to local stresses caused by stones or other hard objects pushing against the rehabilitated PVC-U sewers. That would result in the occurrence of dents in rehabilitation liners. When dents are found, the number of methods that can be applied for the pipe renewal decreases considerably. In such a case, it is possible to

- replace the damaged pipes using the open-cut method, or trenchless technologies such as Berstlining, Hydros, Pipe-Eating or other methods,
- rehabilitate them using the sliplining method, which involves installing pipes with smaller diameter into the host pipe, and then grouting the annular space between these pipes with cement mortar, provided that a partial reduction in the flow capacity is acceptable.

5 CONCLUSIONS

1. CCTV surveys of newly laid PVC-U sewers showed the occurrence of defects typical of visco-elastic pipes, such as excessive deflections, buckling and dents.
2. In PVC-U pipes, different dents were found, namely single point, multiple point, linear and surface ones. The frequency of their occurrence was equal to 0.69 pieces/100 m.
3. Due to the fact that dents are not included in sewer defects acc. European Standard EN 13508-2, dent classification into five groups (from G1 to G5) was proposed. The criterion adopted was dent size defined as the ratio of the dent depth to the sewer diameter, expressed as percentage of the total. G1 group comprises the smallest, and G5 the greatest dent depths.
4. CCTV surveys showed that 14.3% of observed dents were classified as belonging to G1 group (pipes with the smallest dents), 41.4% to G2, 26.6% to G3, 16.4% to G4 and 1.3% to G5 group (pipes with the greatest dents).
5. Dents limit flow capacity of sewers to only a small extent. In the case, the dent grows and the indented piece becomes separated from the pipe, the sewer becomes leaky. Depending on groundwater table position, groundwater with soil particles can infiltrate into the sewer, or sewage can exfiltrate into soil and groundwater. Both cases pose a hazard to structural integrity of the sewer and environmental safety of the neighbouring areas.
6. Dents reduce the application of trenchless methods for sewer rehabilitation. Due to a change in circular shape of sewers, it will not be possible to employ, e.g. CIPP or other close-fit methods.

REFERENCES

Allouti, M., Schmitt, c., Pluvinage, G., 2014. Assessment of a gouge and dent defect in a pipeline by a combined criterion. Engineering Failure Analysis 36, 1–13. http://www.sciencedirect.com/science/article/pii/S1350630713003129.

Allouti, M., Schmitt, C., Pluvinage, G., Gilgert, J., Hariri S., 2012. Study of the influence of dent depth on the critical pressure of pipeline. Engineering Failure Analysis 21, 40–51. http://www.sciencedirect.com/science/article/pii/S1350630711002780.

EN 13508-2, 2003. Conditions of drain and sewer systems outside buildings—Part 2. Visual inspection coding system. Brussels.

Fachverband der Kunstoffrohr—Industrie, 2000. Plastic Pipe Manual. 4 th ed. Essen. (in German).

Janson L.-E. 1999. Plastics Pipes for Water Supply and Sewage Disposal, Borealis, Swen Axelsson AB/Fäldts Grafiska AB, Stockholm.

Kuliczkowska E., 2015, Analysis of defects with a proposal of the method of establishing structural failure probability categories for concrete sewers, Archive of Civil and Mechanical Engineering, 15 (4), 1078–1084, http://www.sciencedirect.com/science/article/pii/S1644966515000175.

Kuliczkowska E., 2016, Risk of structural failure in concrete sewers due to internal corrosion, Engineering Failure Analysis, 66, 110–119, http://sciencedirect.nk022.com/science/article/pii/S1350630716301972.

Kuliczkowska E., Gierczak M., 2013, Buckling failure numerical analysis of HDPE pipes used for the trenchless rehabilitation of a reinforced concrete sewer, Engineering Failure Analysis, 32, 106–112. http://www.sciencedirect.com/science/article/pii/S1350630713001076.

Kuliczkowska E., Zwierzchowska A., 2016, An analysis of early defects in PVC-U sewers and the limitations of their trenchless rehabilitation, Tunneling and Underground Space Technology, 56, 202–210, http://www.sciencedirect.com/science/article/pii/S0886779816301936.

Kunstoffrohrverband e.V. 2000. Kunstoffrohr Handbuch, 4. Auflage, Vulkan-Verlag, Essen.

Macdonald, K.A., Cosham, A., Alexander, C.R., Hopkins, P., 2007. Assessing mechanical damage in offshore pipelines—Two case studies. Engineering Failure Analysis 14, 1667–79. http://www.sciencedirect.com/science/article/pii/S1350630706002196.

Merah, N., 2007. Natural weathering effects on some properties of CPVC pipe material. Journal of Materials Processing Technology 191, 198–201. http://www.sciencedirect.com/science/article/pii/S0924013607002683.

Ostberg, J., Martinsson, M., Stal, O., Fransson, A.M., 2012. Risk of root intrusion by tree and shrub species into sewer pipes in Swedish urban areas. Urban Forestry & Urban Greening. 11, 65–71. http://www.sciencedirect.com/science/article/pii/S161886671100094X.

Rostum J. 2000: Statistical modeling of pipe failures in water networks. PhD Dissertation, Norwegian University of Science and Technology, Trondheim.

Scott, J.F., Davis, P., Beale, D.J., Marlow, D.R., 2013. Failure analysis of a PVC sewer pipeline by fractography and materials characterization. Engineering Failure Analysis 34, 41–50. http://www.sciencedirect.com/science/article/pii/S1350630713002379.

Stein D. 2005: Trenchless Technology for Installation of Cables and Pipelines, Stein & Partner GmbH, Bochum.

Underground Infrastructure of Urban Areas 4 – Madryas et al. (Eds)
 ISBN 978-1-138-55953-0

Development of renewal of water supply networks in Poland in years 2011–2015

M. Kwietniewski, K. Miszta-Kruk & J. Szmulewicz
Faculty of Building Services, Hydro and Environmental Engineering, Warsaw University of Technology, Warsaw, Poland

ABSTRACT: This article presents results of the survey that involves nearly 50% of the length of networks operating in Polish cities. Analysis of the collected data had allowed determining the type of renewal technology and the scope of its application to modernize and rebuild water supply networks. Furthermore, the scope of development of water supply networks has been defined in breakdown by arterial (A) and distributive (D) pipelines as well as water supply connections. The research results achieved in the specified time had also been referred to the state of progress of the renewal of water supply networks in the previous period including about 20 years of pipelines' operation. The research shows that renewal technologies used may be split into four groups consisting of traditional open trench replacement, trenchless replacement, reconstruction and renovation. All the technologies were applied regardless of the type of material of the old pipelines. However, the favorite technology used was traditional open trench technology.

1 INTRODUCTION

The most recent data [Municipal, 2016] indicate that in Poland in 2015 ca. 300 000 km of water supply networks were operated. Despite the fact that the availability of the collective water supply systems is significant (approximately 91,8% of the citizens use the systems) a further systematic development of water supply networks has been observed. In the years 2003–2005, an average annual growth was ca. 5000 km/year (Fig. 1). A careful analysis of network development in the given period allows to observe that it relates mainly to distributive pipelines dominating in water supply networks (Fig. 2); whereas the growth of main pipelines has being noticed only after 2009 (break of the trend—Fig. 3). Until that time, the length of main pipelines in Poland was decreasing by around 900 km/year on average, which could mean that there were not large investments carried out. Only after the year 2009 there have been a slight upward trend to be observed (about 150 km/year on average) in the area of building of arterial networks (Fig. 2). Steady development of distributive pipelines with smaller diameters is mainly associated with development of housing and services buildings in Poland.

Material structure of water supply networks is currently much more diverse than in the past, when dominant were mainly steel, grey cast iron and in large part asbestos cement [Guidelines, 1964]. Since 1990 also other materials have been extensively used to build networks including mainly thermoplastics such as PVC and PE and ductile cast iron (Fig. 4) [Kwietniewski, 2011].

Research on the scope of use of various technologies of the renewal of water supply networks, material structure, failure rate and reliability of the networks has been conducted in Poland in 5-year cycles for almost 25 years by the Department of Water Supply and Wastewater Management of the Warsaw University of Technology.

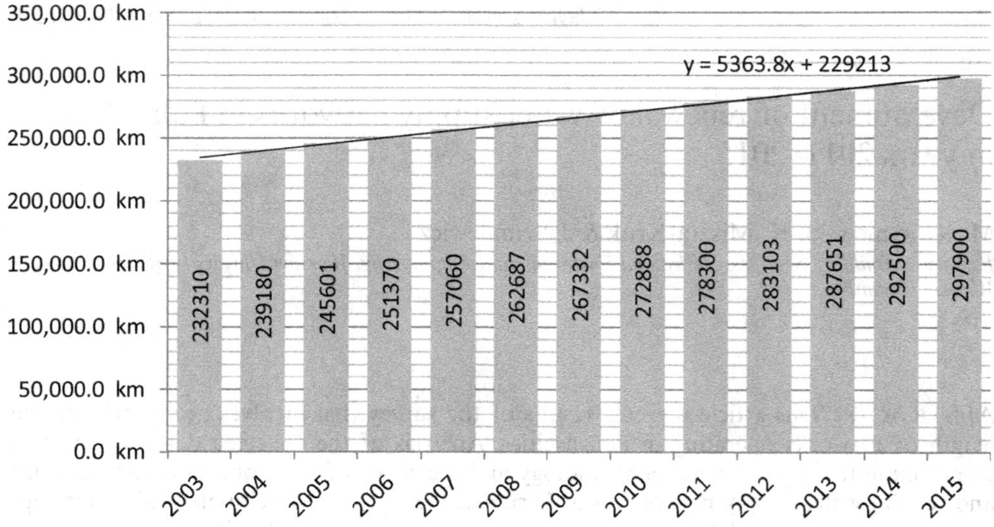

Figure 1. Development of water supply networks (A + D) in Poland in the years 2003–2015 [Municipal, 2016].

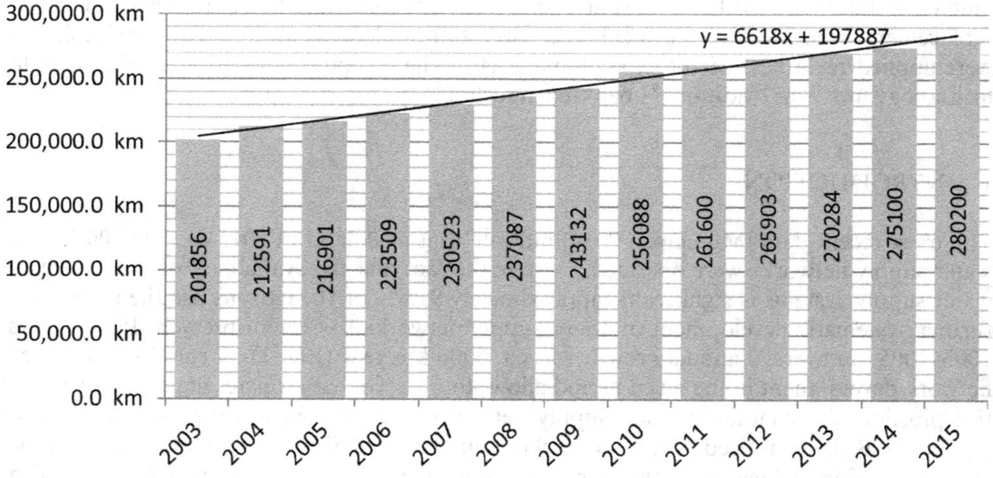

Figure 2. Development of distributive pipelines (D) in Poland in the years 2003–2015 [Municipal, 2016].

This article presents the most important results of the research conducted in the years 2011–2015 and relating mainly to the scope of using various technologies of renewal of water supply systems.

2 RESEARCH SUBJECT AND METHODOLOGY

The tests were mainly aimed to diagnose the following:

- the scope of using various technologies to renew water supply pipelines
- material structure of water supply networks and its changes in the last 5 years.

The basis for the research was the survey data obtained from 105 companies, among which were the largest, operating water supply networks in cities with a population over 100 000.

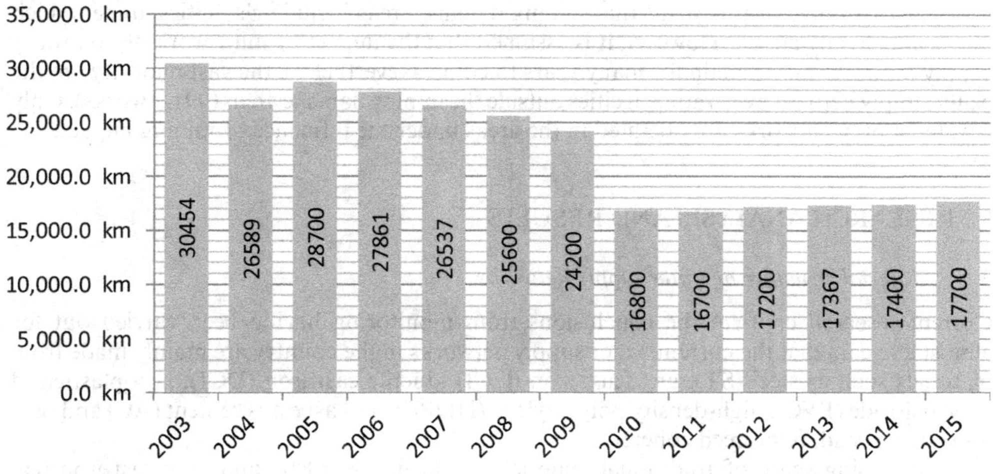

Figure 3. Development of main pipelines in Poland in the years 2003–2015 [Municipal, 2016].

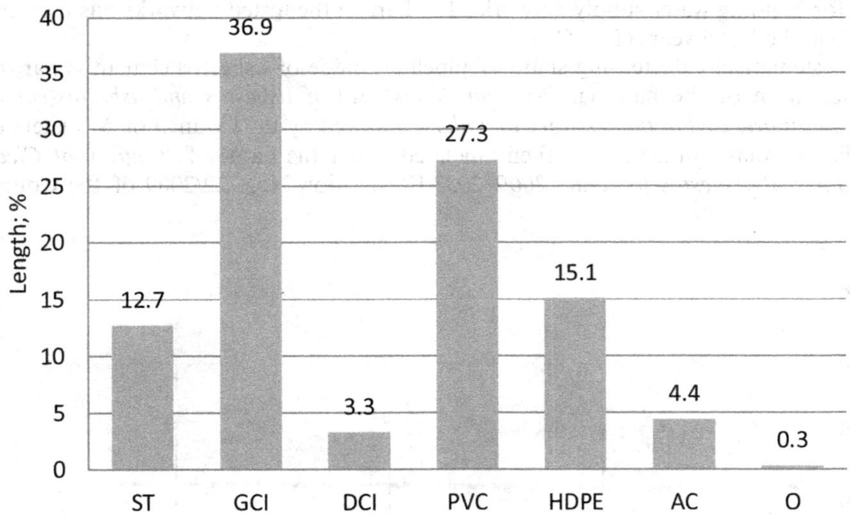

Figure 4. Material structure of examined water supply networks in Poland in 2008 [Kwietniewski, 2011]; Where: steel (ST), grey cast iron (GCI), ductile cast iron (DCI), non-plasticized vinyl chloride (PVC), high-density polyethylene (HDPE), and asbestos cement (AC) and others (O).

The research covered above 48,6% of the length of water supply networks that operate nearly 50% of the population of Polish cities. Therefore, it could be assumed that the gathered research material was a very good base for an assessment of the scope of using renewal and of the material structure of water supply networks in Polish cities.

The main concern in collecting the data were gaps resulting mainly from the fact that the GIS data base has been not implemented yet in full or at all in many Polish companies. The process of GIS implementation, being monitored currently, gives a chance to significantly increase the possibility of acquiring the data for various kinds of research in the future.

Analysis and evaluation of the scope of the renewal of water supply networks were carried out taking into account various technologies used for that purpose. Furthermore, materials of old pipelines being renewed and pipelines already renewed were identified, and also pipelines that were operating in mining damage areas had been distinguished.

It must be noted at this point that mining damage areas include also cities, under which mines are already closed. However, it is assumed that the impact of mine workings on water supply pipelines may remain for many years to come. Nevertheless, the vast majority are the water supply networks operating in cities outside the mining damage areas (99 networks). Only 6 water supply networks are operated in the areas under the influence of mining impact.

3 RESEARCH ANALYSIS AND RESULTS

3.1 *Material structure of water supply networks*

Current research confirm the conclusions from monitoring having been carried out for almost 20 years that the current water supply networks in the country are mainly made from materials such as steel (ST), grey cast iron (GCI), ductile cast iron (DCI), non-plasticized vinyl chloride (PVC), high-density polyethylene (HDPE), and asbestos cement (AC) and others (O), e.g. lead, reinforced concrete.

A decreasing share of traditional materials [i.e. steel (ca. 1,8%) and grey cast iron (ca. 1,0%)] in water supply networks could still be observed in recent years 2011–2015. Moreover, a decreasing share of pipelines made of PVC (ca. 0,7%) and further reduction of the share of AC pipelines (by ca. 0,2%) were also noticed in the examined networks. However, PE is used largely for building water supply networks. Its share in the tested networks has increased by ca. 4,9% in the last 5 years (Fig. 5).

The systematically decreasing share of pipelines made of asbestos cement is part of the implementation of the national *Program of disposal of asbestos and asbestos-containing products that are used in the territory of Poland* adopted by the Council of Ministers of the Republic of Poland in 2002, and then amended under the name of *Program of Cleansing the Country of Asbestos for years 2009–2032* [Resolution No. 122/2009 of the Council of

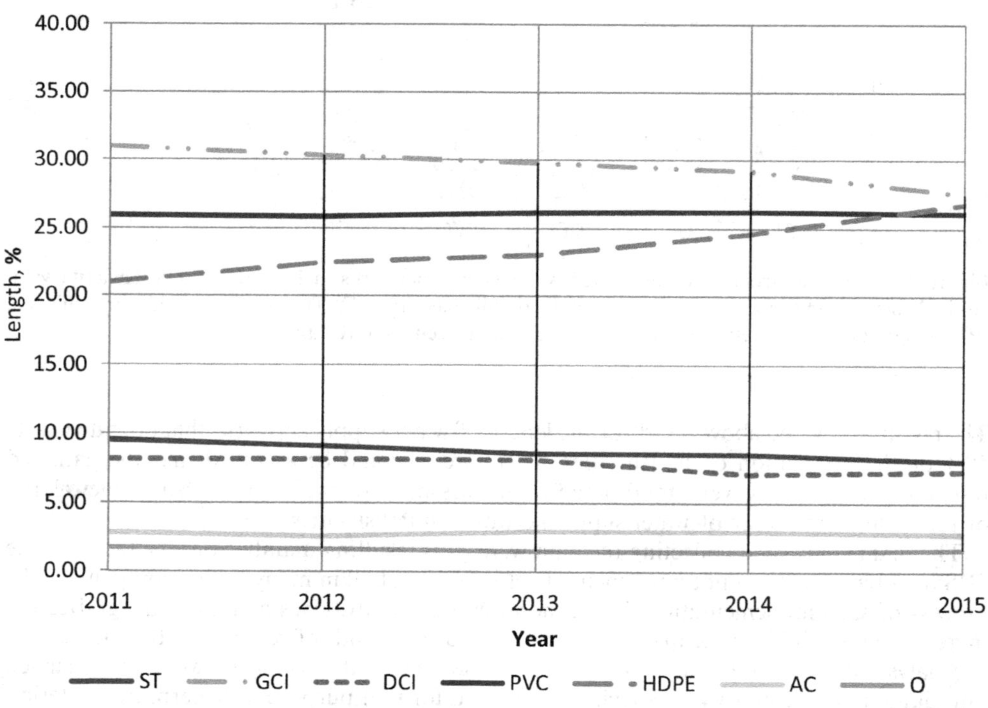

Figure 5. Material structure of tested water supply networks outside the mining damage areas in the years 2011–2015.

Ministers of July 14, 2009]. Based on the current implementation of that program; it is difficult to notice a significant progress in the removal of asbestos-containing products, including water supply pipelines; in the country.

Similar analysis were carried out for water supply networks laid in the mining damage areas. The result is Fig. 6.

Similarly to the case of pipelines from outside the mining damage areas; a downward trend in the shares of steel, ductile cast iron, asbestos cement and PVC is observed in the last five-years period; the share of steel being decreased to the greatest extent (by ca. 4,6%) in the last 5 years. However, the opposite trend is shown by PE, which contributed to an increase in the length of network by 6,3% accordingly. Furthermore, it is worth noting that the PE distinguishes itself by the biggest share in the material structure of the networks operating in mining areas throughout the whole period of the research. Material structure of water supply networks from the end of the year 2015 is shown in the Fig. 7 (pipelines from outside the mining damage areas) and in Fig. 8 (pipelines in the mining damage areas).

Figures 7 and 8 show clearly that both in mining damage areas and outside, water supply networks are mostly built of HDPE. In areas not affected by damage are gray cast iron and PVC pipes are dominated.

3.2 *Renewal of water supply networks*

The technologies applied to renew water supply pipelines in the analyzed period are divided into four groups; namely, traditional trench replacement, trenchless replacement, renovation and reconstruction. The analysis of received results demonstrates clearly that throughout the period considered the most frequently used pipeline renewal technology was the traditional open trench replacement. The use of that technology is showing an upward trend during the period considered and shall decide for the most part on a growing trend in the field of

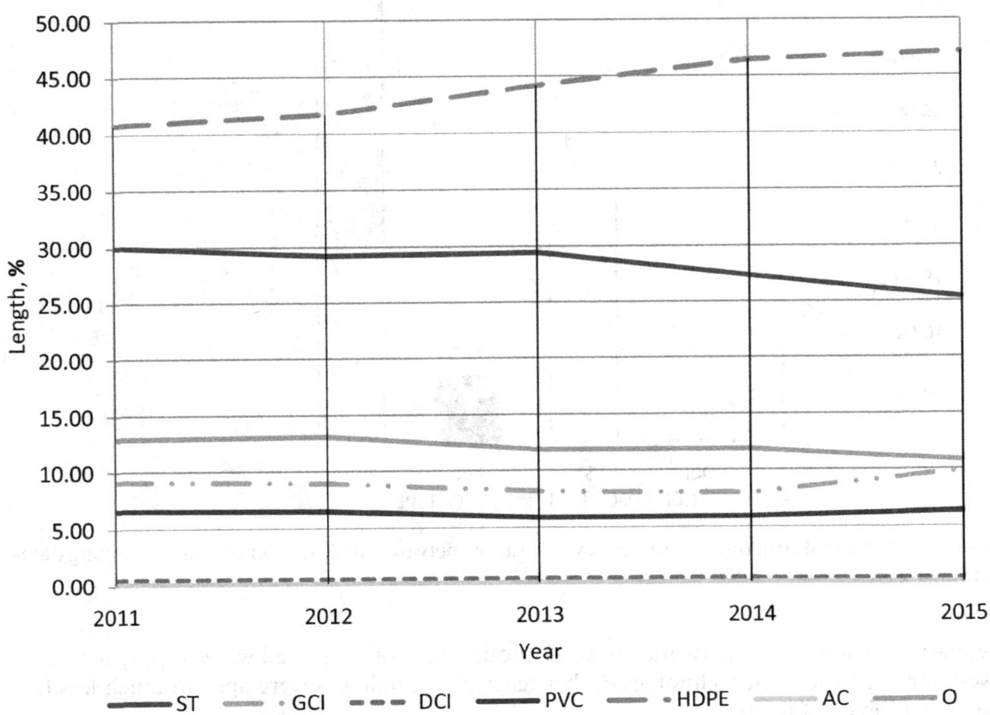

Figure 6. Material structure (%) of water supply network in the mining damage areas in years 2011–2015.

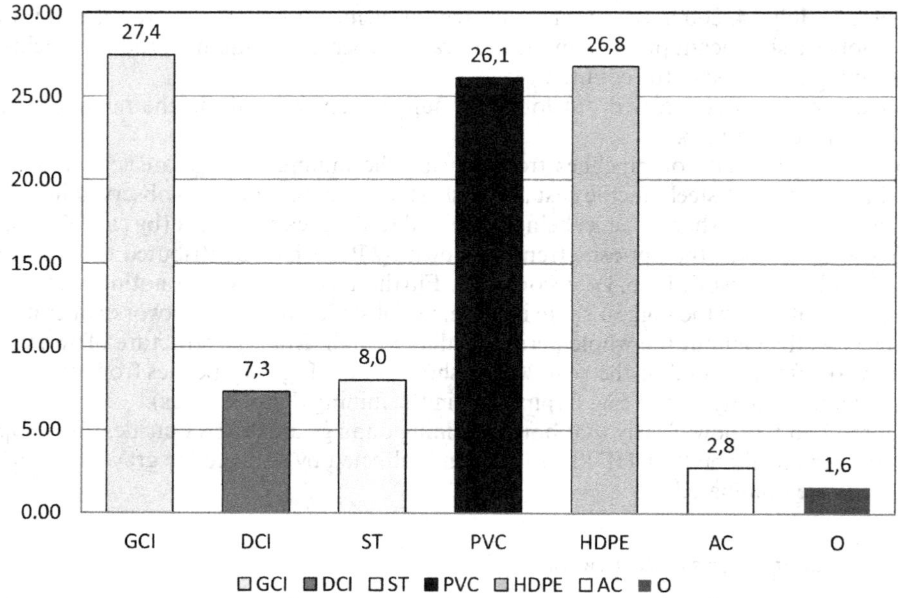

Figure 7. Material structure of water supply networks in 2015 operating outside the mining damage areas.

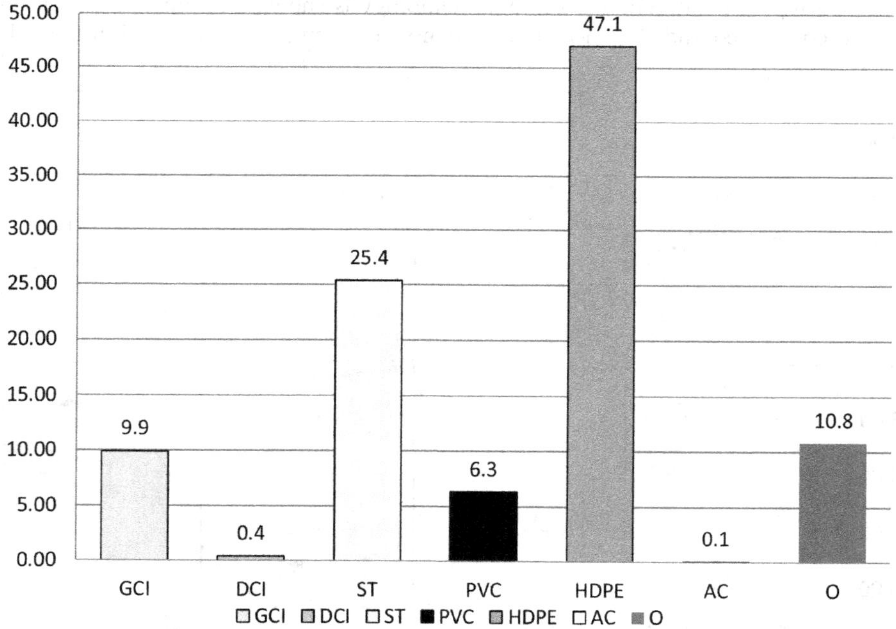

Figure 8. Material structure of examined water supply networks in 2015 operating in the mining damage areas.

renewal of water supply networks. In total, around 80% of examined water supply networks were renewed with that technology. Other renewal technologies were applied much less frequently (Table 1, Fig. 9).

In general, 1,9% of the examined water supply pipelines was renewed in the last 5 years including traditional open trench method (1,4%), trenchless methods (0,3%) and renowation (0,2%).

Table 1. The scope of renewal of examined water supply pipelines laid out outside mining damage areas in the years 2011–2015.

Renewal technology	% of length of renewed pipelines
A. Traditional trench replacement	79,6
B. Trenchless replacement	13,8
including:	
Pipe bursting	1,3
Others	12,5
C. Renowation	4,2
including:	
Relining with a flexible sleeve	0,2
Other (e.g. cementing)	4,0
D. Reconstruction	
e.g. lining a pipeline with a compressed pipe liner	2,4
Total	100,00

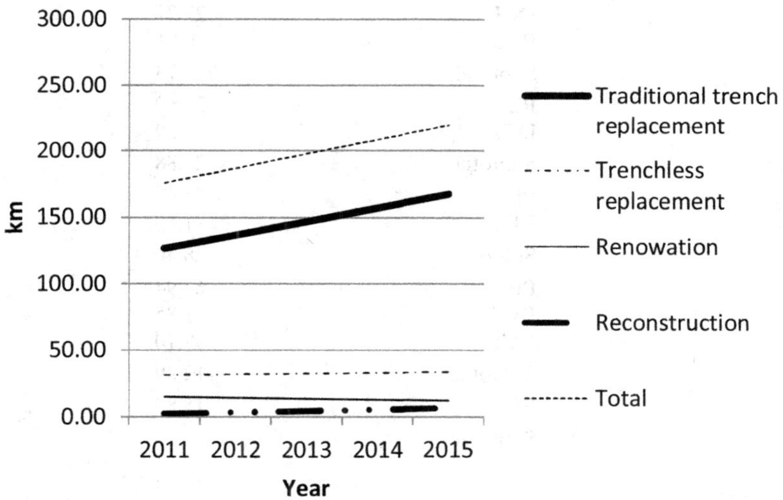

Figure 9. Development of technologies of examined water supply pipelines laid out outside mining damage areas in the years 2011–2015.

By analyzing material structure of old pipelines being renewed and pipelines after renewal (Table 2) it may be noticed that most frequently renewed pipelines are the pipelines built of grey cast iron and of steel. This fact is closely linked to a high failure rate of pipelines built of those materials. They are replaced mainly by pipes built of PE, which is characterised by low failure rate in comparison to other materials. Likewise, asbestos cement pipes are often replaced, not only when broken, but also in accordance with the strategy of disposal of asbestos-containing products from the market. It is usually replaced by PE and ductile cast iron pipes.

PE and PVC pipelines are the rarest pipelines renewed. Pipelines made of those materials are most often used for building and renewal of water supply networks.

Given the dominant share held by open trench technology, material structure of pipelines undergoing renewal and pipelines renewed with that technology has been analyzed. (Table 3 i 4).

Conclusions drawn from the data in the Tables 3 and 4 confirm the above observations, namely: the traditional method was used mainly to replace pipelines of grey cast iron (above 40%) and steel (around 32%). They were being replaced preferably with PE pipes (ca. 72%) and, to a lesser extent, with pipes made of ductile cast iron (ca. 27%).

Table 2. Material structure of old pipelines undergoing renewal and laid out outside mining damage areas in the years 2011–2015.

Pipeline material	Length of renewed pipelines		
Before renewal	After renewal	km	%
GCI	PE	194,81	56,06
	PVC	6,31	1,81
	DCI	146,28	42,10
	GCI	0,08	0,02
	Subtotal:	347,47	100,00
DCI	PE	47,78	86,09
	Steel	0,03	0,05
	DCI	7,59	13,68
	GCI	0,10	0,18
	Subtotal:	55,50	100,00
ST	PE	241,00	89,02
	ST	0,01	0,00
	DCI	29,71	10,97
	PVC	0,01	0,00
	Subtotal:	270,73	100,00
PVC	PE	27,49	95,19
	DCI	1,39	4,81
	Subtotal:	28,88	100,00
PE	PE	26,66	93,68
	ST	1,80	6,32
	Subtotal:	28,46	100,00
AC	PE	63,94	56,64
	PVC	1,85	1,64
	DCI	47,10	41,72
	Subtotal:	112,89	100,00
O	PE	15,80	100,00
	Subtotal	15,80	100,00
Total		859,73	100,00

Table 3. Material structure of water supply pipelines, undergoing renewal with traditional open trench method, laid out outside mining damage areas in the years 2011–2015.

Material of pipeline undegoing renewal	Length, km	% of total length of renewed pipelines
GCI	347,47	40,42
ST	270,73	31,49
AC	112,89	13,13
DCI	55,50	6,46
PVC	28,88	3,36
PE	28,46	3,31
Total	859,73	100,00

3.3 *Renewal of networks operating in the areas of mining damage*

Similar analysis have been conducted not only for pipelines located in the mining damage areas, but also for pipelines operating outside those areas. Unfortunately, in the latter case the collected data are much more modest. After all, the data has been obtained only from 5 companies. Therefore, the results of analysis of these data may only be informative. They are presented in the Tables 5–7 and Fig. 10.

Table 4. Material structure of water supply pipelines renewed with traditional method and laid out outside mining damage areas in the years 2011–2015.

Material of pipelines renewed	Length, km	% of total length of pipelines renewed
PE	617,48	71,82
DCI	232,07	26,99
PVC	8,17	0,95
ST	1,84	0,21
GCI	0,18	0,02
Total	859,73	100,00

Table 5. Material structure of water supply pipelines undergoing renewal in the mining damage areas in the years 2011–2015.

Renewal technology	% of length of renewed pipelines
A. Traditional trench replacement	93,39
B. Trenchless replacement*	2,88
C. Renowation*	3,73
Total	100,00

*There are no data available on the type of technology.

Table 6. Material structure of water supply pipelines undergoing renewal by traditional trench method in the mining damage areas in the years 2011–2015.

Material of pipe under renewal	Length, km	% of total length of renewed pipelines
Steel	137,84	85,93
Ductile cast iron	12,24	7,63
Grey cast iron	5,56	3,47
PE	3,28	2,04
PVC	1,49	0,93
Total	160,41	100,00

Table 7. Material structure of pipelines after renewal by traditional open trench method in the mining damage areas in the years 2011–2015.

Material of pipeline after renewal	Length, km	% total length of pipelines after renewal
PE	160,41	100,00

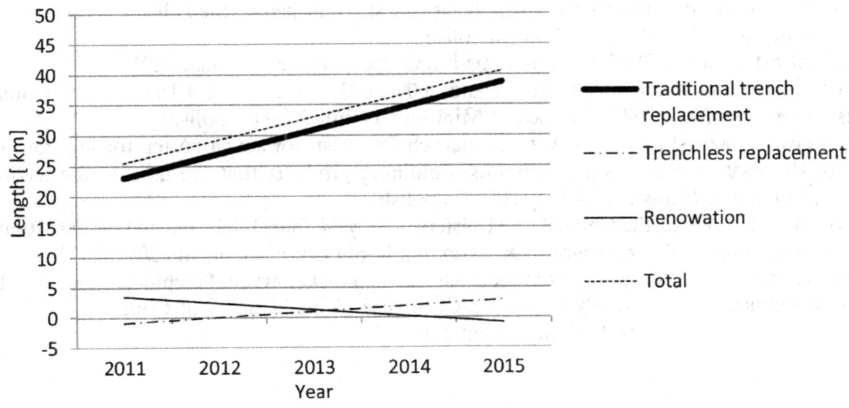

Figure 10. Material structure of old pipelines undergoing renewal and laid out outside mining damage areas in the years 2011–2015.

In the same way as in the case of pipelines from outside mining damage areas, also inside those areas, the most widely used is the traditional open trench method. Trenchless method and renovation were applied to 2,88% and 3,73% of all renewed pipelines, respectively. A significant increase in the implementation renewal with both open trench and trenchless method was observed in 2015 as compared with 2014. There is an upward trend in the field of application of trenchless methods. The most frequently replaced pipelines were the pipelines made of steel (having the highest failure rate). All the water supply pipelines in the mining damage areas had been replaced by pipelines made of PE, material that has spread successfully in those areas.

4 CONCLUSIONS

Pipelines made of traditional materials, namely steel and grey cast iron steel, dominate in municipal water supply networks in Poland. These pipelines are still nearly 35% of the network. Furthermore, pipelines made of HDPE and PVC account for a 53% of examined networks, respectively.

There is a systematic increase in length of networks built of PE in recent years. Over 6% growth of the length of networks built of this material was reported for the period of 2011–2015.

Old pipelines made of traditional materials (grey cast iron and steel) are replaced mainly by PE pipelines. All the pipelines under process of renewal that are located in the mining damage areas are being replaced by this material.

Current research confirm the continuing trends observed for nearly 25 years [Kwietniewski, 2012] on the development of renewal technology of water supply networks in Polish cities as well as their material structure.

In conclusion, it is worth noting that currently many products (pipes, pipe fittings) made of different materials designed for building of water supply networks are available on the Polish market. Therefore, there is an issue of choice of such material, which in given circumstances would best meet requirements of the users of water. A proposal for a methodology for such a choice is presented in the work [Kwietniewski, 2011].

REFERENCES

Kwietniewski M., Miszta-Kruk K., 2012: Trends in application of materials to construction and rehabilitation of water supply networks in Poland after 1990. W: Madras C., Nienartowicz B., Szot A. (Eds), Underground Infrastructure in Urban Areas 2 First Edition, pages 203–211, (Hbk + CD-ROM), CRC Press/Balkema Taylor & Francis Group, London.

Kwietniewski M., Tłoczek M., Wysocki L., 2011: Zasady doboru rozwiązań materiałowo-konstrukcyjnych do budowy przewodów wodociągowych. [The principle of selection of material and structural solutions for building water supply pipelines], Published by the Polish Economic Chamber "Polish Waterworks", Bydgoszcz, 2011 (in polish).

Municipal infrastructure in 2015, Central Statistical Office of Poland, Warsaw, 2016.

Program Oczyszczania Kraju z Azbestu na lata 2009–2032, [Program of Cleansing the Country of Asbestos for years 2009–2032], Council of Ministers, 14 July 2009 (in polish).

Program usuwania azbestu i wyrobów zawierających azbest stosowanych na terytorium Polski. [Program of disposal of asbestos and asbestos-containing products that are used in the territory of Poland], Council of Ministers, 14 May 2002 (in polish).

Resolution No. 122/2009 of the Council of Ministers of July 14, 2009 [Uchwała nr 122/2009 Rady Ministrów z dnia 14 lipca 2009 r. Program Oczyszczania Kraju z Azbestu na lata 2009–2032].

Wytyczne techniczne projektowania komunalnych sieci wodociągowych. [Technical guidelines for the design of municipal water supply networks], Ordinance of the Ministry of Communal Administration of 17 January 1964, (Dz. Bud. no 8, item 26).

Underground Infrastructure of Urban Areas 4 – Madryas et al. (Eds)
© 2018 Taylor & Francis Group, London, ISBN 978-1-138-55953-0

Evaluation of the effect of ribbed road plate foundation conditions on subgrade durability

P. Mackiewicz, Cz. Machelski & A. Szydło
Institute of Civil Engineering, Wroclaw University of Science and Technology, Wroclaw, Poland

ABSTRACT: The road plate presented in this paper is a universal structure designed for excavation works on building sites and on roads and streets under repair. In cases of failures of urban buried utilities and extended repairs the plate ensures proper operational conditions and vehicular traffic safety. It enables continuous heavy goods vehicle traffic without limiting the carriageway width. This paper presents numerical finite element analysis of the effect of road plate foundation conditions on the durability of the subgrade. The displacements and deformations affecting the durability and longevity of such a system are analysed depending on the location of the load on the plate and the extent to which the plate rests on the excavation shelves. Two extreme cases of plate foundation: 1) foundation on weak subsoil and 2) foundation on flexible road pavement are considered.

1 INTRODUCTION

Today failures of buried utilities in strongly urbanized areas are a frequent occurrence. In such cases, extended access to the underground network and its elements is usually required to carry out repairs, inspections and follow-up repairs.

On the initiative of the Municipal Water and Sewerage Company in Wroclaw the Wroclaw Road Plate Project was conceived in order to make the use of roads possible during works connected with the removal of failures of buried utilities. Thanks to its design the road plate enables the passage of vehicles with an unlimited load (permissible on roads by the traffic regulations) over up to 2 m wide excavations running both along and across the roadway.

As regards traffic lane occupancy the use of such plates on general access public roads is agreed with the road authority. An application for the works, with the traffic lane occupancy marked and the way of protecting and signposting specified, is submitted. Thanks to the pre-agreed typical substitute traffic organization schemes road plates can be promptly installed on the construction site, whereby failures can be quickly responded to and repair work can begin immediately. In order to ensure proper safety it is recommended to limit the vehicle traffic speed on the crossover deck to 30 km/h (road sign B-33) and provide information about the conducted works (road sign A-14) and any other dangers. The exemplary use of the road plate on Wroclaw streets is shown in Fig. 1.

2 DESCRIPTION OF ROAD PLATE STRUCTURE

The load-bearing structure of the road plate is made up of the following two main components: component A (riffled deck plate) and component B (381 × 140 × 5.5 corrugated steel sheet). Component B consists of two corrugated sheet parts—elements B1 and B2. The 9 mm thick deck plate has been designed to directly carry traffic. Its top surface is riffled to ensure proper traction for vehicle wheels. A schematic of the road plate is shown in Fig. 2.

As regards road plate type series, four plate shapes are distinguished: a middle plate, a left plate, a right plate and a stand-alone plate. In each of the versions the plates equally well

Figure 1. Exemplary use of road plate on Wroclaw streets.

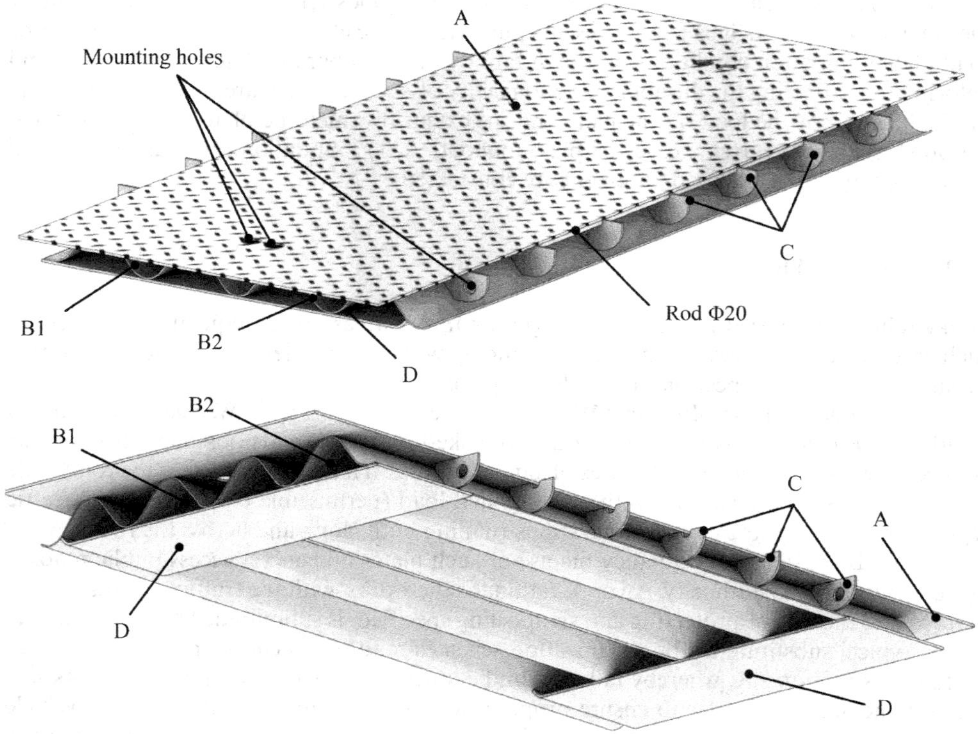

Figure 2. Schematic of road plate: top view and bottom view.

perform their structural and load-bearing functions. The plates only slightly differ in their shape whereby they can be joined together and the edges of the whole structural system covering an excavation can be appropriately finished. Any number of middle plates can be connected together. The plate edges occur in two different forms, which ensures their proper interlocking during assembly. At its ends a continuous system of middle plates is completed with a left plate and a right plate. By joining plates of the different types together one can obtain variously long plate systems. Systems made of two plates (left plate+right plate), three plates (left plate+middle plate+right plate) and more plates (left plate+any number of middle plates+right plate) can be distinguished. The stand-alone plate is used to cover smaller excavations. It does not require any connections (all its edges are properly rounded).

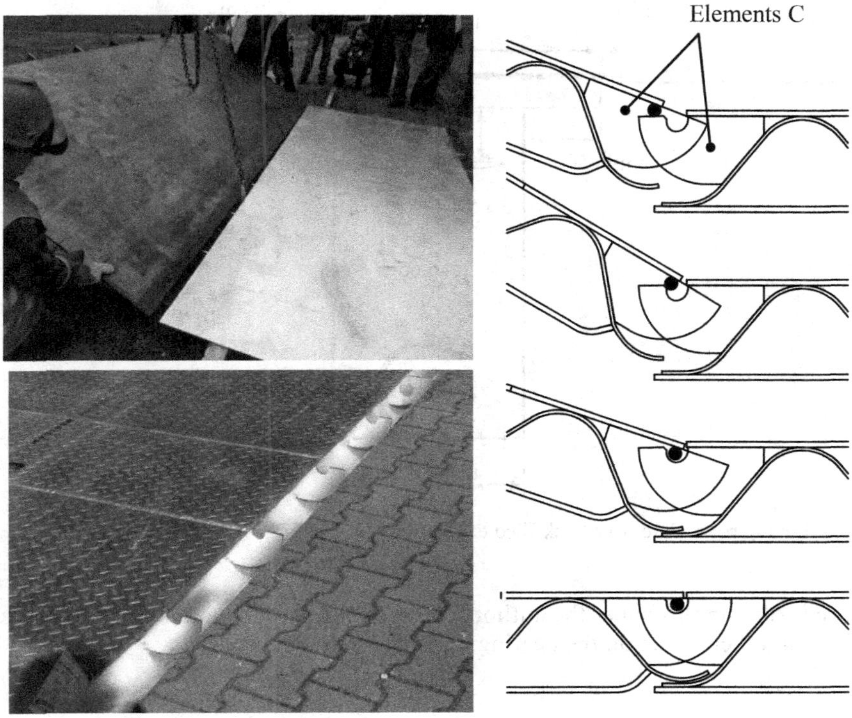

Figure 3. Assembly of road plate.

The plates can be connected by means of elements C called ribs. On one side elements C have "open" round holes for interlocking the plates during assembly. Elements C also perform a structural function in a segment, bracing the supports of the deck plate and the bottom part of the corrugated sheet. The outermost ribs with "closed" holes in them are used for transporting the segments. Along component A between elements C on one side of the plate there is a rod 20 mm in diameter used for interlocking the plates during assembly. Its interlocking takes place in elements C with open holes. The way in which the plates are joined together is shown in Fig. 3.

3 RANGE OF ROAD PLATE APPLICATION

The road plate is intended for laying over a backfilled excavation (Fig. 4). The width of the excavation at its bottom cannot exceed 200 cm. In the top part of the excavation the pavement should be removed to a depth of about 15 cm and to a width of max. 50 cm. This is the so-called shelf on which the road plate rests. The dimensions pertaining to resting the plate on the excavation edges (min. 50 cm) and on the pavement (20 cm) should be complied with. Flat sheet D transfers the vehicle traffic load to the subgrade. The sheet is fillet welded to the corrugated sheet at the segment's end. The dimensions of element D are: $1525 \times 500 \times 9$ mm and those of element A: $3500 \times 1519 \times 9$ mm. Thus the maximum width of the support under element D amounts to 50 cm while the width of the support under the protruding part of element A amounts to 25 cm.

The compactness of the soil in the excavation should amount to about 0.99. The shelves on which the plates (especially plate C) rest should be level and compacted to the index of 0.99. However, in practice it may happen that the road plate will rest unevenly on improperly prepared surfaces and the requirements concerning soil compaction and contact area between the plate's structural elements, the subgrade and the pavement will not be complied with.

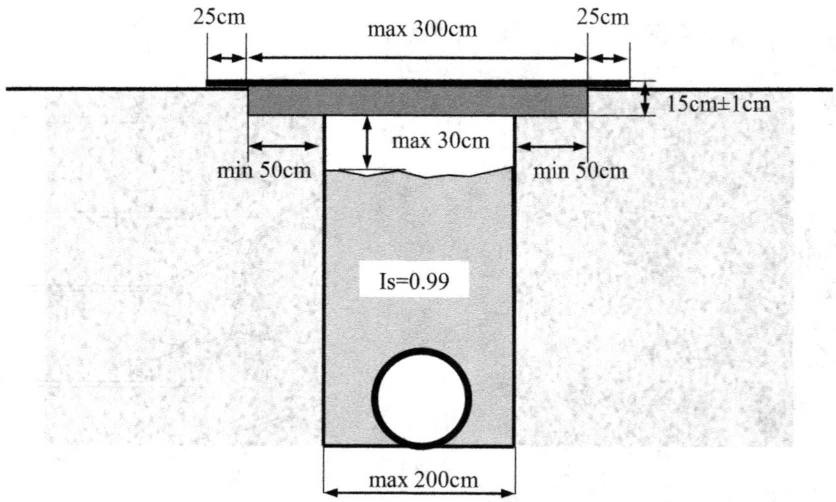

Figure 4. Laying road plate over backfilled excavation.

Therefore further in this paper the authors consider different soil and road plate foundation conditions and their effect on the bearing capacity of the subgrade.

4 NUMERICAL PLATE & SUBGRADE MODEL

Two plate & subgrade configurations were subjected to analysis. In the first configuration the plate is founded on an excavation made in week soil and backfilled. The second configuration is a typical configuration used in practice today: the plate is founded on a flexible road pavement with a properly prepared and compacted subgrade. The finite element method was used for model computations. Linear elastic models characterized by elastic modulus E and Poisson's ratio ν were assumed for the steel plate and the road pavement. The Drucker-Prager elastic-plastic model was used to describe the nonlinear properties of the soil. The soil is described by the modulus of elasticity, internal friction angle ϕ and cohesion c. The latter two parameters are used to determine the yield criterion. In the space of the principal stress this criterion has the shape of a cone of revolution with its axis coinciding with the axis of isotropic stress. The yield surface is described by the relation:

$$F = a \cdot I_1 + \sqrt{J_2} - b \tag{1}$$

where: I_1 = the first stress tensor invariant; I_2 = the second stress tensor invariant; a, b = coefficients dependent on the material parameters,

$$a = \frac{2 \cdot \sin\phi}{\sqrt{3} \cdot (3 - \sin\phi)}, \; b = \frac{6 \cdot c \cdot \cos\phi}{\sqrt{3} \cdot (3 - \sin\phi)}.$$

The load in the loading diagram was assumed as a model of a wheel load of 57.5 kN corresponding to an axle load of 115 kN (Fig. 5). The load pulse was 0.02 s. Different variants of the load simulating the crossing of the excavation by a vehicle were analysed, whereby the most adverse load conditions on the soil shelves were taken into account. Three load locations relative to the road plate's edge, described by variable k: k = 0.00 m, k = 0.25 m and k = 0.50 m, were considered.

The model parameters assumed for the plate & weak soil configuration are listed in Table 1 while the ones for the road pavement-stronger subgrade configuration are shown in Table 2.

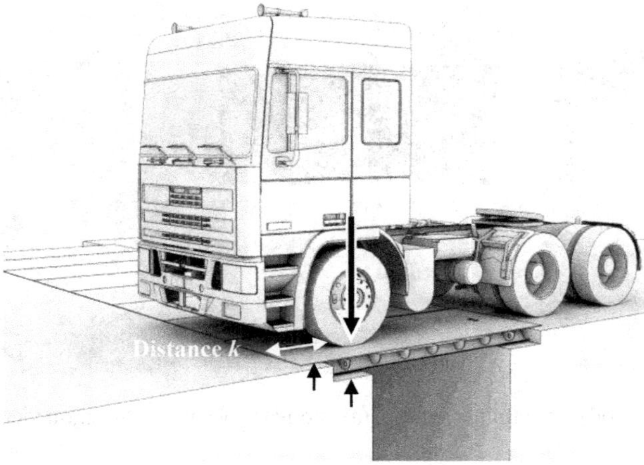

Figure 5. Plate & subgrade system loading scheme.

Table 1. Structural parameters for configuration 1 (excavation in weak soil).

Thickness [m]	Material parameters	Name of layer
0.168	$E = 210$ GPa, $\nu = 0.3$	Steel plate
0.15	$E = 60$ MPa, $\nu = 0.35$, $\phi = 15°$, c = 3 kPa	Top shelf-subgrade G2
2.00	$E = 60$ MPa, $\nu = 0.35$, $\phi = 15°$, c = 3 kPa	Subgrade G2
2.00	$E = 40$ MPa, $\nu = 0.35$, $\phi = 15°$, c = 3 kPa	Backfill-subgrade G3

Table 2. Structural parameters for configuration 2 (excavation in road pavement).

Thickness [m]	Material parameters	Name of layer
0.168	$E = 210$ GPa, $\nu = 0.3$	Steel plate
0.15	$E = 5000$ MPa, $\nu = 0.3$	Top shelf-flexible pavement
2.00	$E = 140$ MPa, $\nu = 0.35$, $\phi = 15°$, c = 3 kPa	Subgrade G1
2.00	$E = 60$ MPa, $\nu = 0.35$, $\phi = 15°$, c = 3 kPa	Backfill-subgrade G2

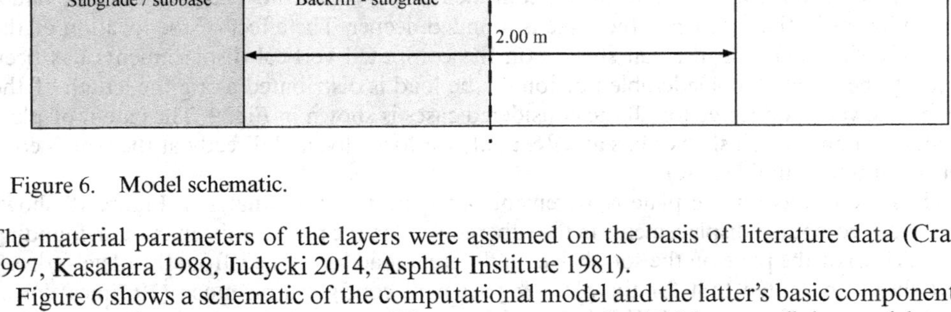

Figure 6. Model schematic.

The material parameters of the layers were assumed on the basis of literature data (Craig 1997, Kasahara 1988, Judycki 2014, Asphalt Institute 1981).

Figure 6 shows a schematic of the computational model and the latter's basic components. The following model boundary conditions were assumed: rigid bottom fixing and lateral

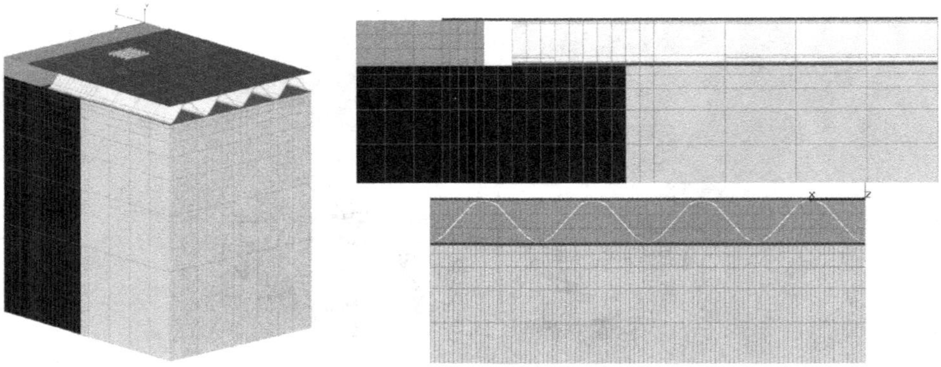

Figure 7. Division of model into finite elements (axonometric view and elevation views).

fixing (possible vertical displacement). Besides the variable load location (k) and the two structural configurations (excavation in weak soil and excavation in road pavement), also the variable plate location (p) on the excavation's shelves was considered. Three setback distances of the steel plate from the top shelf face were considered: $p = 0.01$ m, $p = 0.10$ m and $p = 0.20$ m. Ultimately, computations were carried out for 18 cases. As a result, a distribution of vertical displacements and vertical deformations in the subgrade (on different shelves) depending on the load was obtained.

Discretization and computations were carried out using the SolidWorks/CosmosM software. The model was divided into over 25 000 volume elements. The proper number of divisions into finite elements was determined on the basis of the convergence of the calculated maximum displacements in the subgrade. Figure 7 shows the division of the model into finite elements. The symmetry of the model relative to the axis of the excavation was taken into account.

5 ANALYSIS OF COMPUTATION RESULTS AND PAVEMENT DURABILITY

Computations were performed for eighteen cases (two types of subgrade and different wheel and plate locations on subgrade shelves). Figure 8 shows exemplary results of the computations for the plate & weak soil configuration. The displacements in the subgrade and in the plate for the maximum founding of the plate on the soil shelves ($p = 0.01$ m) at different distances k of the load from the plate's edge are shown.

It should be noted that as the distance of the load from the plate's edge increases, the bottom shelf begins to take part in load transmission and the maximum displacement decreases from 4.4 to 07 mm ($k = 0.50$ m). Similar values were obtained for the cases when only a small part of the plate rests on the shelves (setback distance $p = 0.20$ m). When the wheel is on the plate's edge, the displacement amounts to 5.1 mm, whereas when the wheel is located at a distance of 0.50 m from the edge, the displacement amounts to 0.7 mm. In the latter case, most of the load is transferred to the bottom shelf and the backfill. The displacement values are similar as in the case when the plate is founded deeper. The effect of the location of the relatively rigid plate on the soil shelves on the computed vertical displacements has been found to be small. A considerable portion of the load is distributed along the length of the plate. The variation range for all the considered cases is shown in Fig. 9. The extent of plate foundation on the soil shelves has an effect only for loads located directly at the plate's edge (the most unfavourable case).

Then the results for the plate & pavement configuration were analysed. Figure 10 shows exemplary computed displacements in the subgrade and in the plate for the maximum founding ($p = 0.01$ m) of the plate on the soil shelves at different load locations k from the plate's edge.

It should be noted that for the plate & pavement configuration much (10 times) lower displacement values were obtained. For the load location at the plate's edge a vertical

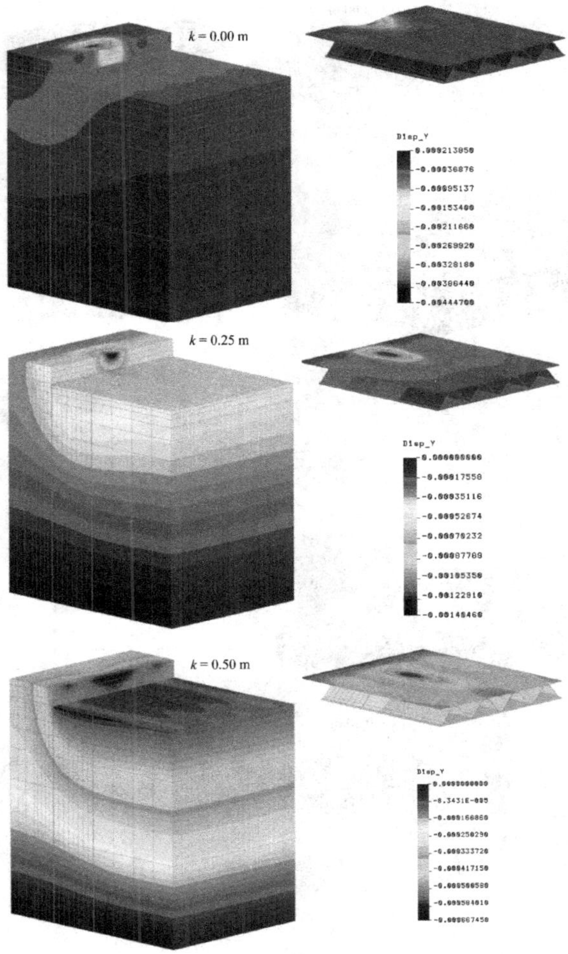

Figure 8. Results of computing vertical displacements uy [m] for $p = 0.01$ m at different distances k of load from plate's edge (plate & weak soil configuration).

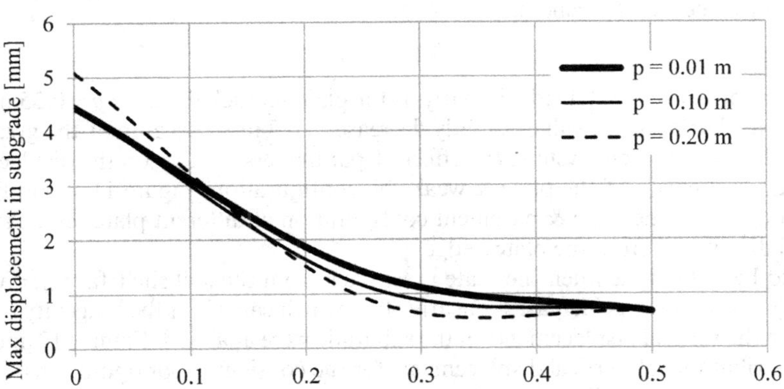

Figure 9. Maximum vertical displacements in subgrade for variable load location and plate foundation on soil shelves (plate & weak soil configuration).

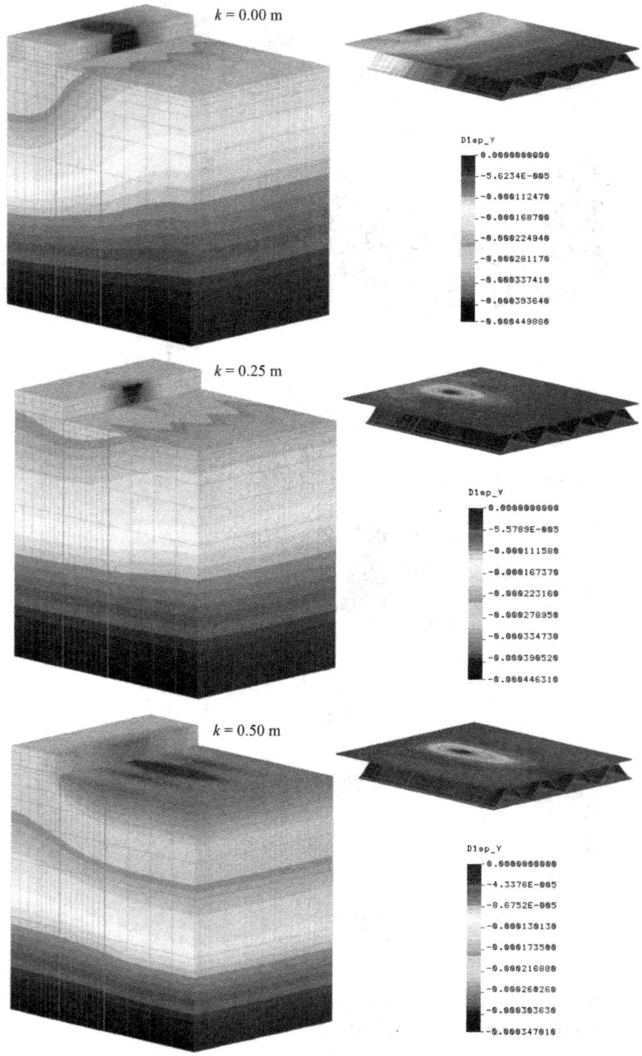

Figure 10. Vertical displacements uy [m] at $p = 0.01$ m for different load locations k relative to plate's edge (plate & pavement configuration).

displacement of 0.4 was obtained, similarly as for plate setback distance $p = 0.25$ m. Only at $p = 0.50$ m the displacement values slightly decrease to 0.3 mm. Because of the greater rigidity of the plate & pavement system, the effect of parameters k and p on the displacements is smaller than in the case of the plate & weak soil configuration. Figure 11 shows the results of computations for the plate & pavement configuration at different plate setback distances p and load distances k from the plate's edge.

It should be noted that when the plate is set back from the soil shelf face ($p = 0.10$ m or $p = 0.20$ m), displacements increase by nearly 50%, which can affect the longevity of the subgrade. Then the vertical displacements in the subgrade were analysed. Figures 12 and 13 show selected distributions of vertical displacements for the considered configurations.

Durabilities for the considered configurations and cases were determined on the basis of the calculated displacement values, using the adopted Asphalt Institute criterion. The subgrade soil structural deformation (structural rutting) criterion interrelates the number of load cycles and the vertical displacements of the subgrade (Bejarano 1999):

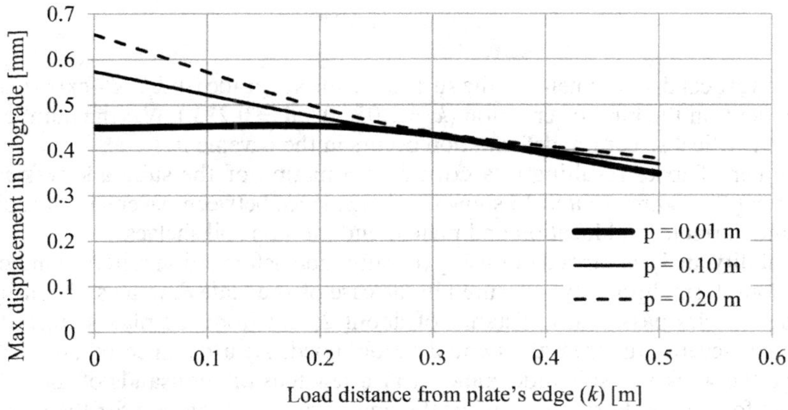

Figure 11. Maximum vertical displacements for different load locations and plate foundation locations on soil shelves.

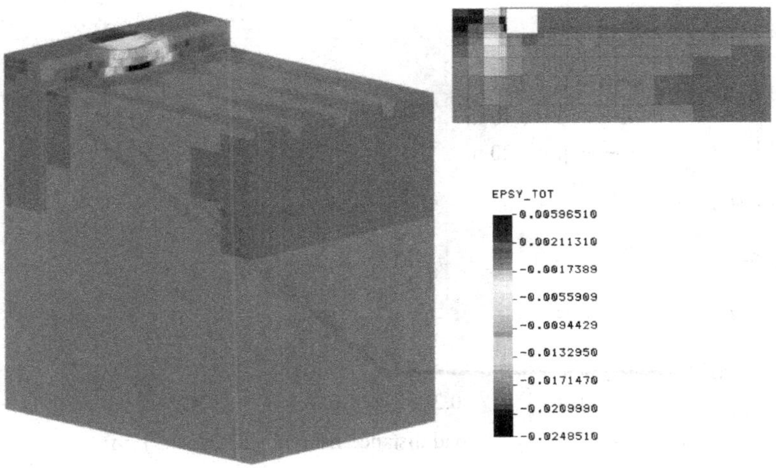

Figure 12. Vertical displacements epsy [–] for $p = 0.20$ m and $k = 0.00$ m (plate & weak soil configuration).

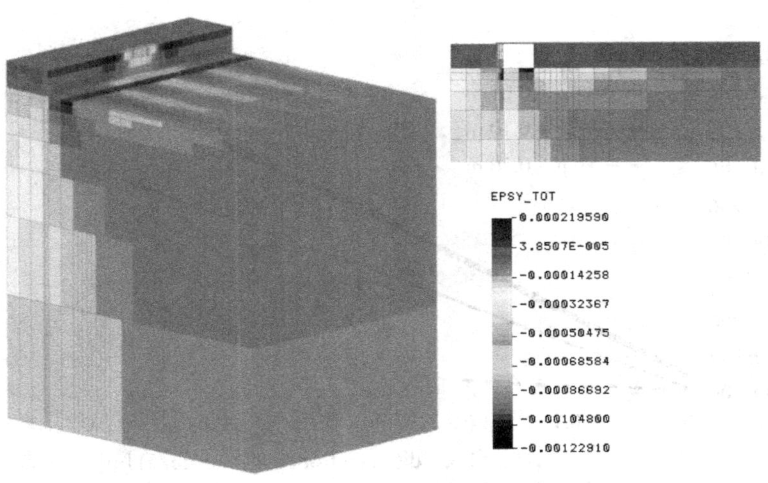

Figure 13. Vertical displacements epsy [–] for p = 0.20 m and k = 0.00 m (plate & pavement configuration).

$$\varepsilon_p = k_p \cdot (1/N)^m \tag{2}$$

where ε_p – a vertical displacement on the surface of the subgrade [-]; k_p, m – experimental criteria dependent on the kind of criterion ($k_p = 1.05\text{e-}02$, $m = 0.223$.); N – the number of load cycles until a critical structural deformation occurs in the pavement.

The number of axles resulting in a critical deformation of the subgrade was calculated from relation (2). Figures 14 and 15 show the dependence between pavement durability degradation and variable load location and plate foundation on soil shelves.

The durability analysis carried out using the subgrade deformation criterion indicates that no proper long-term durability is ensured in the case of the plate & weak soil configuration. Only if the vehicles passed at a distance of about 25 cm from the plate's edge, durability amounting to several hundred axles could be obtained. At a distance of 50 cm from the plate's edge the weak soil subgrade could carry a few tens of thousands of axles. This can be sufficient for only a short period of trafficking during construction for the lowest traffic class. For the plate & weak soil configuration the results only slightly differ depending on the plate setback distance (parameter p) on the soil shelves.

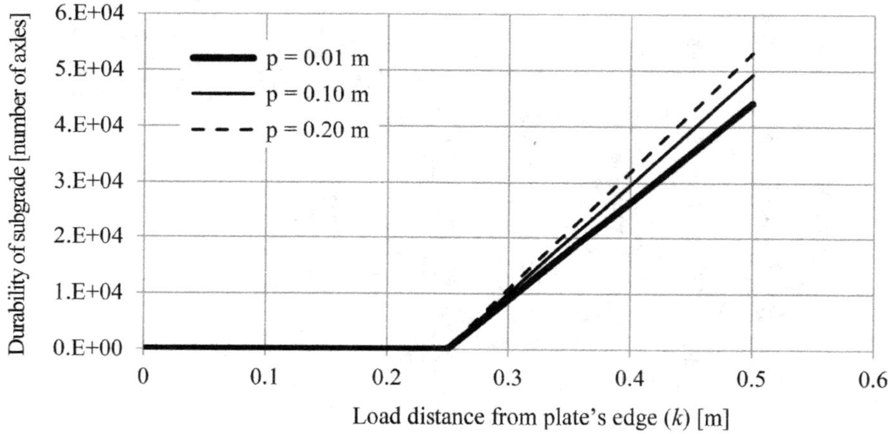

Figure 14. Durabilities of subgrade for variable load location and plate foundation on soil shelves (plate & weak soil configuration).

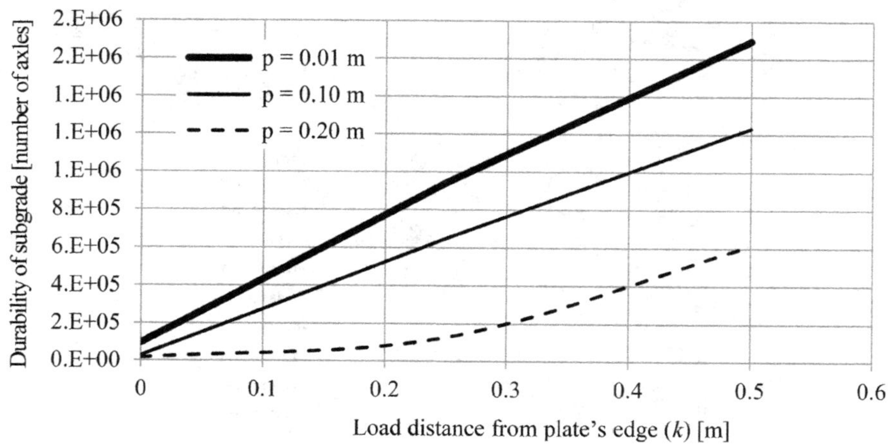

Figure 15. Durabilities of subgrade for variable load location and plate foundation on soil shelves (plate & pavement configuration).

In the case of the plate & flexible pavement configuration there is a marked effect of the location of the plate on the soil shelves. Clearly, the highest durabilities are obtained when the plate rests fully on the shelves ($p = 0.01$ m). Durability increases with the distance of the load from the plate's edge. For the extreme (most unfavourable) case—the load located on the plate's edge ($k = 0.00$ m) – the durability ranged from about 15 000 ($p = 0.20$ m) to 95 000 axles ($p = 0.01$ m). The results indicate that even if the plate is improperly founded on the soil shelves, short-term goods vehicle traffic can be permitted. Class KR1 traffic can be carried if the plate is set back from the shelf face by 10 cm. When the plate fully rests on the shelf, KR2 traffic can be carried. Durabilities will be much higher in cases when vehicles do not pass close to the plate's edge.

6 CONCLUSION

By carrying the numerical analysis the effect of the foundation of the steel road plate on the durability of the subgrade has been evaluated. The computation results indicate that the road plate is suitable for carrying vehicular traffic over road pavement excavations, ensuring the required durability in the repair period. In the case of excavations backfilled with only soil, passenger vehicle traffic and sporadic lorry traffic can permitted. For excavations running along the direction of traffic (along the roadway) the traffic should be organized in such a way that vehicle wheels do not pass close to the plate's edge (it would be best if they passed at a distance of more than 25 cm from the plate's edge).

It should be noted that the weak subsoil conditions analysed here rarely occur in practice. It has been found that in the case of soil-only excavations the extent to which the plate rests on the shelves has no significant effect since the plate also rests on the backfill. In the case of the plate & road pavement configuration, it is absolutely necessary to properly found the plate on the shelves. The maximum plate setback of 10 cm should not be exceeded. The satisfaction of this condition will ensure the safe and long use of the excavation.

The developed road plate systems are an excellent solution in cases of failures of urban buried utilities. If an excavation is made, it can be quickly covered to enable vehicular traffic. The main advantage of the road plate is that it can be quickly and easily transported to the site and quickly assembled and disassembled there.

REFERENCES

Asphalt Institute. 1981. *Thickness design-Asphalt Pavements for Highways and Streets*. MS-1.

Bejarano, M.O. 1999. *Subgrade Soil Evaluation for the Design of Airport Flexible Pavements*. University of Illinois at Urbana-Champaign.

Craig R.F. 1997. *Soil mechanics*. Sixth edition. Spon press, Taylor & Francis Group, London and New York.

Judycki J. 2014. *Catalogue of typical flexible and semi-rigid pavement structures* (in Polish). GDDKiA, Warsaw.

Kasahara, A., Matsuno S. 1988. *Estimation of Apparent Elastic Modulus of Concrete Block Layer. Proc. 3rd Int. Conf. On CBP*. Tokyo.

Underground Infrastructure of Urban Areas 4 – Madryas et al. (Eds)
© 2018 Taylor & Francis Group, London, ISBN 978-1-138-55953-0

The assessment of the durability of a post-tensioned reinforced concrete tank

C. Madryas, A. Moczko, R. Wróblewski & L. Wysocki
Faculty of Civil Engineering, Wroclaw University of Science and Technology, Wroclaw, Poland

ABSTRACT: The paper presents the results of detailed tests of the technical condition of the construction of a reinforced concrete post-tensioned drinking water tank that is embedded in the ground. The study included the determination of the actual mechanical and chemical parameters of the concrete and also verification of the reinforcement diameters, reinforcement spacing and post-tensioned tendons. The tests were performed on the internal and external surfaces of the tank. In order to carry out the tests on the external surfaces, a trench with a depth of approximately 6 m was executed. Structural and strength analysis was conducted with the use of the obtained results of the concrete and reinforcement tests. The results of the tests and structural-strength analysis were used to formulate general conclusions regarding the durability of the construction of post-tensioned concrete tanks that are used to store water that is intended for human consumption.

1 INTRODUCTION

Compressed reinforced concrete tanks provide a greater guarantee of a structure's tightness than comparable reinforced concrete tanks that are constructed without compression. However, steel post-tensioned tendons, due to their high carbon content, have a reduced corrosion resistance, which could endanger the durability of a structure. Tanks that are immersed in the ground, which are intended for storing drinking water, are exposed to both internal and external corrosion. Groundwater may pose a leaching aggressiveness towards concrete. The leaching process of concrete involves the dissolving and eluting of calcium hydroxide and, to a lesser extent, other components of the hardened cement grout. As a result of this process, the structure of concrete is weakened and its porosity increases. This process is particularly intense in the case of the presence of soft waters (e.g. rainwater).

Leaching is a diffusion process and therefore the influence of the environment, which affects the rate of dissolution and leaching of components, is particularly important. The leaching process is greatly accelerated in the case of one-sided water pressure. Moreover, groundwater can cause acid and carbonic acid aggression (groundwater can contain humus acids, inorganic acids and strong acid salts—acid rain).

In turn, a structure is exposed to water from the inside with an increased chlorine content. The current assessment of the technical condition of a construction usually only covers the visual inspection of the structure from the inside. Structural tightness tests are only performed in justified cases. The condition of the concrete cover of post-tensioned tendons, which determines its protective ability and therefore the durability of a tank's structure, is usually not evaluated due to a lack of accessibility. This article describes the results of comprehensive research of the technical condition of a reinforced concrete post-tensioned tank that is immersed in the ground and has been used for about 40 years. Based on the results of conducted tests and analysis, conclusions that concern predicting the durability of the structure of such tanks operating under similar conditions were drawn.

2 GENERAL DESCRIPTION OF THE TANK

A reinforced concrete post-tensioned tank with a capacity of 5000 m^3, which is covered with a floor slab, partially immersed in the ground and partly covered with soil, is the subject of the tests. The geometric dimensions of the tank are as follows:

- total construction height of about 9 m,
- internal diameter equal to 28.5 m.
- thickness of external walls equal to 0.15 m
- thickness of the bottom slab equal to 0.15 m,
- an additional overlay of concrete of approx. 0.05 m that is laid on the bottom slab,
- bottom slab supported on the ring foundation bench with a height of 0.6 m.

The walls of the tank are made of prefabricated concrete slabs with a width of about 1 m and a thickness of 0.15 m. The walls were made using vacuum technology. A hinged support of wall slabs on the bottom slab was used. The joints between the slabs (gaps of about 0.2 m wide) were concreted after the slabs were installed. The tank was compressed with tendons of 5 mm in diameter. The tendon spacing varies from 30 mm in the lower part of the structure to about 60 mm in the upper part of the structure.

The tank was placed in a poorly permeable soil (clay) with a pH index of 5.8 to 6.0, which shows weak aggressiveness to concrete.

No corrosion protection was executed on the outside of the tank. The post-tensioned tendons were covered with a layer of spray concrete. In turn, a coating of epoxy resin was applied inside the tank and a coating of glass fiber and epoxy resin was applied in the areas of joints between the wall slabs. The project assumed a life service of the tank of about 50 years and therefore after 40 years of operation it was decided to carry out a control study. Figure 1 shows the stages of the assembly process of the tank.

Figure 1. View of the assembly process of the tank: The bottom pictures present the mounting of the ceiling on the wall slabs, while the top pictures show the assembled tank in the final compression stage (winding device of post-tensioned tendons).

3 DESCRIPTION OF CONDUCTED TESTS

3.1 *Review of the technical condition of the structure*

A detailed review of the technical condition of the tank construction only showed some local damage to the insulating coatings on the interior surfaces. The damage involved the detachment of a resin coating on a small area (up to about 25 cm^2). There was no visible damage to the spray concrete protecting the post-tensioned tendons.

Wall thicknesses were also verified during the inspection. Due to the fact that the construction was only accessible from the inside, the Impact-Echo method was used. The measurement of thickness using this method is based on determining the propagation time of the excited pulse wave that passes between the accessible and inaccessible surfaces (Δt). By knowing the speed of propagation wave in concrete ($\mathbf{V_P}$), the thickness of the examined element can be determined. However, because the interpretation of test results, which is based on their analysis as a function of time is labour intensive, a fast Fourier Transform (FFT) was used. The FFT enables the recorded time waveforms to be expressed as a function of frequency. The frequency value, which is associated with amplitude extremity, is referred to as the expected frequency ($f_o = f_{extr}$). This frequency, for the known P-wave propagation velocity ($\mathbf{V_P}$) in the tested concrete, corresponds to the thickness of the tested element (T) according to relation (1):

$$T = V_P/2 f_{extr} \tag{1}$$

where T = the thickness of the tested element, V_p = propagation velocity.

The wall thickness measurements were made using a DOCter measuring system from the German Instruments Company. The results of the conducted measurements confirmed that the thickness of the sidewalls is consistent with the thickness from the project and ranges from 151 to 154 mm (150 mm thickness was assumed in the project).

3.2 *Tests of the concrete cover*

The determination of the thickness of the concrete cover in the construction of the tested tank was performed using the non-destructive electromagnetic testing method and the Cover-Master-type CM52 device. It was found that the thickness of the reinforcement cover on the inside of the structure varies from 10 to 35 mm and the average thickness of the concrete cover is equal to about 26 mm. The thickness of the spray concrete protective coating layer varies from 15 to 30 mm. According to the assumptions of the project, the thickness of the spray-coating layer should not be less than 20 mm.

The extent of the carbonation process of the subsurface concrete layer was assessed using the Rainbow-Test and the phenolphthalein test. It was found that the depth of the carbonation of concrete from inside the tank does not exceed 3 mm. The average carbonation depth from the outside (spray concrete) is also equal to about 3 mm, but reaches 6 mm locally. Taking into account the lifetime of the tank (about 40 years), the rate of carbonation of concrete should be considered as very slow. Such a slow carbonation rate is due to the constant high density of the concrete's structure, which makes it difficult for carbon dioxide to penetrate the concrete.

The chloride content was investigated in the concrete cover of the reinforcements and it was found that it did not exceed 0.1% from inside the tank and 0.04% from outside. The permissible content of chloride for reinforced concrete structures should not exceed 0.4% in relation to the weight of cement [1].

3.3 *Tests of the concrete cover*

In order to obtain information about the structure of the tested concrete and its strength parameters, 100 mm diameter core samples were extracted from the selected sections of the walls using a coring drilling machine. The samples were taken according to standard [5]. Strength tests were carried out in accordance with standard [7]. The results are summarized in Table 1.

On the basis of the obtained test results, the concrete class was estimated as C35/45 according to standards [3, 6]. This class is higher than that assumed in the project documentation, which was C30/37 concrete class.

The concrete in the extracted core samples was characterized with a good homogeneity over the entire height of the sample. It was found that the grain size of the aggregate varies up to 16 mm and the concrete has a small porosity.

Regardless of the core sample testing, the compressive strength of the concrete was determined on the basis of measurements obtained directly on site using the pull-out method in 9 selected measuring areas located around the tank. The "CAPO-Test" measuring system from the German Instruments Company was used for this purpose. The tests were carried out in accordance with standard [5] and their results confirmed the concrete class that was previously determined on the basis of core sample testing.

3.4 *Testing of the tensile strength of concrete using the pull-off method*

The tensile strength tests were carried out using the pull-off method according to standard [8]. They were conducted with the use of the Proceq "DYNA" measuring system. The results are summarized in Table 2.

The high average tensile strength of concrete indicated that the tank construction is made of high quality concrete. Concrete substrates with such parameters allow a structure to be repaired using any modern PCC materials.

3.5 *Testing of the water tightness of concrete*

The current standard [4] does not provide any requirements regarding the water tightness of concrete. During the designing process and the execution of the examined tank, standard [14] was obligatory. This standard required that the construction of a tank must be made of concrete of a certain water tightness, which depends on the height of the liquid column acting on the structure and the thickness of this structure. The assessed tank, according to the requirements of this standard,

Table 1. Test results of the core samples taken from the walls.

Sample designation	Cross-section area F [mm^2]	Force P [kN]	Compressive strength [MPa]
Sample O-1	7010	365	52.1
Sample O-2	6995	321	45.9
Sample O-3	7003	308	44.0
Sample O-4	7020	315	44.9
Sample O-5	7000	336	48.0
Sample O-6	6990	358	51.2
Sample O-7	7000	310	44.3
Sample O-8	7035	342	48.6
Sample O-9	7010	303	43.2

Table 2. Results of the tensile strength tests.

Measuring point number	Tensile strength [MPa]	Average tensile strength [MPa]
1	2.93	2.87
2	2.35	
3	2.89	
4	2.45	
5	3.23	
6	2.89	
7	2.78	
8	3.15	
9	3.09	
10	2.97	

should be made of concrete with water tightness of no lower than W8. The assessment of the water tightness of concrete was carried out on the inside surface of the tank using the non-invasive "GWT" method in order to avoid the risk of structural leakage. The obtained results showed that the tested concrete has a high water tightness that is not lower than W8 according to standard [14].

3.6 *The determination of concrete absorbability*

Concrete absorbability tests were performed on core samples that were cut out from the walls of the examined tank. They were carried out in accordance with standard [14]. The concrete absorbance in the individual samples varied from 4.4% to 5.1%. The absorption is therefore in line with standard [14], which was in force during the designing and construction phase of the tank (this standard recognizes absorbability of up to 5% as permissible).

4 STATIC-STRENGTH CALCULATIONS

4.1 *Basic assumptions for the calculations*

Verifying static-strength calculations were performed using the results of the construction tests and data from the project documentation. The following input data was assumed for the calculations:

- post-tensioned reinforcement consisting of 5 mm diameter tendons that were made of grade II steel (according to the grades of steel that were in force during the time of executing the tank) with f_{pk} strength equal to 1471 MPa and f_{pd} computational strength equal to 1260 MPa,
- C35/45 concrete class (according to the results of the conducted tests),
- thickness of tank walls equal to 150 mm,
- concrete deformation modulus E_{cm} equal to 34000 kPa,
- Poisson coefficient of concrete ν equal to 0.2,
- concrete density ρ equal to 2548.4 kg/m^3,
- volume water mass γ equal to 10.0 kN/m^3,
- tank filled to a level of 8.00 m,
- volume weight of soil γ equal to 20.0 kN/m^3,
- angle of internal friction of soil ϕ' equal to 12°.

The following computational factors were assumed:

- $\gamma_F = 1.35/1.00$ for the self-weight of the structure,
- $\gamma_F = 1.20/1.00$ for liquid pressure,
- $\gamma_F = 1.50/1.00$ for ground loads

Calculations were performed for the following characteristic loads:

- self-weight of the construction (constant load),
- loads caused by ceiling weight (constant load),
- loads caused by liquid pressure acting on the walls and bottom (variable load),
- loads caused by ground pressure (variable load).

Calculations were made using the finite element method (FEM). They were performed for small values of displacements and deformations with the assumption of a linear-elastic material model. Three-node finite elements of a rotary-symmetrical shell were used. On the basis of the static calculations, calculations of peripheral reinforcement load bearing capacity, the distribution of post-tensioned tendons and the values of the post-tensioned forces that determine the safety of the construction were performed.

4.2 *Calculations of post-tensioning and load-bearing capacity*

Calculations of post-tensioned losses were performed with the assumption that due to the fact that the structure of the tank is prefabricated, the impact of concrete shrinkage on the internal forces may be omitted.

All strength calculations were performed according to standards [9], [10], [11] and [13]. Calculations were performed assuming that the initial post-tensioned force with a value of 16.46 kN would cause stresses of $0,57f_{pk}$. The maximum stresses after all losses were limited to $0,55f_{pk}$ due to the regulations that were in force when designing the object. For the assumed spacing between the tendons *s*, the post-tensioned losses were calculated and the load-bearing capacity was verified due to the influence of liquid pressure and the self-weight of the structure. The conducted calculations showed that in the applied post-tensioned technology (winding), the tension forces in tendons compensate for the losses caused by friction.

Therefore, the load-bearing capacity of the tank is only assured by the post-tensioned reinforcement.

4.3 *Results of static and strength calculations*

Based on the results of the static-strength calculations, the following was stated:

- tensile stresses occurred in the wall structures due to the most possible unfavourable load scheme—during the tank tightness tests (conducted without the tank being covered with soil). Due to this, some water leaks may occur between the level of 0.50 and 5.70 m above the top foundation surface. In turn, for the quasi-fixed load combination (when the tank is filled with water and covered with soil) there are only compressive stresses, which ensure the tightness of the structure.
- during the tank tightness tests (conducted without the tank being covered with soil), between the levels of 1.30–2.90 m above the top surface of the foundation, the load-bearing capacity of the structure is not assured at the appropriate level. The maximum load-bearing capacity deficiency is equal to 11%. In turn, for a constant and transient load combination (when the tank is filled with water and covered with soil), the load-bearing capacity of the structure is provided.
- the results of the verifying static-strength calculations confirmed that the tank can be safely operated. However, it is not advisable to check the tank's tightness when the walls are not covered with soil.

5 SUMMARY

A construction of a tank that is intended for drinking water must meet the following criteria:

- a lack of a possibility of adverse effects on the health and taste parameters of water,
- the safe transfer of all operating loads,
- a tight construction that prevents against water exfiltration and also the infiltration of groundwater into the interior of a tank,
- appropriate durability.

The structure of the tank is made of concrete and the insulation of the inner surface is made of epoxy resin and glass fabric. Concrete is proven to be the best known and most commonly used material in building objects that are used for storing drinking water (concrete was even used in the Roman Empire). It has never been found to have a negative effect on water and is generally considered as a completely safe material. Epoxy resins are also considered as safe and inert to water intended for consumption. A threat for maintaining the required quality parameters of water can only be caused by water leaking through a tank's construction.

The conducted verifying static-strength calculations showed that the structure of the tank can safely carry all operating loads and that the state of safety is not currently threatened. However, performing the tightness tests of the tank when its walls are not covered in soil is not permitted.

The tests performed on part of the exposed wall showed that the thickness of the cover of the post-tensioned tendons ranged between 15 and 30 mm and the neutralization depth was equal to about 5 mm. The depth of concrete neutralization L_p during the expected life service of a tank is determined using dependence (2):

$$L_p = L_u\,(t_l/t_o)^{1/2} \tag{2}$$

where L_u = depth of the damaged concrete in the examined construction [cm], t_l = total expected operating life [in years] and t_o = the period in which a structure operates until the moment of assessment [in years].

Using the above, the following depth of concrete neutralization was obtained:

$$L_p = 0.5(65/30)^{1/2} = 13 \text{ mm}$$

After 100 years of operation, the depth of neutralization of the spray concrete in the cover would be equal to about 13 mm. This means that there should not be any corrosion of the post-tensioned steel in the next 60 years.

The results of the detailed tests and calculations of the tank's construction confirmed its relatively good technical condition and the possibility of its long-term operation.

The tested tank has been operating in average conditions and despite some assembly errors, the most important of which is too thin a concrete cover of the post-tensioned tendons, it has confirmed its suitability and high durability. It is worth emphasizing the fact that the structure is fully water tight despite its very thin wall thickness.

In the case of reinforced concrete tanks that are not post-tensioned, the tightness of the structure is often a major problem. The construction of such a tank requires the use of tight concrete, which in turn requires a relatively high cement content and a low w/c ratio. The high content of cement causes increased shrinkage, which can lead to the occurrence of cracks and leaks. The low w/c ratio makes it difficult to thicken the concrete, which in turn can lead to local discontinuities of the structure and local leakage. A post-tensioned structure does not have such faults, which was confirmed by the investigation of the tank. The good technical condition of the tested tank confirms that in any average ground-water conditions, corrosion-resistant insulations applied on the ground side of the structure are not necessary. In the case of structural durability, internal insulations are also unnecessary. Internal insulations may only be needed in the case of storing soft water, which has a corrosive influence against concrete.

REFERENCES

[1] Czarnecki L., Emmons P. 2003. The repair and protection of concrete structures. Polski Cement.

[2] Project of fresh water tanks located in Morask from May 1975 by the "Budownictwo Komunalne" Polish Construction Office (ul. Piekary 14/15, 60–967 Poznan).

[3] PN-B-03264: 2002 Concrete, reinforced concrete and prestressed concrete structures. Static calculations and design.

[4] PN-EN 206-1 Concrete, Part 1: Requirements, Properties, Manufacturing and Compatibility.

[5] PN-EN 12504-1: 2001 Testing of concrete in constructions. Part 1: Core drilling. Cutting, evaluation and testing of compressive strength.

[6] PN-EN 13791: 2008 Evaluation of compressive strength in constructions and prefabricated concrete products.

[7] PN-EN 12390-3: 2001 Concrete testing—Part 3: Compressive strength of specimens for testing.

[8] PN-EN 1542: 2000: Products and systems for the protection and repair of concrete structures. Test methods. Measurements of pull-off adhesion.

[9] PN-EN 1992-1-1: 2008. Eurocode 2. Design of concrete structures. Part 1-1: General rules and regulations for buildings.

[10] PN-EN 1992-3: 2008. Designing of concrete structures. Part 3: Siloes and tanks for liquids.

[11] PN-EN 1990: 2004/A1: 2008. Eurocode: Fundamentals of structural design.

[12] EN 1991-1-3. Eurocode 1. Impacts on constructions. Part 1-1: General impacts—Volume weight, self-weight, operational loads in buildings.

[13] EN 1991-1-3. Eurocode 1. Impacts on constructions. Part 1-5: General impacts—Thermal effects.

[14] PN-88/B-06250. Ordinary concrete.

Underground Infrastructure of Urban Areas 4 – Madryas et al. (Eds)
© 2018 Taylor & Francis Group, London, ISBN 978-1-138-55953-0

On designing underground extensions in existing heritage-listed buildings

H. Michalak & K. Kościńska-Grabowska
Warsaw University of Technology, Warsaw, Poland

ABSTRACT: The advances of civilization are increasingly forcing the need for the adaptation or modernization of works in heritage-listed buildings. Modern technologies for deep foundation construction allow and facilitate adaptation and modernization of works in a way that is safe for existing buildings. With this kind of works, it is possible to introduce new functional and spatial solutions which allow for a more effective use or exposure of heritage-listed buildings.

1 INTRODUCTION

Apart from adaptation works altering their original function, existing buildings, including listed ones, often require modernization works. The main reason for the works is to improve the conditions for the use of the building by adapting existing facilities to modern standards of living as well as to the requirements of current legislation and standards. The transformation of an existing building may involve, for example, adding underground floors to it or lowering its basement.

The planned conversion of existing buildings should be considered individually as a multifaceted issue carried out in collaboration with designers, architects, geotechnicians, installers and, in the case of listed buildings, specialists in the field of conservation and archaeology.

The construction of additional underground floors should be preceded by an interdisciplinary diagnostic work of both the building subject to extension within the underground section, as well as the buildings located within the effected area of the new investment, including architectural, historical, conservation, structural, geotechnical and urban conditions. The diagnosis should lead to defining the scope and technology of the works and the need for the possible introduction of permanent or temporary protection of existing buildings and facilities located in the affected area.

Numerical modelling is an important tool to assist forecasting the movements of the ground and buildings erected on it. The developed numerical models should be calibrated by taking into account the actual measurement results and experience arising from the construction of buildings in similar soil and water conditions and ground cover loading.

In order to calibrate numerical models, it is essential to adopt correct assumptions in the model. The calibration process can be improved by specifying the main factors and their impact on the effectiveness of modelling.

The most important factor influencing the effectiveness of numerical modelling systems "subsoil—underpinned building—adjacent buildings" is the appropriate determination of the elastic modulus E_0, the ground loosening zone and the ground cover loading.

2 STRUCTURAL SOLUTIONS FOR UNDERGROUND PARTS

The construction of additional underground parts under existing buildings, including heritage-listed buildings, or lowering their basements usually requires interference with the foundation area. Selection of the method to perform this type of work depends on numerous factors, including above all the technical condition of the existing foundations, depth of the underpinning (deepening) and the soil and water conditions (Michalak et al. 1998; Michalak 2009).

If an existing foundation must be underpinned due to a change in the level of the floor of basement or by making new basement rooms, the numerical analyses must take into account the factors that may cause occurrence of movements, i.e. increasing the stress in the areas of adjacent "stands" of the underpinned section, as a result of making an excavation and its protection, destress in the area of the underpinned wall. The numerical analyses must also take into account the area of loosened ground, destressing the bottom of excavation, accuracy of the underpinning at the joint point of the new foundation with the old one (Karczmarczyk 2005).

Two numerical analyses of the systems (subsoil—underpinned building—adjacent buildings) that were carried out, using the method of inverse analysis, i.e. the method using computer simulation consisting in execution of computational models constructed using the real measurement results, are discussed below in light of the analysed investments—the Museum of Praga and the Historical Museum of the City of Warsaw.

2.1 *Museum of Praga*

In order to increase the height of basement rooms and thus to adapt them to the current requirements, the foundations of the two existing buildings A and B were underpinned, and new underground parts for the needs of the planned museum were designed. The existing real estate and the interior courtyard of the museum are entered in the register of historic monuments (Fig. 1).

Wall paintings located in one of the existing buildings—a former Jewish House of Prayer from 1934 are entered in the register of movable heritage—they are the only such relic in Warsaw (Fig. 2).

The foundations of the basement walls in the buildings were underpinned using a traditional method (Fig. 3a and b). The length of the section of underpinning with the application of waterproof concrete of C25/30 class and reinforcement made of steel of A-IIIN class, was limited to 1.20 m, and its width and height to 0.70÷1.0 m. Underpinning of the foundations was preceded by making rim beams and structures protecting the walls of the building.

2.2 *The Historical Museum of the City of Warsaw*

The Historical Museum of the City of Warsaw consists of eight old tenement buildings that form the northern street frontage of the Old Town Square, and three tenement buildings located on Nowomiejska Street (Fig. 4).

Figure 1. Parts of elevation of the building of the Museum of Praga under construction.

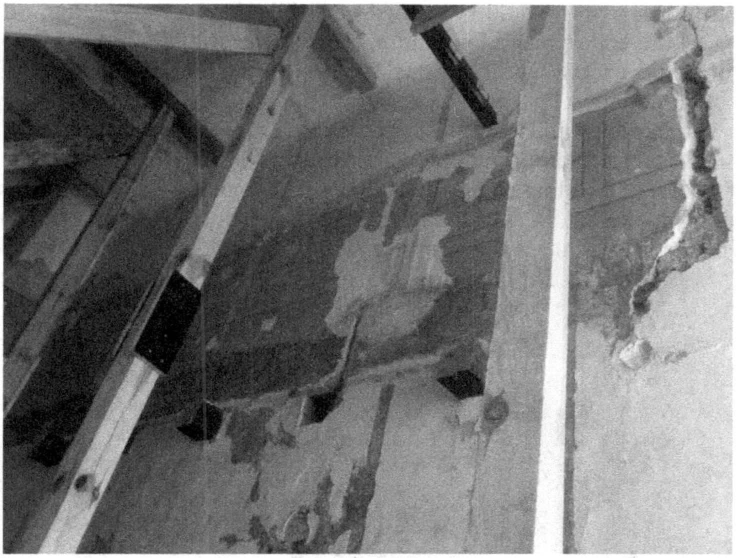

Figure 2. Historic wall paintings in the building of the former House of Prayers. Visible temporary protection of the ceiling with a wooden structure.

a) b)

Figure 3. The building of the Museum of Praga: a) making of underpinning, b) the view of the executed underpinned walls and their bracing.

The tenement buildings are located within the area that is included in a list of sites of the UNESCO World Heritage Sites (Fig. 5).

In order to adapt the basement rooms to their new functions, it was necessary to underpin the existing foundations (Fig. 6). A traditional method was applied, using sections (stands) of 0.8÷1.2 m width, with access to the foundation from both sides. A working distance allowing for simultaneous execution of several stands of at least 4.0 m was assumed together with the principle that the works would be carried out while digging out no more than 20% of the projection of the underpinned foundations. Special regard was given to the necessity to protect and brace the excavation of the working stands made, and to seal the joint point of the new and old foundation by making pressurised injections.

The works within each stand included in particular: excavating the ground to a level below the underpinned wall; protecting and bracing the excavation; clearing the bottom of the wall; the laying of a level of bedding concrete of C8/10 class and thickness 0.10 m; the laying of reinforcement of one stand (by driving in bars into the ground to the length of works equalling 0.40 m in order to connect the reinforcement with another stand); laying concrete to

Figure 4. The view of the facade of the eight old tenement buildings forming the northern frontage of the Old Town Square.

Figure 5. An interesting architectural detail of the facade of the tenement buildings.

a)

b)

Figure 6. A part of the interior of the Historical Museum of the City of Warsaw: a) making the underpinning, b) apparent difference of the levels of basement floors and foundation walls underpinned with the use of a traditional method. The photographs provided by Stołeczny Zarząd Rozbudowy Miasta [Warsaw City Development Authority], www.szrm.pl.

the footing with concrete of C25/30 class, with the water resistance ratio W8. At the footing that was made, insulation of torch-on felt was applied. The foundation walls of full ceramic bricks of 15 MPa class, on gauged mortar of M10 class were designed.

The final result of underpinning the existing walls can now be seen in the museum due to glass floors in some of the rooms (Fig. 7).

Figure 7. A part of the interior of the Historical Museum of the City of Warsaw with a glass floor. The photographs provided by Stołeczny Zarząd Rozbudowy Miasta [Warsaw City Development Authority], www.szrm.pl.

3 NUMERICAL MODELS

3.1 *General information*

In the analyses of numerical systems (subsoil—underpinned building—adjacent buildings), the information from pre-design and design works was used in order to determine the base model. Then from execution and geodetic measurements carried out during the works to calibrate the models in the particular stages, the final model was conceived with consistent values of movement, taking into account the geodetic survey double error values.

In the case of each of the discussed realizations, the results of geodetic surveys of movements of characteristic points (bench marks) located in the adopted computational cross-section of the investment area, and in particular phases of execution were analysed. Then, computational models were developed and calibrated through inverse analysis. Computer simulations were made assuming a panel model.

Three stages of calibration of numerical models were assumed. In the first stage, the data concerning the subsoil in accordance with geodetic-engineering documentation was adopted, assuming geotechnical parameters given there of the separate layers of soil (primary ground deformation modulus E_0, the Poisson number ν, weight density γ; Figs. 8 and 9). The results of the analyses were characterised by a significant difference of vertical movements of the subsoil in relation to the results of geodetic surveys performed during the execution.

In the second stage, the change of the primary ground deformation modulus E_0 and its influence on the results of subsoil movements were analysed. A few dozen of computer simulations were carried out, taking into account the change of the ground deformation modulus in the area below the foundation level; the remaining ground parameters were adopted in accordance with geodetic-engineering documentation. Results obtained by increasing the modulus by 50, 100, 150, 200 and 300% in relation to the base parameters—depending on the depth of their deposition—were analysed. Finally, in the second stage, the change of the modulus by 50% was adopted in the case of soils deposited below the level of foundation in the area of thickness of approximately 10 m, i.e. to the depth of 16 m below the ground level and increasing it four times below this depth. The results of the analyses were closer to the results of real ground movements, but they were characterised by an error which exceeded the double movement measurement error several times.

In the third stage, increasing primary ground deformation modulus E_0 by 50% was adopted in the case of all layers of ground deposited below the fill that occur to the depth of 16 m below the ground level. With regard to deeper layers, increasing the modulus four times was adopted. The results of these numerical calculations were consistent with the results of the actually measured vertical movements (Fig. 10a and b).

During the model calibration, factors such as (1) ground loosening in the area of the excavated area and during execution of stands of a given section, (2) inaccuracy of execution of joint point of the underpinned element with the underpinning (the presence of low strength mortar was taken into consideration by introducing layers of finite elements, approx. 1 cm thick, and modulus $E_0 = 10$ MPa) and (3) the presence of asphalt pavement on the street were

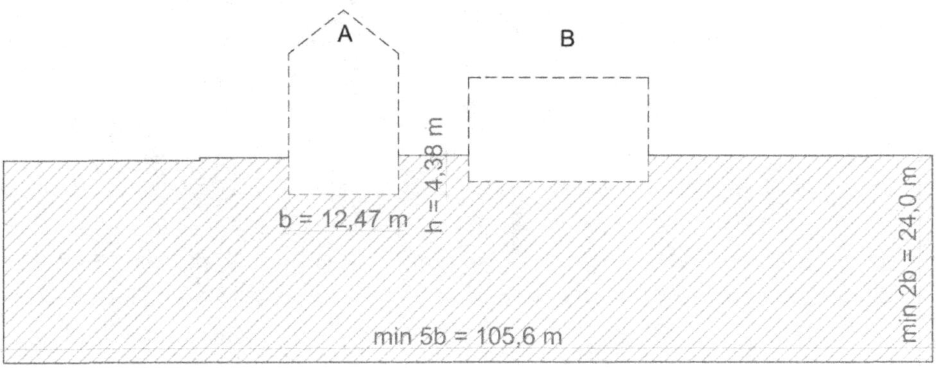

Figure 8. The outline of the ground panel assumed for numerical analyses. Vertical cross-section through buildings A and B and the area of ground panel marked with hatching.

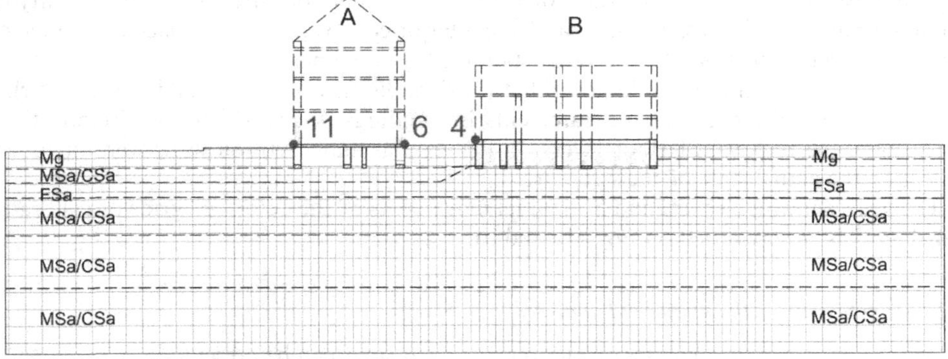

Figure 9. The outline of computational base model together with the adopted ground parameters and cross-sections of buildings A and B. The hatching marks the area of ground panel. Points 11, 6, 4—are bench marks where the geodetic measurements have been made.

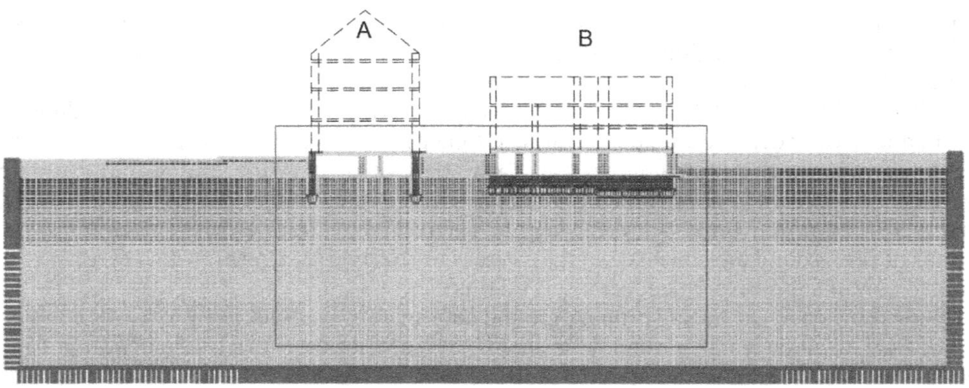

a)

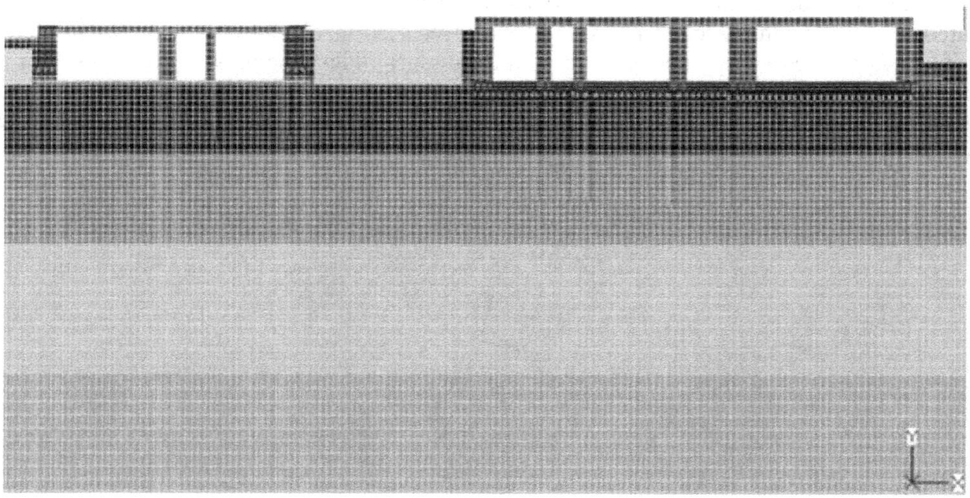

b)

Figure 10. The outline of the calibrated computational model (a) and the enlarged area of this model presenting the underpinning zone of the foundations of the buildings (b).

analysed as well. Also the manner of loading placement and its value in the analysed buildings were verified, as well as the value of the placed load of soil surcharge in the vicinity of the underpinned buildings, as the area of the interior courtyard was the place of storage of materials and the location of the construction facility containers.

It was also taken into consideration that the demolition works were taking place during the initial vertical movement measurements in buildings A and B. In the discretisations performed, the removal of floor layers on the ceilings and partition walls were included, reflecting the actual state of the structure.

The grids showing vertical and horizontal movements of the deformed grids of calibrated panel models are presented on Fig. 11a and b).

3.2 *Conclusions of the numerical analyses*

Based on the numerical analyses, the following conclusions can be presented:

The main factor determining the model calibration was the change of ground deformation moduli E_0. Geotechnical parameters determined according to standard PN-B-03020:1981 (now: PN-EN 1997-1. Eurokod 7, PN-EN 1997-2. Eurokod 7) are characterised by the con-

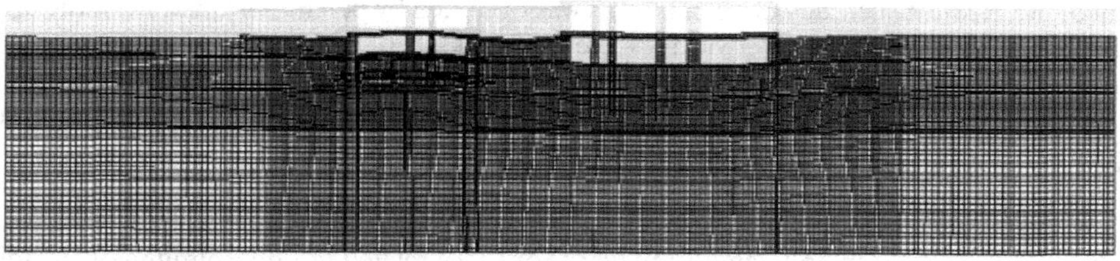

a)

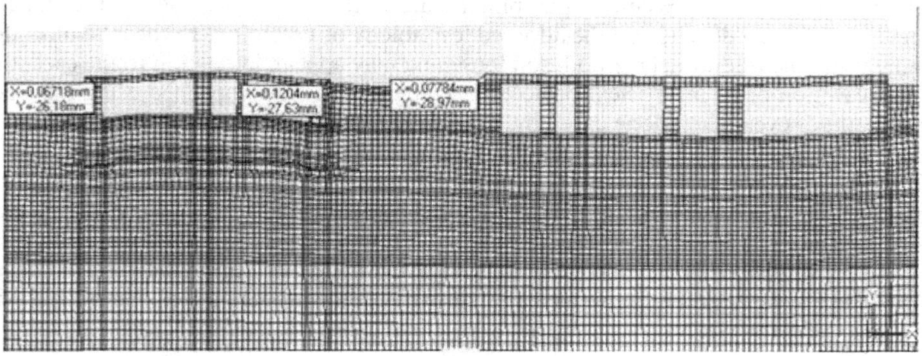

b)

Figure 11. The grids of vertical and horizontal movements of the calibrated panel model: a) entire model, b) enlarged part.

stant deformation modulus E_0 depending on the condition and type of soil, whereas the influence of stress condition of the "history" of the subsoil load (preconsolidated soil) and the scope of minor deformations is omitted. From the literature (Janowski 2010, Jastrzębska et al. 2010, Łupieżowiec 2009, Michalak 2006, Popielski 2007, Truty 2008, Zaczek-Peplinska et al. 2011) it appears that the subsidence obtained when taking into account the standard parameters significantly differ from the actual (measured) vertical movements of the buildings. It must be pointed out that in the case of minor deformations ($10^{-5} \div 10^{-3}$), the deformation modulus is characterised by nonlinearity, and the modulus values obtained from the standard correlations are up to several times smaller from the values in the case of very small deformations (less than 10^{-5}). In accordance with the results of the analyses (Jastrzębska et al. 2010, Michalak 2006) with the larger deformation modulus of the subsoil layers located deeper, the vertical movements obtained are similar to the actual ones. In the work (Jastrzębska et al. 2010) it was found that: *In their view, the subsoil deformations under the working load (of 150–200 kPa) are completely within the scope of minor deformations stated at the beginning* (Jastrzębska et al. 2010). In the numerical analyses (Łupieżowiec 2009, Zaczek-Peplinska et al. 2011), a significant increase of the deformation modulus was introduced in a considerable part of the ground model, excluding the layers of compressible ground.

The computer simulations presented, including the inverse analysis, allowed for determination of deformation modulus values assumed similar to the computational models. The principles adopted, taking into account the change of the subsoil modulus, may be considered effective.

A significant element while preparing the computational models is to recognize the existing buildings in the aspect of its conditions, including architectural and historical, conservation, urban, geotechnical and structural conditions. On this basis, it is possible to select parameters of the computational model, discreticise the layout, adopt the layout of the soil layers and their parameters, and to estimate the loads affecting the structures and subsoil.

The manner of placement of the load has a significant influence on the values of obtained movements. In the case of the analysed buildings for which the foundations were underpinned, the greatest accuracy of calculations was obtained by placing the load from the upper floors of a given building to the width of the wall (in kN/m).

The construction, which uses the area of the investment as a building site, including storage of materials, is characterised with the continuity of the works and overlapping of subsequent stages of execution. It is an additional difficulty while determining completion of particular stages, to which the movement measurements are assigned in the used method of inverse analysis.

During the excavation and demolition works or those related to underpinning of the foundations, soil loosening occurs. It should be remembered that estimating the sizes of the area of soil weakening and effectiveness of the executed post-tensioning in the joint points of the old foundation with the executed underpinning are difficult to be assessed at the stage of modelling. It largely depends on the adopted technology of execution. In the case of underpinning using the traditional method, soil loosening may appear, in relation to the execution of an excavation or its insufficient protection. Underpinning accomplished using the traditional method, even when done carefully, may cause significant movements, ranging up to 25 mm (Karczmarczyk 2005). These movements may cause damage to the buildings.

Based on the results of the numerical analyses carried out, it can be concluded that the adopted stages of proceeding and general assumptions for modelling and calibration were adopted correctly and may constitute a procedure that may be used in subsequent analyses of the systems "subsoil—underpinned building—adjacent buildings" for predicting movements. The values of vertical ground movements obtained from the model showed compliance with the actual, measured values.

3.3 *Analysis of numerical modelling effectiveness*

In the process of calibration of the numerical model concerning the Museum of Praga, five basic factors that constitute the numerical model were distinguished, including: ground parameters (deformation modulus E_0); values of load of equipment and the stored construction materials (during execution of the investment); values of load of the underpinned buildings; change of ground parameters resulting from loosening of soil as a result of the works carried out; the manner of placing the load to the building in the immediate vicinity. The results obtained from the numerical analyses of the models are the following.

The lack of account that is taken from increasing the deformation modulus E_0 results in the creation of movements different from the consistent movements in the final model. The error determined as a result of comparison of models at the stage of execution was up to 57%.

In the case of underpinning, the estimation error of the soil surcharge within the area of execution and in the immediate vicinity, from the construction works and material and equipment storage, was up to approximately 30%. Therefore, in the analysed model, a value decreased by 50% from the construction execution in relation to the loads assumed in the corrected final model was adopted.

Due to the simultaneous foundations underpinning and disassembly of pieces of the structure of the part of the building above the ground, the value of subsoil loads was decreased from that building by 20%. In the case of the discussed tenement building with three storeys above the ground where demolition works were made during underpinning, leaving only the floor beams as bracing for the structure, the 20% decrease results in the creation of movements several times smaller than the consistent movements. The load in the model was placed as evenly distributed on the width of the wall. It means that underestimation of the loads in the discussed tenement building may result in more than five times smaller values of vertical movements (subsidence).

Loosening of soil as a result of underpinning the foundations using the traditional method was not taken into account. The adopted deformation modulus E_0 of fine sands was 70 MPa. In the final model (calibrated), cutting of soil structures was taken into consideration assuming the modulus E_0 in the zone of the underpinned foundations equalling 40 MPa. The vertical movements that occurred were several times smaller than the consistent movements obtained in the final model.

Originally, the manner of load placement to building A as a concentrated load placed in the wall axis was adopted, and eventually, in the final model (calibrated) as a load evenly distributed across the width of the wall. The movements differing from the consistent movements in the final model were obtained, and the error resulting from comparing the models was 1÷5%.

4 FINAL CONCLUSIONS

In the case of the proposed expansion of the underground part of the existing building by additional floors, it is necessary to perform diagnostic tests to determine its technical condition. The diagnostics of the historic buildings should be extended to include historical, archaeological and conservatory research. On this basis, the history of the monument, original form, the scope of subsequent transformations, material stress-strain properties and technologies used during its construction are recognised (Janowski 2010).

Adoption of correct assumptions for numerical modelling depends on comprehensiveness of diagnostic tests of the building subject to underpinning and extension in the scope of the underground part as well as on the existing buildings located in the impact zone of this investment, properties of subsoil, existing infrastructure, etc.

The effectiveness of modelling (calibration) of the systems "subsoil—underpinned building—adjacent buildings" in the analysed cases was mostly influenced by (Kościńska-Grabowska & Michalak 2017):

- Determination of gradation of changes with the depth of the deformation modulus – the lack of consideration of these changes resulted in occurrence of up to 110% error;
- Identification of the area of possible soil loosening—failure to take this fact into account resulted in occurrence of more than 100% error;
- Determination of the load from the structures of the existing buildings undergoing expansion with additional underground floors—underestimating these loads resulted in an error of approximately 50%;
- Determination of the load values from storing the construction materials, equipment, and construction containers—incomplete diagnosis in this scope resulted in occurrence of up to 30% error;
- Determination of the manner of placement of loads from the structures of the existing buildings subject to underpinning and extension with additional underground floors, including assumption of the load as concentrated, resulted in up to 12% error. Therefore, this load was modelled as evenly distributed across the width of the structural walls.

In the light of the presented results of the numerical analyses, it may be concluded that the most significant factors that affect the effectiveness of numerical modelling of the models "subsoil—underpinned building—adjacent buildings" are appropriate determination of the deformation modulus E_0 and the area of soil loosening.

The developed numerical models may constitute a tool for forecasting vertical movements of subsoil and structures founded on it. Computer simulations, taking into account the inverse analysis (Michalak 2006), allow for determination of close to actual values of deformation moduli adopted to computational models and may be used in the analyses of similar systems "subsoil—underpinned building—adjacent buildings" for forecasting vertical subsoil movements.

REFERENCES

Collective work edited by
Janowski Z. 2010. Diagnostyka, modernizacja i rewitalizacja budowli zabytkowych. Kielce-Krynica: LVI Konferencja Naukowa KILiW PAN i KN PZITB.

Jastrzębska M., Łupieżowiec M. 2010. Osiadanie fundamentów z uwzględnieniem zmienności modułów odkształcenia. Kielce-Krynica: LVI Konferencja Naukowa KILiW PAN i KN PZITB.

Kościńska-Grabowska K., Michalak H. 2017. O projektowaniu rozbudowy części podziemnych w budynkach zabytkowych. "Inżynieria i Budownictwo", nr 3.
Karczmarczyk S. 2005. Wzmocnienie fundamentów w budynkach zabytkowych. Wisła-Ustroń: XX Ogólnopolska Konferencja Warsztat Pracy Projektanta Konstrukcji.
Łupieżowiec M. 2009. Analiza posadowienia obiektów wielkowymiarowych z uwzględnieniem silnej zmienności sztywności gruntu w zakresie małych odkształceń. "Czasopismo Techniczne", zeszyt 4-Ś.
Michalak H., Pęski S., Pyrak S., Szulborski K. 1998. O diagnostyce zabudowy usytuowanej w sąsiedztwie wykopów głębokich. "Inżynieria i Budownictwo", nr 6.
Michalak H. 2009. Garaże wielostanowiskowe. Projektowanie i realizacja. Warszawa: Wydawnictwo Arkady.
Michalak H. 2006. Kształtowanie konstrukcyjno-przestrzenne garaży podziemnych na terenach silnie zurbanizowanych. Warszawa: Oficyna Wydawnicza Politechniki Warszawskiej.
Popielski P. 2007. Wyznaczanie wzmocnionych parametrów podłoża. Analiza wstecz w praktycznych zagadnieniach posadowienia budowli przy wykorzystaniu pakietu HYDRO-GEO. "Nowoczesne Budownictwo Inżynieryjne", nr 9-10.
Truty A. 2008. Sztywność gruntów w zakresie małych odkształceń. Aspekty modelowania numerycznego. "Czasopismo Techniczne", zeszyt 3-Ś.
Zaczek-Peplinska J., Popielski P. 2011. Wykorzystanie monitoringu geodezyjnego do weryfikacji modeli numerycznych oddziaływania realizowanej inwestycji na obiekty sąsiednie. "Czasopismo Techniczne", zeszyt 3-Ś.

Underground Infrastructure of Urban Areas 4 – Madryas et al. (Eds)
 ISBN 978-1-138-55953-0

Sewer damage and its consequences with regard to issues relating to plastic sewers

B. Przybyla
Wrocław University of Science and Technology, Wrocław, Poland

ABSTRACT: This paper presents issues relating to damage in pipeline networks with a particular focus on consequences of the damage for the environment. Closer attention was devoted to the damage to plastic sewers along with its effects with regard to the behaviour of plastic sewers in time.

1 INTRODUCTION

Research carried out in Poland indicates that the technical condition of sewage networks is not satisfactory, and the data concerning this problem is still incomplete (Kwietniewski 2011, Kuliczkowska 2008). Other European countries face similar situation, and even in these countries that inspect the technical condition of sewers regularly and are aware of the need to conduct technical rehabilitation, a large part of networks require short or medium term interventions—see results in (Berger et al. 2016).

It is emphasized in (Kwietniewski 2011) that previous results of failure tests concerning sewage networks are limited to sewers mainly made of traditional materials: stoneware, concrete, reinforced concrete, and to a lesser degree to sewers made of plastic. Therefore, additional research is called for in order to identify types of damage in plastic sewers (mainly PE, PVC, PP, GRP), their consequences and the level of risk.

Study (Madryas 1993) includes the following definition: sewer damage is a change from the condition of serviceability to non-serviceability, when the sewer is no longer able to meet expected requirements. An old German instruction (ATV M 143, 1998) presents a similar definition: "sewer damage in terms of proper operability of the network is a condition in which the sewer loses its ability to operate properly". The aforementioned definitions refer to the sewer condition, which is understood as a conventional level of its ability to function (Wieczysty 1990).

As we know, the sewer carries a particular amount of sewage to the collector at a given time, provided it is leak tight. As a result of sewer damage, the sewer cannot function properly. If the sewer stops performing its duties, it means that the sewer lost its operability. If the sewer only performs a part of its duties against its specification, it means that the sewer partly lost its operability. That is why, relevant literature (e.g. Madryas et al. 2010) differentiates between partial and total sewer damage in terms of its functioning (performing the duties). The sewer damage refers to a pipeline of a specific length; it is accepted that it is at least the section between inspection chambers (manholes).

It is necessary to draw a line between the partial and total damage due to the specificity of sewer functioning: damage resulting in complete inoperability occurs very rarely. In most common situations (in particular in sewers without proper maintenance) the sewer functions at a lower standard, which inflict various losses (environmental, social and other). It is only possible to determine the moment in which the sewer stops functioning completely, when the sewage is no longer carried. In a majority of cases, the sewer functions in a limited scope, causing great losses. For example, the sewer loses its operability, but it manages to carry the sewage due to a bypass in the ground that prevents the sewer from being completely blocked. However, such a structure is of course unstable; accidental external forces may bring about a collapse of the ground and a complete blockage of the flow.

The assessment of the sewer condition is contingent on prior appropriate investigations. Inspections of sewer functioning are usually conducted by TV inspections, when the measurements of observable changes are taken (e.g. deformations) and (though rarely in Poland) additional pressure tightness tests are carried out.

Due to the specificity of TV inspections (visual observations), the registered images constitute a set of observable (noticeable) changes related by their cause, and thus possible to be distinguished from among the others that are located nearby. These include longitudinal cracks, infiltrations, losses of materials in the sewer walls, exposed reinforcement, etc. Generally these sets of changes are called sewer damage. In sewers a limited set of damage is found, but related features in a separate sewer may expand the admissible range to a various degree. Consequently, types of damage have different intensity. Moreover, owing to various conditions in which the sewer functions, e.g.: the ground, the level of ground water, the land development above the sewer etc., similar damage may have different negative consequences.

The condition of sewer damage and the very damage should be carefully differentiated. The sewer damage condition reflects the destruction as a whole, following the assessment (analysis) carried out by an expert who examines the damage that was registered during the inspection. The sewer under investigation may have (and usually has) a considerable number of flaws; however, they do not constitute one set of data sufficient to determine the state of the sewer. The knowledge about one or a few flaws in one sewer may be insufficient to determine the condition about the whole sewer. All the more, not all changes in the sewer can be noticed during a TV inspection, for example, changes in the material structure that occur in time as a result of natural aging and corrosion. Although the signs of corrosion are registered during inspections of concrete sewers, the actual level of risk may be determined after additional investigations that require collecting samples for destructive tests.

If it is not feasible to observe any damage, on principle the natural limitations of human sight (equipped with a device, e.g.: TV camera) and the scope of damage are to blame. Such situation occurs if changes of material features occur slowly; the damage is then referred to as parametric damage. The degradation of the sewer is slow, and the damage condition is not immediately noticeable. A classic model of a prolonged degradation of concrete sewer affected with corrosion is presented in Figure 1. These phenomena may also be of significance

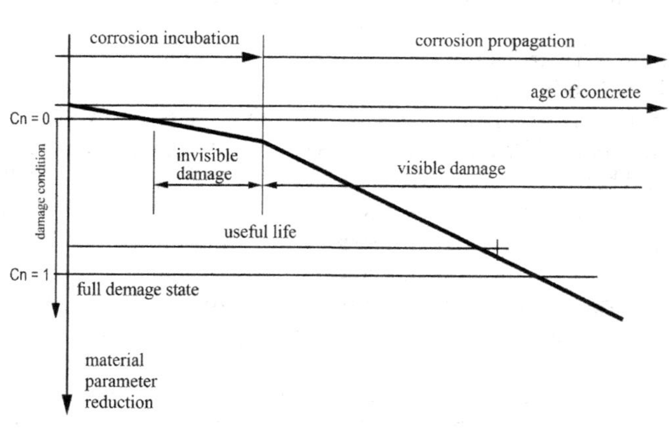

Figure 1. Degradation model of a concrete sewer as a result of corrosion (based on (Madryas et al. 2010)).

for plastic sewers, for which long-term and short-term material parameters are taken into consideration at the designing stage.

It should also be emphasized that the assessment of the sewer technical condition need to focus on a given criterion, for example, maintaining the capacity, hydraulic performance, effects on the environment. Overall, the starting point for the assessment of the sewer condition may be a statement from [8]: properly functioning sewer carries an expected amount of sewage, remaining neutral for the environment in which it is located and at the lowest possible (to the best of current knowledge) nuisance when operated by network managers.

2 DAMAGE, CAUSES AND CONSEQUENCES OF DAMAGE

There are numerous ways of classifying the sewer damage. An exemplary classification may include:

a. the cause of the damage,
b. the level (degree) of damage advancement (the degree of sewer degradation),
c. the pace of changes in the environment,
d. the level of risk (consequences) the damage incurs,
e. the possibility of damage removal,
other ...

On the basis of the major universal classifications presented in (Madryas 1993, Wieczysty 1990) and taking into account the pace of changes in the environment, the damage is divided into gradual damage (parametric) and sudden damage (usually identified with catastrophic damage) (Figure 2). Moreover, independent damage (or in other words: primary), which have various causes except for another damage, and dependent damage (in other words: secondary), which was initiated by another damage in the sewer or which was propagated by the other damage. Additionally, if the possibility of damage removal is taken into consideration, researchers differentiate permanent damage (it can mainly be eliminated by technical rehabilitation) and reversible damage, which can be removed by maintenance works, e.g.: by cleaning the sewer, removing the roots etc.

The causes behind sewer damage are traditionally divided into the following groups:

A(0) – structural errors,
B(0) – technological and material errors,
C(t) – exploitation errors,
D(t) – aging and corrosion processes,
E(t) – disturbances (loads).

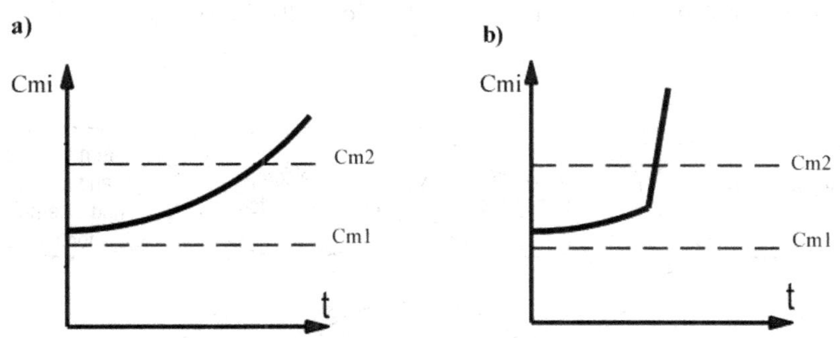

Figure 2. Conceptual diagram of gradual damage (parametric) (a): and sudden (catastrophic) damage (b): t – time, Cmi – sewer characteristics (features), (Cm1,Cm2) – admissible scope of changes in the sewer characteristics.

If the causes of damage result from improper decisions made by any participants of the construction and maintenance processes, they are referred to as errors. Structural and technological and material errors are not dependent on time counted from the moment the sewer was put into operation. The errors arise from improper construction works, material defects, or an incorrect design. They may lead to damage even before the sewer is put into operation, or they may become visible after a time, during the sewer operation.

In order to analyze the causes behind the sewer damage, a set of forces that affect the sewer operation must be determined. The forces may be divided into:

a. a subset of forces that result from the sewer functioning (internal forces),
b. a subset of forces that result from environmental influences (external forces).

Internal forces, which include: chemical aggressiveness of sewage, the contents of abrasive fractions, sewage temperature, flow velocity and its output, and chemical aggressiveness over the sewage level are the main cause of aging and/or corrosion processes D(t) in the sewer, usually impossible to be avoided. When designing the sewer and planning the sewer exploitation, the designers seek to obtain the greatest resistance of the sewer to similar forces and consequently slow down the aging processes. Any errors in designs, during construction works or usage accelerate the aging process of the sewer and bring about the state of damage.

The subset of external forces include mechanical loads (static or dynamic), chemical aggressiveness of the ground, destructive effect of other constructions, in particular during the construction works or in the event of failure, destructive effect of roots and other factors. The external forces occur at random; their another feature is the variation in time.

The damage mechanism is finally related to the determined function and processes (designing, construction, and usage) and the course and intensity of random events (failures, external forces, aging processes). It is presented schematically in Figure 3 and by the following equation based on (Przybyła 1999):

$$\begin{aligned}&\{D(t) \vee [D(t) \wedge A(0) \wedge B(0) \wedge C(t)] \vee [D(t) \wedge E(t)] \vee [D(t) \wedge A(0) \wedge B(0) \wedge C(t) \wedge E(t)]\} \Rightarrow \\ &\Rightarrow \text{ state of sewer damage}\end{aligned} \tag{1}$$

Once the changes (in comparison to an ideal condition and proper operation) in the sewer have arisen, they intensify the destructive processes. Not necessarily does the set of internal and external forces that affect the sewer need to change; it is enough that the conditions change in which particular set elements affect the sewer. The damage then propagates, which means the existing damage increases and new primary and secondary damage arises.

Figure 4 demonstrates my concept of damage classification for the purposes of study (Przybyła 1999). In Poland at present we commonly use the damage classification (description) system compliant with Norm EN 13508 and (still) the system developed by DWA

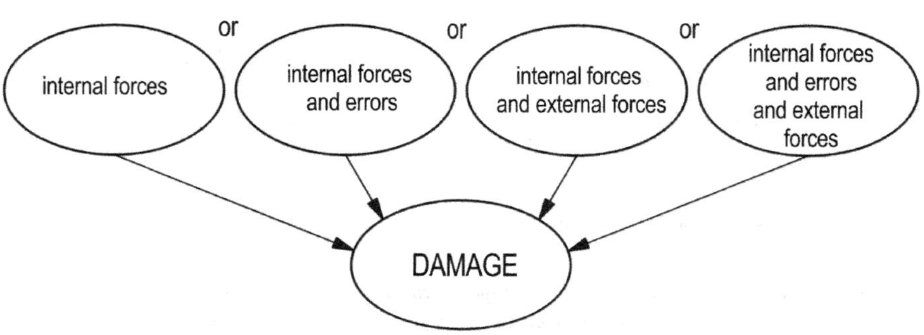

Figure 3. Damage mechanism in an operating sewer (Przybyła 1999).

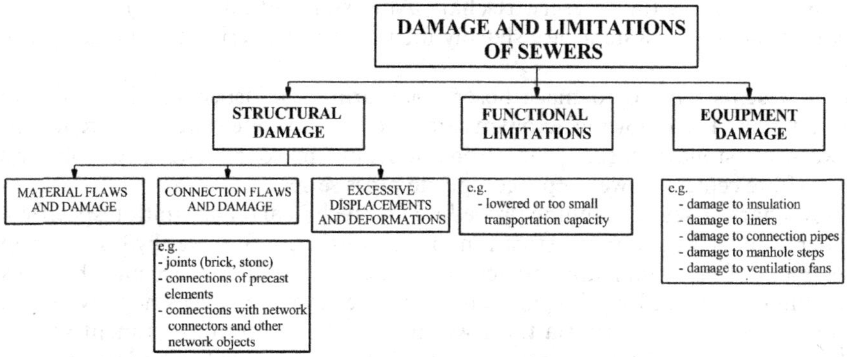

Figure 4. Concept of sewer damage classification (Przybyła 1999).

Table 1. Types of damage in sewers made of various materials.

No.	TYPE OF DAMAGE	c	rc	st	ir	m	ps	e	cp
1.	leakage in pipe connections, or in welds in sewer constructions	+	+	+	+	+	+	+	+
2.	leakage in walls of pipes or sewer constructions	+	+			+		+	
3.	leakage in connections between pipes and manholes or other network objects	+	+	+	+	+	+	+	+
4.	other obstructions	+	+	+	+	+	+	+	+
5.	sediments	+	+	+	+	+	+	+	+
6.	hardened sediment	+	+	+	+	+	+	+	+
7.	protruding sealing	+	+	+	+	+	+	+	+
8.	incrustations	+	+			+			
9.	protruding obstructions in cross-section of the sewer	+	+	+	+	+	+	+	+
10.	root ingrowth	+	+	+	+	+	+	+	+
11.	horizontal position deviations	+	+	+	+		+	+	+
12.	vertical position deviations	+	+	+	+		+	+	+
13.	position deviations in the sewer axis	+	+	+	+		+	+	+
14.	position deviations: deflection	+	+	+	+	+	+	+	+
15.	mechanical abrasion	+	+	+	+	+	+	+	+
16.	corrosion in connection zone	+	+	+	+	+	+	+	+
17.	mortar corrosion in joints					+			
18.	external corrosion	+	+	+	+	+	+	+	+
19.	internal corrosion	+	+	+	+	+	+	+	+
20.	deformation: over the admissible value for flexible pipes						+	+	
21.	cracks in connection zone	+	+	+	+		+	+	+
22.	longitudinal cracks	+	+	+	+	+	+	+	+
23.	transverse cracks	+	+	+		+		+	+
24.	cracks from one point	+	+	+		+	+	+	+
25.	local cut areas	+	+	+		+		+	+
26.	lack of a wall part	+	+	+	+	+		+	+
27.	lack of a wall part in connection zone	+	+	+	+			+	+
28.	cut areas in the whole cross-section	+	+	+				+	+
29.	sewer collapse	+	+	+	+	+		+	+
30.	blocked branch	+	+	+	+	+	+	+	+
31.	rupture in branch zone	+	+	+	+	+	+	+	+
32.	branch with improper structure	+	+	+	+	+	+	+	+
33.	protruding connection pipe	+	+	+	+	+	+	+	+
34.	connection pipe with improper structure	+	+	+	+	+	+	+	+
35.	rupture in connection pipe	+	+	+	+	+	+	+	+
36.	sewage backup	+	+	+	+	+	+	+	+

(Deutsche Vereinigung für Wasserwirtschaft, Abwasser und Abfall e.V.), German technical organisation. These systems are sharply focused on registering the damage during TV inspections.

Table 1 presents types of damage observable during TV inspections in sewers made of various materials. The following acronyms are used: c – concrete sewers, rc – reinforced concrete sewers, st – stonework sewers, ir – iron sewers, m – masonry sewers, ps – plastic sewers, e – eternit (fibre cement) sewers, cp – composite pipes sewers

The flow chart of sewer planning procedure and technical rehabilitation specified in EN 752: 2008 redirects the sewer investigation to the assessment of sewer hydraulic properties, the assessment of environmental impact, the assessment of endurance and the assessment of operating deficits. A long-lasting damage in the sewer produces negative effects that can be classified by basic criteria for sewer technical condition assessment as presented in the flow chart. Therefore, the damage is analysed with regard to its impact on current load capacity and expected sewer duration, sewer tightness or possible limitation of sewer hydraulic capacity. It corresponds with the analysis of technical condition in terms of (respectively): endurance criterion, environmental criterion, hydraulic criterion shown in Figure 5.

The complexity of damage mechanisms, the interdependence between particular type of damage in the sewer and the relations between the sewer and its environment make it difficult to determine how serious the effects of particular type of damage is with regard to specified criteria. For example, for a given type of damage (e.g. a separated connection) both exfiltration and infiltration is possible, depending on the level of ground water around the sewer.

Different consequences of deteriorated sewer condition in relation to assessment criteria are presented in Table 2.

The consequences of deteriorated sewer condition presented in Table 2 comprise negative effects for:

- economy and living conditions,
- natural environment,
- sewerage system (sewer network and waste treatment plant) as a complex of technical devices,
- system managers.

Figure 6 shows the diagram of aforementioned consequences of damaged sewer operation.

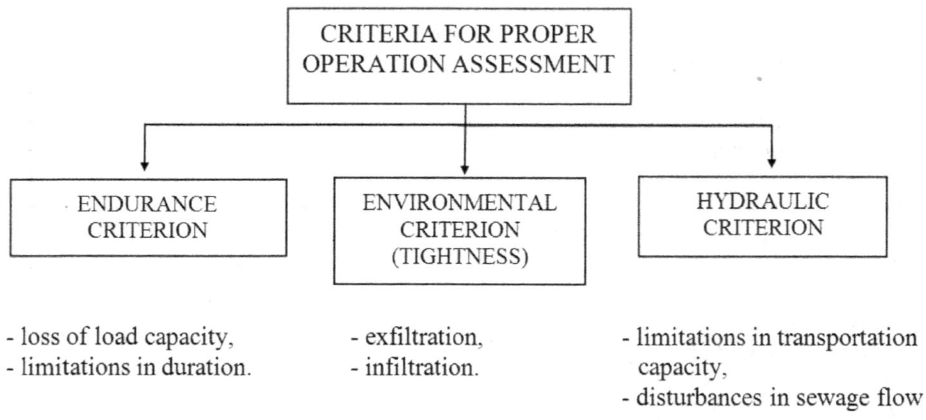

Figure 5. Assessment criteria for proper sewer operation with selected factors directing the analysis of damage influence.

Table 2. Consequences of deteriorated sewer condition (Przybyła 1999).

CRITERIA	DETERMINANTS	CONSEQUENCES
ENDURANCE CRITERION	LOSS OF LOAD CAPACITY	LOSS OF THE ABILITY TO CARRY LOAD—destruction of the sewer (construction catastrophe) PARTIAL OR TOTAL FLOW BLOCKAGE LOOSENING OR COLLAPSING OF GROUND OVER THE SEWER—consequences in the form of damage to surrounding constructions LOSS OF LEAK TIGHT PROPERTIES
	LIMITATIONS IN DURATION	ACCELERATION OF LOSS OF LOAD CAPACITY NECESSITY OF CONDUCTING MORE FREQUENT INSPECTIONS LOSS OF LEAK TIGHT PROPERTIES
ENVIRONMENTAL CRITERION	EXFILTRATION	SOIL CONTAMINATION—(environment degradation, land exclusion from agricultural production, lowered recreational values, health hazard) GROUND WATER CONAMINATIONS—(environment degradation, threat to water sources for various purposes, lowered recreational values, health hazard), impact of aggressive water on constructional objects, including the sewer INCREASED LEVEL OF GROUND WATER—turning lands into marshes, damage to nearby constructions (buoyancy, flooding rooms, changes of ground geotechnical parameters) LOSSENED GROUND AROUND THE SEWER—uneven subsidence of the sewer (cunette deformation, additional loss of leak tight properties, lowered capacity), change of load distribution (loss of load capacity and shorter sewer durability), consequences in the form of damage to surrounding constructions
	INFILTRATION	LOWERED LEVEL OF GROUND WATER—land drying (environment degradation, destruction of urban green spaces), damage to nearby constructions (changes of ground geotechnical parameters: subsidence) ADDITIONAL HYDRAULIC SEWER LOAD—lowered sewer capacity, overloaded pump rooms and waste treatment plants etc. (increased network maintenance costs, contaminated collector water), accelerated erosion in the sewer (shorter sewer durability) TRANSFER OF GROUND TO THE SEWER—overloaded pump rooms and waste treatment plants etc. (increased network maintenance costs, contaminated collector water), lowered sewer capacity due to siltation, increased abrasion intensity (shorter sewer durability) LOOSENING OF GROUND, EMPTY SPACES, COLLAPSING OF GROUND AROUND AND OVER THE SEWER—change of load distribution (loss of load capacity and shorter sewer durability), consequences in the form of damage to surrounding constructions

(*Continued*)

Table 2. (*Continued*)

CRITERIA	DETERMINANTS	CONSEQUENCES
HYDRAULIC CRITERION	LIMITATIONS IN TRANSPORTATION CAPACITY	PERIODICAL OR PERMANENT SEWAGE BACKUP—sewer operating under pressure which results in losing leak tight properties, flooding rooms on bottom floors (health hazard, damage to buildings, losses of property value) IMPOSSIBILITY OF COLLECTING SEWAGE FROM USERS—periodical or permanent exclusions of sanitary equipment from use (worsening of hygiene and health conditions, various organisational problems, losses in production and service industry) IMPOSSIBILITY OF COLLECTING RAINWATER—flooding street and pavements—communication disturbances
	DISTURBANCE IN SEWAGE FLOW	SEWAGE SPLASHING—creation of aggressive environment bringing about corrosion (shorter sewer durability) CAVITATION (shorter sewer durability) INCREASED FLOW RESISTANCE

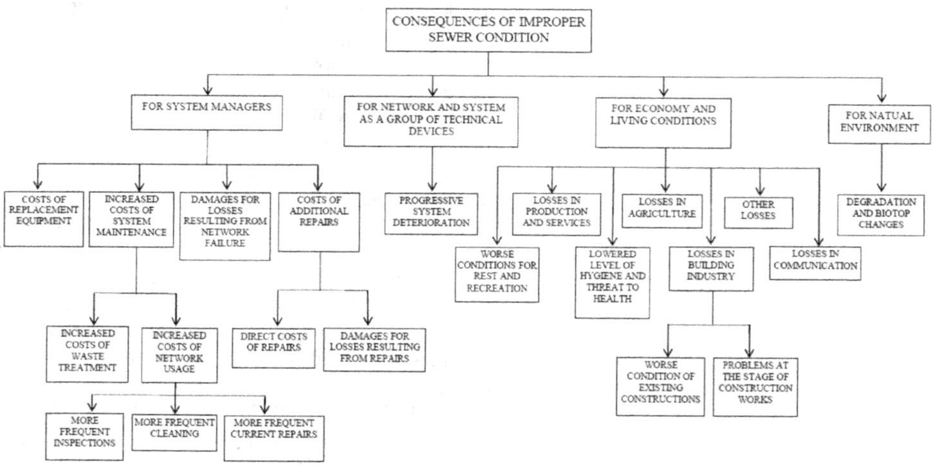

Figure 6. Consequences of deteriorated sewer condition.

3 DAMAGE TO PLASTIC SEWERS

The monograph (Kuliczkowska 2008) presents the results of extensive research (TV inspections) into sewers carried out over years by the Kielce University of Technology in cooperation with enterprises that manage sewerage systems in selected towns in Poland. Among the sewers that were analysed in detail there were sewers made of PVC, which allowed comparison of damage to them with the damage to concrete and stonework sewers. Those inspections as well as other TV inspections of operating sewers (e.g. Eckert 2012, Cierpiał et al. 2014) reveal a limited set of damage types to plastic sewers; cf. Table 1 and the percentage of plastic sewers. The plastic sewers (flexible) have their own types of damage that are not to be found in rigid sewers. They include:

- pipe transverse deflation (ovalisation),
- loss of wall stability,
- local deformations (dents e.g. made by stones),
- damage (cracks) of the pipe external surface,
- pipe longitudinal bending.

Examples of those types of damage are presented in Figure 7. The damage to the pipe external surface damage is related to the construction of pipelines using trenchless methods, where the greatest threat accompanies the application of HDD technologies and during the technical rehabilitation with the cracking (burst lining) technology. Since it is impossible to access the sewer, one cannot register the damage from inside and the assessment of the actual scope of damage and its propagation is limited. The pipe deflection is determined when the inspection reveals that the sewer walls are not straight in their longitudinal axis, both in the horizontal and vertical plane, which also occurs outside the pipe connection zone in contrast to rigid sewers and it is not necessarily related to the wall structure damage (like ruptures in rigid pipes). These changes result from improper manufacturing, in particular from errors in condensing the backfilling, and, as a secondary damage, they result from ground loosening nearby the sewer (e.g. following infiltration).

Moreover, during regular inspections it is difficult to register cases of damage to plastic pipes that results from illegal discharge of sewage that has a high temperature and is chemically aggressive to the plastic e.g. petroleum products to the polyethylene. This type of damage brings about local deterioration of material parameters and further local deformation (high temperature) or losses in wall thickness, loss of flexibility and tendency to crack (influence of chemical compounds).

Interesting results of investigations into the frequency of various types of damage in PVC pipes are presented in (Kuliczkowska 2008). The results are demonstrated, using original names of damage descriptions, in Table 3 (the investigations did not take into account the lack of straightness and incorrect slopes). The authors of the aforementioned study also attempted to classify damage by types of threat within the framework of three standard

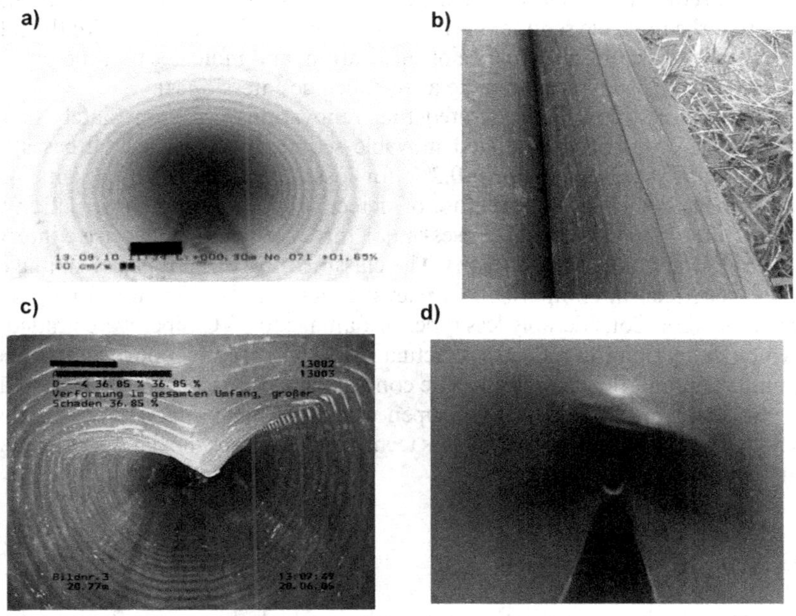

Figure 7. Examples of typical plastic pipe damage: ovalisation (a), cracks in the external surface (b), loss of stability (c), local deformation (d) (my own images and Internet sources).

Table 3. Frequency of PVC sewer damage (based on (Kuliczkowska 2008)).

Linear damage [% of the length of inspected sewers]	
movable sediment	42.3%
fixed sediment	2.2
abrasion of sides and bottom of the sewer	0.02
deflection of the sewer top	6.7
Non-linear damage [items/100 m]	
household connections, sealings	0.1/100 m
improper connections	0.01
infiltration	0.17
sealing protruding into the sewer inside	0.03
build-up after infiltration	0.06
chipped pipes at the connections	0.04
longitudinal displacement	3.1
transverse cracks and ruptures	0.01
diagonal cracks and ruptures	0.03
losses of coating fragments	0.06
local dents e.g. made by stones	0.7

assessment criteria (hydraulic and usage criterion, structure safety criterion, environmental criterion); they graded a given damage type from I to IV (where I means the sewer needs immediate renovation, class IV means the sewer will need renovation in a longer term).

The most frequent types of damage include movable or fixed sediment, deflection of the sewer top (as linear damage) and longitudinal displacement at the connections, local dents made by stones, observed infiltration, improper household connections resulting in leakage (as non-linear damage). These types of damage, both primary and secondary, first of all result from incorrect construction works (unless the design contains errors, i.e. unclassified ground in the filling zone, which is rather unlikely) and "economical" usage when large quantities of movable sediment are left in the inspected sewer without being removed. Simultaneously, most often the movable sediment is the soil carried inside the sewer with the infiltrating water; it proves the scale and significance of infiltration, and indicates that the pipe profile is not correct (sediments usually accumulate at so-called counterslopes).

The damage classification demonstrated that renovations are immediately needed in a small number of cases: for accumulated movable sediment: 0.04% (of all cases); for longitudinal displacement of connections: 0.3%; for losses of pipe wall fragments: 43%. This last type of damage, resulting from the use of defective pipes during construction works or their subsequent mechanical damage, poses major threat and requires decisive interventions, though it rarely occurs (0.06 items/100 m). The classification of all types of damage to PVC pipes is more beneficial in comparison to relevant damage classification for rigid concrete and stonework sewers. Considerably less types of damage to PVC pipes were graded I, II and III class, which in my opinion reflects the actual risk level posed by poor technical condition of functioning rigid and flexible sewers. The conclusion is also valid for those damage types typical of plastic sewers assembled in an open trench. However, it may not necessarily be valid for sewers made by means of trenchless technologies, where the damage to external pipe surface can be essential.

4 CONCLUSIONS

The sewage networks, in particular these networks in which the sewage flows under gravity, can function for a long period of time despite their partial damage. The hydraulic capacity of given pipes may be lowered, the construction may be weakened without being revealed, at least between inspections that are carried out relatively seldom (on average once in 5 or 10 years). Such situations pose a serious threat, and various consequences may ensue (Table 2). Relevant literature rather focuses on the damage theory for rigid sewers, whereas little can be found about the behaviour of flexible plastic sewers. One of the reasons behind it is that flexible sewers have been used for a shorter period of time, and they are of smaller diameters, which causes less problems, and consequently, the knowledge about them was not so urgently needed. However, a smaller number of registered failures in plastic pipes not necessarily proves their advantage over rigid ones, which still play an important role in new sewage networks.

Plastic sewers are resistant to short-term overloads, but if the overloads are long-lasting, rheological processes take place and the plastic properties deteriorate. Pipe designs take into account long-term parameters; moreover, over longer periods of time the loads are usually distributed evenly along the pipe due to the pressure redistribution in the ground. However, pressure points and longitudinal cracks on external surface of pipes present a threat for the pipeline, in particular if improper polyethylenes are used, which further leads to a phenomenon called slow crack growth. That is why, the manufacturers recommend that the products should be adjusted to possible loads and threats related to the way of installation and operating conditions (multi-layer pipes, polyethylenes grade 80 or higher, and/or cross-linked RC).

Most often plastic pipes seem to face the problem of leakage and its effects. These effects are indirectly related to all the types of damage that require immediate repairs in compliance with the study specified in Chapter 3. The loss of leak tight properties with the phenomena of exfiltration and/or infiltration cause the loosening of the ground around the sewer, which on principle pose a threat due to the fact that the capacity of flexible sewer depends on the behaviour of the ground (it is taken into consideration in calculations for the pipe-ground system, e.g. in compliance with directive ATV DVWK A 127). Even if the load capacity is not lost, the pipes deflect towards the loosened ground and the pipes lose its straightness. The soil carried inside the sewer as sediment limits the sewer's transportation capacity (hydraulic capacity); other environmental effects of leakage are presented in Chapter 2.

The loss of leak tight properties directly results from:

- excessive transverse deformation (ovalisation) in the connection zone,
- connection displacement,
- damage of seals, incorrect installation of seals in connections (protruding seals) or (less frequently) their chemical destruction during usage,
- lack of proper welding of elements by means of electrofusion sockets or butt welding,
- discontinuities in pipes (various types of ruptures, losses and chipping, in particular in PVC pipes and composite pipes).

Deformations around pipe connections (usually in the form of ovalisation) may be lower than the long-term threshold for a pipe (e.g. 15% or 9% depending on the recommendations), yet the pipe may lose its leak tight properties; it is the properties of the connection that matter.

The sewer leakage are repaired by means of methods of pipeline technical renovations or local repairs (liners, injections, resin spraying, sealing packers) that have been used on the market since many years. It should be emphasised that a part of widely used methods cannot be applied to the technical rehabilitation of plastic sewers. They include, first of all, methods of local repairs, which are made by connecting (gluing) a sealing element (packer, cap liner, etc.) with the wall of plastic pipe or the method with centrifugally cast sealant. The system manufacturers declare that their solutions can be applied after the substrate is adequately prepared; however, it turns out to be difficult for non-passage ducts, even though cutting robots are used.

These considerations do not cover all the issues relating to plastic sewer damage. The author aimed first of all to draw attention to the specificity of these issues, despite a small number of relevant studies.

REFERENCES

ATV M 143 (Regelwerk Abwasser-Abfall). 1998. *Inspektion, Instandsetzung, Sanierung und Erneuerung von Entwässerungskanälen und—leitungen*. Hannef: Abwassertechnische Vereinigung e.V.

Berger, C. & Falk, C. & Hetzel, F. & Pinnekamp, J. & Roder, S. & Ruppelt, J. 2016. Zustand der Kanalisation in Deutschland. Ergebnisse der DWA-Umfrage 2015. *KA Korrespondenz Abwasser, Abfall*, 63(6): 498–509.

Cierpiał, Z. & Kuliński, E. 2014. Awaryjność przewodów wodociągowych I kanalizacyjnych eksploatowanych w Przedsiębiorstwie Wodociągów i Kanalizacji Okręgu Częstochowskiego S.A. *VII Ogólnopolska Konferencja Techniczna PRiK, Sieci kanalizacyjne i wodociągowe z tworzyw sztucznych*; *Proc. symp., Zawiercie, 27–28 November 2014*: 27–37.

Eckert, R. 2012. PE Sewer Piping Systems, *KA Korrespondenz Abwasser, Abfall · International Edition*, no 5: 25–31.

Kuliczkowska, E. 2008. *Kryteria planowania bezwykopowej odnowy nieprzełazowych przewodów kanalizacyjnych*. Kielce: Wydawnictwo Politechniki Świętokrzyskiej.

Kwietniewski, M. 2011. Awaryjność infrastruktury wodociągowej i kanalizacyjnej w Polsce w świetle badań eksploatacyjnych. *XXV Konferencja Naukowo-Techniczna, Awarie Budowlane 2011; Proc. intern symp., Międzyzdroje, 24–27 May 2011*: 128–140.

Madryas, C. 1993. Odnowa przewodów kanalizacyjnych, *Prace Naukowe Instytutu Inżynierii Lądowej Politechniki Wrocławskiej*. Wrocław: Wydawnictwo Politechniki Wrocławskiej.

Madryas, C. & Przybyła, B. & Wysocki, L. 2010. *Badania i ocena stanu technicznego przewodów kanalizacyjnych*. Wrocław: Dolnośląskie Wydawnictwo Edukacyjne.

Przybyła, B. 1999. Ocena i kształtowanie konstrukcji przewodów kanalizacyjnych w ujęciu teorii niezawodności. *Prace Naukowe Instytutu Inżynierii Lądowej Politechniki Wrocławskiej*. Wrocław: Wydawnictwo Politechniki Wrocławskiej.

Wieczysty, A. 1990. *Niezawodność systemów wodociągowych i kanalizacyjnych*. Kraków: Wydawnictwo Politechniki Krakowskiej.

Underground Infrastructure of Urban Areas 4 – Madryas et al. (Eds)
© 2018 Taylor & Francis Group, London, ISBN 978-1-138-55953-0

Three-parameter metering method for diversification of water supply

J. Rak & K. Boryczko
Faculty of Civil and Environmental Engineering and Architecture, Rzeszow University of Technology, Rzeszow, Poland

ABSTRACT: The subject of the article is the methodology for calculating the degree of water supply diversification in water supply systems. The presented methodology proposed three parameters affecting the degree of water supply diversification: the maximum production capacity of water intakes, the volume of water tanks, flowability of pressure pipelines of the second degree pumping station. Dimensionless simpson ratio were calculated and the degree of diversification of the given collective water supply system, as well as compare even very different water supply systems was determined. The paper presents the calculation of simpson diversification indexes for selected collective water supply systems.

1 INTRODUCTION

Agglomeration (lat. Agglomeratio—accumulation) is an intensive built up area, characterized by a significant number of people staying in the area. Monocentric agglomerations are the large urban centres with satellite towns and urbanized villages around them. Polycentric agglomeration is created by several interconnected large urban areas. Located close to each other agglomerations in time transform into a unit with a high level of urbanization, referred to as megalopolis. Located close to each other dynamically developing agglomerations can create metropolitan areas, which include a few/several cities around the major urban centre and related with it functionally and economically.

Critical elements of municipal infrastructure conditioning the smooth functioning of large urban areas are the following systems:

- energy supply, energy raw materials and fuel,
- communication,
- ICT networks,
- financial,
- food supply
- water supply,
- health care,
- transport,
- rescue,
- ensuring the continuity of public administration activity,
- production, storage, storage and use of chemical and radioactive substances, including pipelines with hazardous substances.

The concept of portfolio diversification, known in economics, is the most popular and described as one of the most effective methods of reducing investment risk. This concept means the division of the portfolio into different types of investments, among others, in terms of the type of market (e.g. raw materials, currency, shares, bonds), trade (in the case of shares) or geographical coverage of given entities (e.g. shares, shares funds of enterprises from a specific region). In the natural sciences diversification is identified with the biodiversity of fauna and flora in different biosytems (Hill 1973, Pielou 1966, Simpson 1949). Diversification is an important issue in the electricity, gas and heat supply of urban agglomerations (Babiarz & Blokus-Roszkowska 2015, Goodman & Evenhuis 2014, Muller et al. 2008, Ohimain 2015, Olivier & Root 2014, Ponisio 2015, Tchórzewska–Cieślak at al. 2016). Each agglomeration for the proper functioning requires good quality water supply in necessary amount. It is particularly important in crisis or emergency situations. The possibility of water supply from several sources clearly increases the diversification of water supply (Boryczko & Rak 2016, Rak 2015). If you shut off the water intake or the water intakes the volume of water stored in water supply network tanks plays a key role (Lee et al. 2016, Rak 2015, Sarbu & Ostafe 2016). An important role is also played by the collective pressure pipelines of the second degree pumping station, whose number and diameter directly affect the reliability of water supply (Bajer 2007, Boryczko & Rak 2016, Farahmandfar et al. 2017, Iwanejko & Bajer 2009, Kutylowska 2015, Kutylowska 2015, Rak 2014, Studziński 2014, Szpak & Tchórzewska-Cieślak 2015, Tchórzewska-Cieślak & Rak 2010, Vieira & Cunha 2017, Zimoch & Paciej 2013). Reducing the water supply or a total lack of water supply can cause threat to water consumers lives or health, as well as financial losses of the recipients and water companies. The operator of the collective water supply system (CWSS) having a high degree of diversification has greater room for maneuver in case of crisis, and his decisions do not have to be taken under pressure of time (Ke et al. 2016, Li et al. 2016).

The subject of the article is the methodology for calculating the degree of water supply diversification in water supply systems. The presented methodology proposed three parameters affecting the degree of water supply diversification: the maximum production capacity of water intakes, the volume of water tanks, flowability of pressure pipelines of the second degree pumping station. On the basis of these data it is possible to calculate the dimensionless Simpson ratio and assess the degree of diversification of the given collective water supply system, as well as compare even very different water supply systems. The paper presents the calculation of diversification indexes for selected CWSS in Poland.

2 METHOD

For the assessment of diversification the following parameters were proposed:

- Q – Parameter associated with the resource of water in the water supply subsystems (WSS) (Rak & Włoch 2015),
- V – Parameter associated with the volume of water in the network water tanks (Rak 2015),
- M – parameter related to flowability of pressure pipelines of the second degree pumping station

Water resources in the individual WSS mean the maximum daily production capacity. The parameter V includes the volume of the network tanks, i.e. all the tanks located in the technological line after the second stage pumping stations. The study used the specific flowability (m^3/s) and taking into account the diameters of pressure pipelines of the second degree pumping station, directly correlated them with the flow. The M parameter takes into account the diameters of the collective pressure pipelines of the second degree pumping station and their roughness. The proposed methodology adopted the same roughness of collective pressure pipelines of the second degree pumping station for the tested systems of public water supply. In further studies it will be possible to take into account the material and age of the pipe (the pipe will have a specific value of roughness).

Diversification degree for parameter Q, V, M were determined according to Simpson formula (Simpson 1951, Simpson 1949):

$$d_Q = 1 - \sum_{i=1}^{m} u_i^2 \tag{1}$$

$$d_v = 1 - \sum_{j=1}^{n} u_j^2 \tag{2}$$

$$d_M = 1 - \sum_{k=1}^{p} u_k^2 \tag{3}$$

where d_Q = the diversification index of water resources (intakes) in the CWSS; d_V = the diversification index of the volume of water in the network water tanks; d_F = the diversification flowability of collective pressure pipelines of the second degree pumping station; u_i = share of maximum daily production capacity of the i-th CWSS (m^3/d) in the total maximum daily capacity of water supply to CWSS; u_j = share of the volume of the j-th tank (m^3) in a total volume of water in the network water tanks; u_k = share of flowability of k-th pressure pipelines of the second degree pumping station in a sum of flowability collective pressure pipelines of the second degree pumping stations; m = number of WSS; n = number of network tanks; p = number of collective pressure pipelines of the second degree pumping stations.

Tables 1–4 presents numerical values diversification index for 2, 3, 4 and 5 elements.

Analyzing tables 1÷4 it is noticed that an increasing m, n or p does not mean that $d_Q/d_V/d_M$ will also increase.

Global diversification degree was calculated according to:

Table 1. The numerical values of diversification index (m, n, p = 2).

m, n, p = 2	$u_1 = 0{,}5$ $u_2 = 0{,}5$	$u_1 = 0{,}6$ $u_2 = 0{,}4$	$u_1 = 0{,}7$ $u_2 = 0{,}3$	$u_1 = 0{,}8$ $u_2 = 0{,}2$	$u_1 = 0{,}9$ $u_2 = 0{,}1$	$u_1 = 0{,}95$ $u_2 = 0{,}05$	$u_1 = 0{,}99$ $u_2 = 0{,}01$
$d_Q/d_V/d_M$	0,50	0,48	0,42	0,32	0,18	0,095	0,0198

Table 2. The numerical values of diversification index (m. n. p = 3).

m. n. p = 3	$u_1 = 0.33$ $u_2 = 0.33$ $u_3 = 0.33$	$u_1 = 0.4$ $u_2 = 0.3$ $u_3 = 0.3$	$u_1 = 0.5$ $u_2 = 0.3$ $u_3 = 0.2$	$u_1 = 0.6$ $u_2 = 0.3$ $u_3 = 0.1$	$u_1 = 0.6$ $u_2 = 0.2$ $u_3 = 0.2$	$u_1 = 0.7$ $u_2 = 0.2$ $u_3 = 0.1$	$u_1 = 0.8$ $u_2 = 0.1$ $u_3 = 0.1$
$d_Q/d_V/d_M$	0.667	0.66	0.62	0.54	0.56	0.46	0.34

Table 3. The numerical values of diversification index (m. n. p = 4).

m. n. p = 4	$u_1 = 0.25$ $u_2 = 0.25$ $u_3 = 0.25$ $u_4 = 0.25$	$u_1 = 0.3$ $u_2 = 0.3$ $u_3 = 0.2$ $u_4 = 0.2$	$u_1 = 0.4$ $u_2 = 0.3$ $u_3 = 0.15$ $u_4 = 0.15$	$u_1 = 0.5$ $u_2 = 0.3$ $u_3 = 0.1$ $u_4 = 0.1$	$u_1 = 0.6$ $u_2 = 0.2$ $u_3 = 0.1$ $u_4 = 0.1$	$u_1 = 0.7$ $u_2 = 0.1$ $u_3 = 0.1$ $u_4 = 0.1$
$d_Q/d_V/d_M$	0.75	0.74	0.705	0.64	0.58	0.48

Table 4. The numerical values of diversification index (m. n. p = 5).

m. n. p = 5	$u_1 = 0.2$ $u_2 = 0.2$ $u_3 = 0.2$ $u_4 = 0.2$ $u_5 = 0.2$	$u_1 = 0.3$ $u_2 = 0.3$ $u_3 = 0.2$ $u_4 = 0.1$ $u_5 = 0.1$	$u_1 = 0.4$ $u_2 = 0.3$ $u_3 = 0.1$ $u_4 = 0.1$ $u_5 = 0.1$	$u_1 = 0.5$ $u_2 = 0.2$ $u_3 = 0.1$ $u_4 = 0.1$ $u_5 = 0.1$	$u_1 = 0.6$ $u_2 = 0.1$ $u_3 = 0.1$ $u_4 = 0.1$ $u_5 = 0.1$	$u_1 = 0.8$ $u_2 = 0.05$ $u_3 = 0.05$ $u_4 = 0.05$ $u_5 = 0.05$
$d_Q/d_V/d_M$	0.80	0.76	0.72	0.68	0.60	0.35

$$d = d_Q + d_V + d_M \tag{4}$$

In the calculations it was assumed that if $u_i = 1.0$, $u_j = 1.0$ or $u_k = 1.0$, the value of the index d_Q, d_V or d_M taken to calculate the index d from the equation (4) is 0.0.

The following standards for the ratio d were adopted:

- lack of diversification $d = 0$
- low diversification $0 < d \le 1.0$
- average diversification $1.0 < d \le 1.5$
- sufficient diversification $1.5 < d \le 2.0$
- very satisfactory diversification $2.0 < d \le 3.0$

3 RESULTS

Calculations of the assessment of the diversification degree for selected CWSS are presented below.

- Rzeszów

– Q – two intakes:
$u_1 = 0.43$
$u_2 = 0.57$
According to (1):
$d_Q = (1 - (0.43^2 + 0.57^2) = 0.49$

– V – twelve tanks:
$u_1 = 0.018$
$u_2 = u_3 = 0.049$
$u_4 = u_5 = u_6 = u_7 = 0.081$
$u_8 = 0.480$
$u_9 = u_{10} = u_{11} = u_{12} = 0.02$
According to (2):
$d_V = (1 - (0.018^2 + 2{\cdot}0.049^2 + 4{\cdot}0.081^2 + 0.480^2 + 4{\cdot}0.02^2) = 0.737$

– M – four collective pressure pipelines of the second degree pumping station:
$u_1 = 0.024$
$u_2 = u_3 = 0.049$
$u_4 = 0.878$
According to (3):
$d_M = (1 - (0.24^2 + 0.49^2 + 0.49^2 + 0.878^2) = 0.224$

According to (4):
$d = 0.49 + 0.737 + 0.224 = 1.451$ – average diversification

The calculations for another CWSSs were shown in Tables 5–10.

Three out of the analysed CWSS, were classified in the category of sufficient diversification. Very satisfactory diversification was found for one CWSS, which is due to balanced tanks volumes and a high value of the index d_M. As for the proposed in the method three parameters, there are always two of them in the given CWSS, i.e. water intakes and pressure pipelines of the second degree pumping station. Analysing the results of the research it was found that the lack of water tanks has an impact on the low value of the index d (Bydgoszcz, $d_V = 0$). With the increase in the number of water intakes, water tanks or collective pressure pipelines of the second degree pumping station the value of the indexes d_Q, d_V, d_M, grows.

Table 5. The calculation of indicator d for Tarnow CWSS.

Q	V	M
$u_1 = 0.168$	$u_1 = 0.088$	$u_1 = u_2 = 0.332$
$u_2 = 0.634$	$u_2 = u_3 = u_4 = 0.070$	$u_3 = u_4 = 0.054$
$u_3 = 0.198$	$u_5 = 0.105$	$u_5 = u_6 = 0.114$
–	$u_6 = 0.007$	–
–	$u_7 = u_8 = 0.282$	–
–	$u_9 = u_{10} = 0.005$	–
–	$u_{11} = u_{12} = u_{13} = u_{14} = 0.004$	–
$d_Q = 0.531$	$d_V = 0.80$	$d_M = 0.748$

d = 2.086 – very satisfactory diversification.

Table 6. The calculation of indicator d for Krosno CWSS.

Q	V	M
$u_1 = 0.60$	$u_1 = u_2 = 0.5$	$u_1 = u_2 = 0.463$
$u_2 = 0.28$	–	$u_3 = 0.074$
$d_Q = 0.552$	$d_V = 0.5$	$d_M = 0.566$

d = 1.618 – sufficient diversification.

Table 7. The calculation of indicator d for Bydgoszcz CWSS.

Q	V	M
$u_1 = 0.704$	–	$u_1 = u_2 = 0.046$
$u_2 = 0.296$	–	$u_3 = u_4 = 0.36$
–	–	$u_5 = u_6 = 0.094$
$d_Q = 0.417$	$d_V = 0.0$	$d_M = 0.719$

d = 1.136 – average diversification.

Table 8. The calculation of indicator d for Gorzow Wielkopolski CWSS.

Q	V	M
$u_1 = 0.23$	$u_1 = 0.667$	$u_1 = 0.673$
$u_2 = 0.176$	$u_2 = 0.111$	$u_2 = 0.177$
$u_3 = 0.594$	$u_3 = 0.222$	$u_3 = u_4 = 0.061$
–	–	$u_5 = 0.028$
$d_Q = 0.564$	$d_V = 0.494$	$d_M = 0.508$

d = 1.566 – average diversification.

Table 9. The calculation of indicator d for Poznan CWSS.

Q	V	M
$u_1 = 0.652$	$u_1 = 0.667$	$u_1 = u_2 = 0.285$
$u_2 = 0.348$	$u_2 = 0.333$	$u_3 = u_4 = 0.149$
–	–	$u_5 = u_6 = 0.061$
–	–	$u_7 = 0.010$
$d_Q = 0.454$	$d_V = 0.444$	$d_M = 0.786$

d = 1.684 – sufficient diversification.

Table 10. The calculation of indicator d for Racibórz CWSS.

Q	V	M
$u_1 = 0.75$	$u_1 = u_2 = u_3 = 0.333$	$u_1 = 0.138$
$u_2 = 0.25$	–	$u_2 = 0.862$
$d_Q = 0.375$	$d_V = 0.667$	$d_M = 0.238$

d = 1.280 – average diversification.

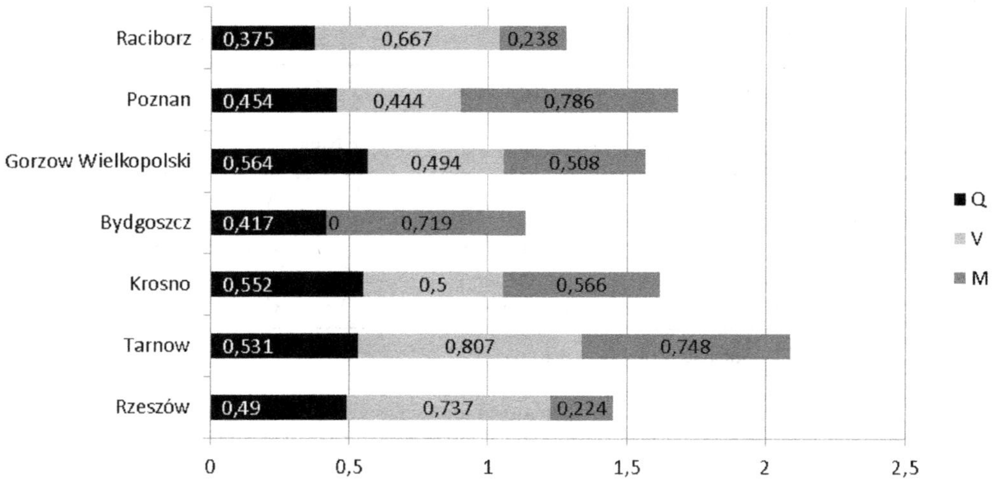

Figure 1. Graphical representation of analysis results.

4 CONCLUSIONS

Diversification degree depends on:

- number of intakes, tanks, collective pressure pipelines of the second degree pumping station, evenness of distribution of intakes productivity, volume of network tanks, flowability of collective pressure pipelines of the second degree pumping station,
- share of the volume of accumulated water in the size of its collection from the water supply system (the development of criteria and standards is in the process of research).

In comparison with other indicators diversification (Boryczko & Rak 2016, Rak 2015) proposed by the Simpson index rewards systems with more intakes, tanks, collective pressure pipelines of the second degree pumping station.

Dimensionless values of the global diversification index predispose it to analyse the degree of diversification of water supply in different CWSS. It is possible to compare even very different systems, with one or several intakes, equipped or not equipped with water supply tanks, with different numbers and different collective pressure pipelines of the second degree pumping station.

The authors are aware that the analyses of the diversification performed in particular cities are result of the technical conditions, the size of water resources in each source, the need for expansion of the CWSS elements. The designers of these systems certainly were not guided by the ability to calculate and assess the degree of diversification because they did not know such concept. We realize that a large number of cities are supplied from a single source (e.g. for Montreal it is a big St. Lawrence river) and the CWSS operate in those cities with satisfactory reliability. The content of the work is a new contribution to the possibility of conscious design of the CWSS expansion, taking into account the diversification as the basis of preliminary reliability.

REFERENCES

Babiarz, B. & Blokus-Roszkowska, A. 2015. Probabilistic model of district heating operation process in changeable external conditions. *Energy and Buildings* 103: 159–165.

Bajer J. 2007. Reliability analysis of variant solutions for water pumping stations, In Dudzińska M.R. Pawłowski L., Pawłowski A. (eds), *Environmental Engineering:* 253–261. New York, Singapore: Taylor & Francis Group.

Boryczko, K. & Rak, J. 2016. Oceny dywersyfikacji zaopatrzenia w wybranych miast w wodę metodą dwuparametryczną z wykorzystaniem wskaźnika Pielou. *Instal* 6: 60–63.

Farahmandfar, Z., Piratla, K.R. & Andrus, R.D. 2017. Resilience Evaluation of Water Supply Networks against Seismic Hazards. *Journal of Pipeline Systems Engineering and Practice* 8: 1

Goodman, K.R., Evenhuis, N.L., Bartosova-Sojkova, P. & O'grady, P.M. 2014. Diversification in Hawaiian long-legged flies (Diptera: Dolichopodidae: Campsicnemus): Biogeographic isolation and ecological adaptation. *Molecular Phylogenetics and Evolution* 81: 232–241.

Hill, M. 1973. Diversity and Evenness: a Unifying Notation and its Consequences. *Ecology* 2: 427–432. 54

Iwanejko, R. & Bajer, J. 2009. Determination of the optimum number of repair units for water distribution systems. *Archives of Civil Engineering* 55: 87–101. 1

Ke, W., Leia, Y., Shaa, J., Zhangb, G., Yana, J., Lind, X., Pana, X. 2016. Dynamic simulation of water resource management focused on water allocation and water reclamation in Chinese mining cities. *Water Policy* 18: 844–861. 4

Kutylowska, M. 2015. Modelling of Failure Rate of Water-pipe Networks. *Periodica Polytechnica-Civil Engineering* 59: 37–43. 1

Kutylowska, M. 2015. Neural network approach for failure rate prediction. *Engineering Failure Analysis* 47: 41–48.

Lee, H.M., Yoo, D.G., Kang, D., Jun, H. & Kim, J.H. 2016. Uncertainty quantification of pressure-driven analysis for water distribution network modeling. *Water Science and Technology-Water Supply* 16: 599–610. 3

Li, F., Wang, W. & Ramírez, L.H.G. 2016. The determinants of two-dimensional service quality in the drinking water sector—evidence from Colombia. *Water Policy* 18: 983–997. 4

Muller, R., Steinert, M. & Teufel, S. 2008. Successful diversification strategies of electricity companies: An explorative empirical study on the success of different diversification strategies of German electricity companies in the wake of the European market liberalization. *Energy Policy* 36: 398–412. 1

Ohimain, E.I. 2015. Diversification of Nigerian Electricity Generation Sources. *Energy Sources Part B-Economics Planning and Policy* 10: 298–305. 3

Olivier, J. & Root D. 2014. The diversification strategies of large South African contractors into southern Africa. *Journal of the South African Institution of Civil Engineering* 56: 88–96. 2

Pielou, E.C. 1966. The measurement of diversity in different types of biological collections. *Journal of Theoretical Biology* Elsevier: 131–144. 13

Ponisio, L.C., M'gonigle L.K., Mace K.C., Palomino J., De Valpine P. & Kremen C. 2015. Diversification practices reduce organic to conventional yield gap. *Proceedings of the Royal Society B-Biological Sciences* 282: 1799

Rak, J. 2014. Metoda oceny stopnia dywersyfikacji dostawy wody dla wybranych miast w Polsce. *Instal* 5: 38–40.

Rak, J. 2015. Propozycja oceny dywersyfikacji objętości wody w sieciowych zbiornikach wodociągowych. *Journal of Civil Engineering, Environment and Architecture* 32: 339–349. 62

Rak, J. & Włoch, A. 2015. Models of level diversification assessment of Water Supply Subsystems, In Kolonko A. Madryas C., Nienartowicz B., Szot A. (eds), *Underground Infrastructure of Urban Areas 3:* 237–244. London: Taylor & Francis Group.

Sarbu, I. & Ostafe G. 2016. Optimal design of urban water supply pipe networks. *Urban Water Journal* 13: 521–535. 5

Simpson, E.H. 1949. Measurement of diversity. *Nature* 163: 688.

Simpson, E.H. 1951. The Interpretation of Interaction in Contingency Tables. *Journal of the Royal Statistical Society, Series B* 13: 238–241.

Studziński, A. 2014. Amount of labour of water conduit repair, In Van Gelder P.H.A.J.M., Steenbergen R.D.J.M., Miraglia S., Vrouwenvelder A.C.W.M. (eds), *Safety, Reliability and Risk Analysis: Beyond the Horizon:* 2081–2084. London: Taylor & Francis Group.

Szpak, D. & Tchorzewska-Cieslak B. 2015. Water producers risk analysis connected with collective water supply system functioning, In Zamojski, W., Mazurkiewicz, J., Sugier, J. et al. (eds), *Dependability Engineering and Complex Systems. Advances in Intelligent Systems and Computing:* 479–489. Switzerland Springer.

Tchórzewska-Cieślak, B. & Rak, J. 2010. Method of identification of operational states of water supply system, In Dudzińska M.R. Pawłowski L., Pawłowski A. (eds), *Environmental Engineering III:* 521–526. London: Taylor & Francis Group.

Tchórzewska–Cieślak, B., Pietrucha-Urbanik, K. & Urbanik, M. 2016. Analysis of the gas network failure and failure prediction using the Monte Carlo simulation method. *Eksploatacja I Niezawodnosc-Maintenance and Reliability* 18: 254–259. 2

Vieira, J. & Cunha, M.C. 2017. Nested Optimization Approach for the Capacity Expansion of Multiquality Water Supply Systems under Uncertainty. *Water Resources Management* 31: 1381–1395. 4

Zimoch, I. & Paciej, J. 2013. Zastosowanie Geograficznych Systemów Informacyjnych w zarządzaniu oraz prowadzeniu kontroli jakości wody przeznaczonej do spożycia przez ludzi. *Instal* 38–41. 1

Underground Infrastructure of Urban Areas 4 – Madryas et al. (Eds)
© 2018 Taylor & Francis Group, London, ISBN 978-1-138-55953-0

The impact of the channel retention before the tank on its retention capacity

M. Starzec & J. Dziopak
Department of Infrastructure and Water Management, Rzeszow University of Technology, Poland

ABSTRACT: The study defines the impact of the inlet channel's bottom fall on the relevance of the use of its storage capacity in order to reduce the required usable volume of a multi-chamber storage reservoir. Tests using hydrodynamic simulations were conducted in three model catchment areas with various shapes and with various systems of channels, using hydrodynamic software SWMM 5.1. The obtained results confirmed the assumptions as to the relevance of including the storage volume of the inlet channel. The demonstrated significant share of channel holding has a measurable application effect because it makes it possible to considerably reduce the required usable volume of the multi-chamber storage reservoir, and even to abandon its construction in extreme cases. It turns out that the effectiveness of the use of the storage volume of the inlet channel determined in the tests is greater when the fall of this channel is lower. This results from an obvious phenomena because when the value of the inlet channel's bottom fall decreases, sewer backwater reaching increasingly further is observed which makes it possible to incorporate a longer section of the sewage network to accumulate the sewage, ensuring its gravitational operation. It was demonstrated in the course of simulation tests on model catchment areas of various shapes and systems of the sewage network that the greatest benefits from the use of channel holding are observed: (1) with a high level of the location of the edge between chambers towards the bottom of the inlet collector, (2) with substantial values of the flow rate β reduction coefficient and (3) with a dense system of channels of the storm water drainage system. Due to the simple and cheap manner of using the storage volume of the inlet channel, it is advisable to include this phenomenon when developing design concepts, regardless of the network's hydraulic parameters and the hydrological parameters of the drained area.

Keywords: multi-chamber storage reservoir; channel holding; design flow; drainage system

1 INTRODUCTION

One of the main consequences of urban areas development is the increase in the amount of wastewater generated within them, that causes the need for effective collection and discharge of this medium. However, wastewater should not be treated only as the source of the problems, because it may find use for various purposes (Kordana 2017, Stec et al. 2017). Sewage discharge systems are the most expensive and the most difficult infrastructural investment projects, especially when they operate in a gravitational system (Benzerra et al. 2012). The construction of sewage systems in a catchment area equipped with other underground and transport infrastructure is particularly difficult and expensive (Gogate et al. 2017, Przybyła et al. 2011). The maintenance of a proper fall of the sewage network channel bottom in order to enable the gravitational flow and to avoid collisions with the existing infrastructure often requires the construction of ducts very deep which, in turn, generates high investment costs.

The storm water drainage system, especially a combined sewage system, is the most capital-intensive type of sewage systems (Zeleňáková et al. 2014). This results from the fact that the channels in these systems are designed for short extreme flows which are characterized by great irregularities of flows (Kaźmierczak & Kotowski 2012). There are a number of conditions which impose technical progress in storm water management. In practice, this is aimed at searching for innovative technologies and solutions which will make it possible to obtain a more efficient operation of drainage systems with minimum financial expenses (Bartoszek et al. 2015, Kordana et al. 2014, Słyś & Kordana 2014).

Drainage systems are equipped with various facilities, especially with the use of multi-chamber storage reservoirs, for many years in order to improve their operation (Szeląg & Mrowiec 2016, Mazurkiewicz et al. 2016, Wiśniowska-Kielian et al. 2013). The beneficial impact of storage reservoirs on the operation of the sewage system, the protection of the receiver water as well as the technical efficiency of a sewage treatment plant has often been confirmed in practice (Dziopak & Starzec 2016, Pochwat 2017, Suligowski & Orłowska-Szostak 2013). However, apart from the unquestionable advantages of such facilities, they also have fundamental disadvantages, including the high cost of such investment projects as well as the lack of available land to build them. These conditions, and others, inspire the search for effective solutions which make it possible to reduce the required usable volume of storage reservoirs without deteriorating the effectiveness of the operation of the entire sewage system (Pochwat et al. 2017, Stec & Słyś 2014).

The inlet channel to the storage reservoir is usually characterized by high geometry. Therefore, one of the vital issues which should be taken into account when dimensioning a sewage tank is the practical use of the storage volume of the inlet collector. The result of using the volume of the inlet channel is a proper reduction of the required usable volume of the storage reservoir by a properly determined volume of channel holding.

Measurable benefits which may be obtained due to the use of channel holding depend on numerous factors. The most significant ones include: (1) the geometry of the inlet channel, the maximum possible impoundment height of sewage in this channel and, first of all, (3) the inlet channel's bottom fall which has a direct impact on the range of backwater, still ensuring a gravitational flow of sewage.

The geometry of the flow chamber of multi-chamber tanks determines the hydraulic conditions for its operation (Dziopak & Słyś 2007, Słyś & Dziopak 2011). The forced maintenance of a high level of sewage in the flow chamber in these tanks leads directly to a beneficial impoundment of sewage in the inlet collector. The maximum assumed level of sewage in the multi-chamber tank results from the location of the level of the edge between chambers through which the sewage overflows from the flow chamber to the accumulation chamber. This phenomenon results in a forced increase in the level of sewage in the inlet channel by a certain amount which results in the use of the channel storage volume of the sewage system ducts covered by the sewage backwater.

The usable value of the inlet collector's storage volume results from the difference between the total surface of its cross-section and the collector's active surface calculated with flow elicited with the determined duration of dependable rain to dimension the tank's geometry at a specific length of the collector covered by the sewage backwater. Therefore, the selection of dependable rain to dimension a specific type of storage reservoir directly affects the manner in which the storage volume of the inlet collector and its value are calculated.

2 OPTIONS ADOPTED IN THE ANALYSIS

The article presents an attempt to determine the impact of the inlet channel's bottom fall on the relevance of the use of its storage capacity to reduce the required usable volume of a multi-chamber storage reservoir, operating in a storm water drainage system. For this purpose three different model catchment areas were adopted for tests by formulating three diverse hydrodynamic models of the operation of a storm water drainage system with diagrams presented in Figure 1, using hydrodynamic software SWMM 5.1. Each diverse drainage system consists of 18 sections of channels, each 250 m long. The ducts of the analyzed networks are arranged with constant falls assuming, respectively, the values 1, 2 or 3 ‰, depending on the analyzed option. The drainage area assigned to the assumed catchment

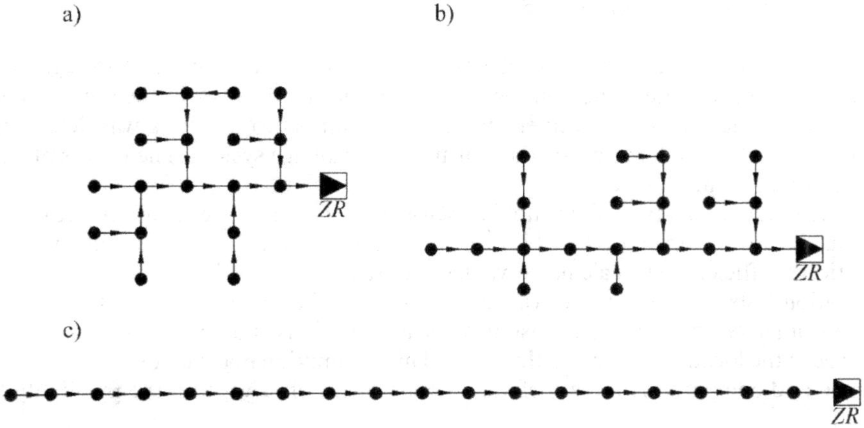

Figure 1. Adopted schematic representation of the sewer networks with the locations of the multi-chamber storage reservoirs.

areas is constant and is each time 144.0 ha. The coefficient of surface runoff is assumed at 0.25. Therefore, the total area of the reduced catchment area from which the storm water is drained to the storm water drainage system is 36.0 ha.

In order to calculate the unit rainfall intensity, Błaszczyk's (1) formula was applied (Błaszczyk et al. 1983). It defines the correlations between rain intensity and duration:

$$q = \frac{6{,}631\sqrt[3]{H^2 \cdot c}}{t_d^{2/3}} \tag{1}$$

where q – unit rain intensity; H – amount of mean annual precipitation; c – precipitation frequency; and t_d – precipitation duration.

For the purpose of this simulation, the amount of mean annual precipitation adopted was of the order of $H = 600$ mm/year, while the precipitation frequency was $c = 2$ years. The conducted tests assumed that the precipitation is of constant intensity throughout the duration of the tests and it starts at the same time in every point of the catchment area.

Due to the lack of a function describing a hydraulic system of a multi-chamber tank in the software SWMM 5.1, the proper representation of its operation was achieved as a result of using two single-chamber tanks described with the Storage function. The first tank was assigned with the function of a flow chamber, while the second tank serves as an accumulation chamber. The flow of sewage over the edge between chambers from the flow chamber to the accumulation chamber was described with the Outlet function, while the discharge of sewage from the accumulation chamber towards the discharge channel—with the Weir function. The tests assumed another assumption that the discharge of sewage from the accumulation chamber takes place only in periods when the level of sewage in this chamber is higher than in the flow chamber.

The tests included seven different levels of the location of the overflow edge: (1) at the height of the inlet channel's bottom (level 0), (2) at one fourth of the inlet channel's diameter (level 1/4 D), (3) at half of the inlet channel's diameter (level 1/2 D), (4) at three fourths of the inlet channel's diameter (level 3/4 D), (5) at the height of the vault of the inlet channel (level D), (6) 0.5 m above the vault of the inlet channel (level D + 0.5 m) and (7) 1.0 m above the level of the vault of the inlet channel (level D + 1.0 m).

The simulations were conducted with changes in steps regarding the value sewage flow rate β reduction coefficient which are: 0.3; 0.4; 0.5; 0.6; 0.7 and 0.8 which results in six different values of the intensity of the runoff of sewage from tanks in each of the assumed catchment areas.

3 ANALYSIS OF TEST RESULTS

In order to determine the maximum values of the intensity of the runoff of sewage from the storage reservoir *ZR*, determined on the basis of the adopted values of the sewage flow rate β reduction coefficient, the peak intensity of sewage supply to this tank was determined in three different adopted diagrams of the storm water drainage system. The results of calculations are presented in Table 1.

The maximum intensity of the runoff of sewage from the storage reservoir *ZR* was determined on the basis of the data listed in Table 1 as well as the adopted values of the flow rate β reduction coefficient. The calculated values are presented in Table 2.

Simulation tests were conducted with the assumption that the maximum runoff of sewage from the storage reservoir *ZR* is a constant value in all options and does not change along with the change in the location of the overflow edge. This assumption made it possible to thoroughly determine and compare the impact of the impoundment of sewage on the possibility to use

Table 1. Maximum intensity of storm water supply to storage reservoir *ZR* based on the determined dependable duration of rain.

	$i = 1$ ‰		$i = 2$ ‰		$i = 3$ ‰	
Catchment	$Q_{d,max}$* dm³/s	T_{dm}** min	$Q_{d,max}$* dm³/s	T_{dm}** min	$Q_{d,max}$* dm³/s	T_{dm}** min
I	2595.10	14	2894.93	12	3206.50	10
II	2237.25	19	2623.56	14	2939.65	12
III	1456.31	40	1723.96	31	1940.11	27

*$Q_{d,max}$ – maximum intensity of storm water supply to storage reservoir *ZR*, dm³/s.
**T_{dm} – dependable duration of rain to dimension the sewage system, min.

Table 2. Maximum intensity of storm water runoff from storage reservoir.

	$i = 1$ ‰		$i = 2$ ‰		$i = 3$ ‰	
Catchment	β* –	$Q_{O,max}$** dm³/s	β* –	$Q_{O,max}$** dm³/s	β* –	$Q_{O,max}$** dm³/s
I	0,3	778.53	0.3	868.48	0.3	961.95
	0.4	1038.04	0.4	1157.97	0.4	1282.60
	0.5	1297.55	0.5	1447.47	0.5	1603.25
	0.6	1557.06	0.6	1736.96	0.6	1923.90
	0.7	1816.57	0.7	2026.45	0.7	2244.55
	0.8	2076.08	0.8	2315.94	0.8	2565.20
II	0.3	671.18	0.3	787.07	0.3	881.90
	0.4	894.90	0.4	1049.42	0.4	1175.86
	0.5	1118.63	0.5	1311.78	0.5	1469.83
	0.6	1342.35	0.6	1574.14	0.6	1763.79
	0.7	1566.08	0.7	1836.49	0.7	2057.76
	0.8	1789.80	0.8	2098.85	0.8	2351.72
III	0.3	436.89	0.3	517.19	0.3	582.03
	0.4	582.52	0.4	689.58	0.4	776.04
	0.5	728.16	0.5	861.98	0.5	970.06
	0.6	873.79	0.6	1034.38	0.6	1164.07
	0.7	1019.42	0.7	1206.77	0.7	1358.08
	0.8	1165.05	0.8	1379.17	0.8	1552.09

*β – sewage flow rate reduction coefficient, –;
**$Q_{O,max}$ – maximum intensity of storm water runoff from storage reservoir, dm³/s.

channel holding in each design option. Seven different levels of the location of the overflow edge between chambers were assumed during the simulations. This level directly affects the height of the sewage level in the storage reservoir as well as the location of the sewage level in the inlet channel during the channel holding process. The higher the crown of the overflow between chambers, the further the backwater caused by the impoundment of sewage in the flow chamber reaches deep into the storm water drainage system, situated above this chamber.

In order to determine the exact impact of channel holding of the inlet channel on the required usable volume of the multi-chamber storage reservoir *ZR*, located on the outlet from the catchment area, four hundred and eighty six cases were examined in total, assuming different values of parameters affecting the examined phenomenon. The values of the required usable volume of the tank are presented in Tables 3 and 4 with the channel's bottom fall $i = 1$ ‰, in Tables 5 and 6 with the channel's bottom fall $i = 2$ ‰ and in Tables 7 and 8 with the channel's bottom fall $i = 3$ ‰, with markings meaning, respectively:

- V – required usable volume of storage reservoir *ZR*, m^3;
- ΔV – difference between usable volumes of storage reservoirs *ZR*, m^3;
- h_k – foundation level of the crown of overflow between chambers *ZR*, –;
- % – difference between usable volumes of storage reservoirs *ZR*, %.

The differences in usable volumes of storage reservoirs ΔV were calculated using the formula (2):

$$\Delta V = V_0 - V_i \tag{2}$$

where: V_0 – usable volume of storage reservoir *ZR* with a crown of overflow between chambers located at the level of the bottom of the inlet channel, m^3; Vi – usable volume of the examined storage reservoir V, m^3.

Table 3. Maximum required usable volumes of storage reservoir *ZR* with collectors of the storm water drainage system constructed with fall $i = 3$ ‰.

		$\beta = 0.30$			$\beta = 0.40$			$\beta = 0.50$		
Catchment	h_k	V [m^3]	ΔV [–]	% [m^3]	V [m^3]	ΔV [–]	% [m^3]	V [m^3]	ΔV [–]	% [m^3]
I	0 D	1370	0	0.00	964	0	0.00	665	0	0.00
	1/4 D	1366	4	0.29	960	4	0.41	661	4	0.60
	1/2 D	1335	35	2.55	939	25	2.59	632	33	4.96
	3/4 D	1282	88	6.42	892	72	7.47	583	82	12.33
	D	1233	137	10.00	844	120	12.45	538	127	19.10
	D + 0.5 m	1004	366	26.72	620	344	35.68	314	351	52.78
	D + 1.0 m	817	553	40.36	416	548	56.85	114	551	82.86
II	0 D	1424	0	0.00	1000	0	0.00	683	0	0.00
	1/4 D	1421	3	0.21	996	4	0.40	679	4	0.59
	1/2 D	1391	33	2.32	969	31	3.10	658	25	3.66
	3/4 D	1338	86	6.04	922	78	7.80	606	77	11.27
	D	1287	137	9.62	875	125	12.50	559	124	18.16
	D + 0.5 m	1059	365	25.63	653	347	34.70	337	346	50.66
	D + 1.0 m	881	543	38.13	450	550	55.00	135	548	80.23
III	0 D	1824	0	0.00	1290	0	0.00	895	0	0.00
	1/4 D	1821	3	0.16	1287	3	0.23	892	3	0.34
	1/2 D	1798	26	1.43	1274	16	1.24	880	15	1.68
	3/4 D	1760	64	3.51	1232	58	4.50	844	51	5.70
	D	1729	95	5.21	1201	89	6.90	813	82	9.16
	D + 0.5 m	1571	253	13.87	1052	238	18.45	662	233	26.03
	D + 1.0 m	1458	366	20.07	938	352	27.29	549	346	38.66

Table 4. Maximum required usable volumes of storage reservoir *ZR* with collectors of the storm water drainage system constructed with fall $i = 3$ ‰.

		$\beta = 0.60$			$\beta = 0.70$			$\beta = 0.80$		
Catchment	h_k	V [m³]	ΔV [–]	% [m³]	V [m³]	ΔV [–]	% [m³]	V [m³]	ΔV [–]	% [m³]
I	0 D	435	0	0.00	267	0	0.00	133	0	0.00
	1/4 D	432	3	0.69	263	4	1.50	129	4	3.01
	1/2 D	428	7	1.61	260	7	2.62	126	7	5.26
	3/4 D	368	67	15.40	202	65	24.34	83	50	37.59
	D	326	109	25.06	157	110	41.20	39	94	70.68
	D + 0.5 m	114	321	73.79	1	266	99.63	0	133	100.00
	D + 1.0 m	1	434	99.77	0	267	100.00	0	133	100.00
II	0 D	447	0	0.00	265	0	0.00	127	0	0.00
	1/4 D	444	3	0.67	262	3	1.13	124	3	2.36
	1/2 D	432	15	3.36	258	7	2.64	120	7	5.51
	3/4 D	374	73	16.33	198	67	25.28	78	49	38.58
	D	329	118	26.40	153	112	42.26	37	90	70.87
	D + 0.5 m	114	333	74.50	0	265	100.00	0	127	100.00
	D + 1.0 m	2	445	99.55	0	265	100.00	0	127	100.00
III	0 D	595	0	0.00	360	0	0.00	178	0	0.00
	1/4 D	592	3	0.50	357	3	0.83	175	3	1.69
	1/2 D	588	7	1.18	354	6	1.67	173	5	2.81
	3/4 D	550	45	7.56	321	39	10.83	146	32	17.98
	D	520	75	12.61	291	69	19.17	116	62	34.83
	D + 0.5 m	373	222	37.31	154	206	57.22	0	178	100.00
	D + 1.0 m	255	340	57.14	40	320	88.89	0	178	100.00

Table 5. Maximum required usable volumes of storage reservoir *ZR* with collectors of the storm water drainage system constructed with fall $i = 2$ ‰.

		$\beta = 0.30$			$\beta = 0.40$			$\beta = 0.50$		
Catchment	h_k	V [m³]	ΔV [–]	% [m³]	V [m³]	ΔV [–]	% [m³]	V [m³]	ΔV [–]	% [m³]
I	0 D	1457	0	0.00	1019	0	0.00	717	0	0.00
	1/4 D	1453	4	0.27	1016	3	0.29	714	3	0.42
	1/2 D	1415	42	2.88	1002	17	1.67	703	14	1.95
	3/4 D	1366	91	6.25	945	74	7.26	641	76	10.60
	D	1218	239	16.40	812	207	20.31	502	215	29.99
	D + 0.5 m	847	610	41.87	432	587	57.61	140	577	80.47
	D + 1.0 m	439	1018	69.87	72	947	92.93	0	717	100.00
II	0 D	1525	0	0.00	1062	0	0.00	731	0	0.00
	1/4 D	1522	3	0.20	1059	3	0.28	727	4	0.55
	1/2 D	1494	31	2.03	1032	30	2.82	714	17	2.33
	3/4 D	1430	95	6.23	980	82	7.72	662	69	9.44
	D	1284	241	15.80	848	214	20.15	535	196	26.81
	D + 0.5 m	946	579	37.97	500	562	52.92	190	541	74.01
	D + 1.0 m	585	940	61.64	168	894	84.18	0	731	100.00
III	0 D	1936	0	0.00	1367	0	0.00	948	0	0.00
	1/4 D	1933	3	0.15	1364	3	0.22	945	3	0.32
	1/2 D	1909	27	1.39	1348	19	1.39	934	14	1.48
	3/4 D	1871	65	3.36	1312	55	4.02	896	52	5.49
	D	1780	156	8.06	1235	132	9.66	828	120	12.66
	D + 0.5 m	1578	358	18.49	1037	330	24.14	618	330	34.81
	D + 1.0 m	1374	562	29.03	822	545	39.87	397	551	58.12

Table 6. Maximum required usable volumes of storage reservoir *ZR* with collectors of the storm water drainage system constructed with fall $i = 2$ ‰.

		$\beta = 0.60$			$\beta = 0.70$			$\beta = 0.80$		
Catchment	h_k	V [m^3]	ΔV [–]	% [m^3]	V [m^3]	ΔV [–]	% [m^3]	V [m^3]	ΔV [–]	% [m^3]
I	0 D	479	0	0.00	290	0	0.00	146	0	0.00
	1/4 D	475	4	0.84	287	3	1.03	142	4	2.74
	1/2 D	472	7	1.46	283	7	2.41	139	7	4.79
	3/4 D	410	69	14.41	231	59	20.34	93	53	36.30
	D	276	203	42.38	120	170	58.62	17	129	88.36
	D + 0.5 m	6	473	98.75	0	290	100.00	0	146	100.00
	D + 1.0 m	0	479	100.00	0	290	100.00	0	146	100.00
II	0 D	482	0	0.00	289	0	0.00	143	0	0.00
	1/4 D	479	3	0.62	285	4	1.38	140	3	2.10
	1/2 D	475	7	1.45	282	7	2.42	136	7	4.90
	3/4 D	419	63	13.07	236	53	18.34	90	53	37.06
	D	296	186	38.59	125	164	56.75	13	130	90.91
	D + 0.5 m	17	465	96.47	0	289	100.00	0	143	100.00
	D + 1.0 m	0	482	100.00	0	289	100.00	0	143	100.00
III	0 D	628	0	0.00	378	0	0.00	186	0	0.00
	1/4 D	625	3	0.48	375	3	0.79	183	3	1.61
	1/2 D	622	6	0.96	373	5	1.32	180	6	3.23
	3/4 D	583	45	7.17	341	37	9.79	153	33	17.74
	D	520	108	17.20	280	98	25.93	94	92	49.46
	D + 0.5 m	302	326	51.91	68	310	82.01	0	186	100.00
	D + 1.0 m	105	523	83.28	0	378	100.00	0	186	100.00

Table 7. Maximum required usable volumes of storage reservoir *ZR* with collectors of the storm water drainage system constructed with fall $i = 1$ ‰.

		$\beta = 0.30$			$\beta = 0.40$			$\beta = 0.50$		
Catchment	h_k	V [m^3]	ΔV [–]	% [m^3]	V [m^3]	ΔV [–]	% [m^3]	V [m^3]	ΔV [–]	% [m^3]
I	0 D	1513	0	0.00	1041	0	0.00	715	0	0.00
	1/4 D	1509	4	0.26	1038	3	0.29	711	4	0.56
	1/2 D	1454	59	3.90	992	49	4.71	661	54	7.55
	3/4 D	1326	187	12.36	900	141	13.54	571	144	20.14
	D	915	598	39.52	503	538	51.68	216	499	69.79
	D + 0.5 m	134	1379	91.14	0	1041	100.00	0	715	100.00
	D + 1.0 m	0	1513	100.00	0	1041	100.00	0	715	100.00
II	0 D	1677	0	0.00	1181	0	0.00	812	0	0.00
	1/4 D	1674	3	0.18	1177	4	0.34	808	4	0.49
	1/2 D	1642	35	2.09	1141	40	3.39	774	38	4.68
	3/4 D	1480	197	11.75	1017	164	13.89	641	171	21.06
	D	1120	557	33.21	654	527	44.62	319	493	60.71
	D + 0.5 m	532	1145	68,28	76	1105	93.56	0	812	100.00
	D + 1.0 m	213	1464	87,30	0	1181	100.00	0	812	100.00
III	0 D	2094	0	0.00	1476	0	0.00	1017	0	0.00
	1/4 D	2091	3	0.14	1473	3	0.20	1014	3	0.29
	1/2 D	2057	37	1.77	1450	26	1.76	993	24	2.36
	3/4 D	1922	172	8.21	1333	143	9.69	894	123	12.09
	D	1747	347	16.57	1168	308	20.87	734	283	27.83
	D + 0.5 m	1353	741	35.39	762	714	48.37	337	680	66.86
	D + 1.0 m	949	1145	54.68	390	1086	73.58	47	970	95.38

Table 8. Maximum required usable volumes of storage reservoir *ZR* with collectors of the storm water drainage system constructed with fall $i = 1$ ‰.

		$\beta = 0.60$			$\beta = 0.70$			$\beta = 0.80$		
Catchment	h_k	V [m³]	ΔV [–]	% [m³]	V [m³]	ΔV [–]	% [m³]	V [m³]	ΔV [–]	% [m³]
I	0 D	466	0	0.00	265	0	0.00	113	0	0.00
	1/4 D	462	4	0.86	261	4	1.51	110	3	2.65
	1/2 D	436	30	6.44	259	6	2.26	106	7	6.19
	3/4 D	339	127	27.25	149	116	43.77	16	97	85.84
	D	55	411	88.20	1	264	99.62	0	113	100.00
	D + 0.5 m	0	466	100.00	0	265	100.00	0	113	100.00
	D + 1.0 m	0	466	100.00	0	265	100.00	0	113	100.00
II	0 D	523	0	0.00	303	0	0.00	136	0	0.00
	1/4 D	520	3	0.57	299	4	1.32	132	4	2.94
	1/2 D	506	17	3.25	296	7	2.31	129	7	5.15
	3/4 D	376	147	28.11	175	128	42.24	39	97	71.32
	D	96	427	81.64	5	298	98.35	0	136	100.00
	D + 0.5 m	0	523	100.00	0	303	100.00	0	136	100.00
	D + 1.0 m	0	523	100.00	0	303	100.00	0	136	100.00
III	0 D	664	0	0.00	389	0	0.00	182	0	0.00
	1/4 D	661	3	0.45	386	3	0.77	179	3	1.65
	1/2 D	636	28	4.22	365	24	6.17	168	14	7.69
	3/4 D	555	109	16.42	290	99	25.45	96	86	47.25
	D	399	265	39.91	142	247	63.50	0	182	100.00
	D + 0.5 m	57	607	91.42	0	389	100.00	0	182	100.00
	D + 1.0 m	0	664	100.00	0	389	100.00	0	182	100.00

In turn, the percentage difference referring to the usable volume of storage reservoirs was determined using the following equation (3):

$$\% = \frac{V_0 - V_i}{V_0} \cdot 100 \tag{3}$$

It was assumed that the case in which the crown of the overflow edge of the overflow between chambers is located at the height of the bottom of the inlet channel to the tank will be the reference level. In this case, it is not possible to use the storage capacity of the inlet channel when the tank is being filled.

The obtained results of simulation test with the use of hydrodynamic modelling clearly proved that the basic parameter that affects the value of channel holding when dimensioning the required storage volume of tank *ZR* is the fall of the bottom of the inlet channel to the tank. It was also determined that an increase in the required storage volume of tank *ZR* is observed at low levels of the location of the overflow edge and with a decrease in the fall of the bottom of the channels of dimensioned storm water drainage systems. This dependence takes place regardless of the adopted scheme of the drainage system ducts and the value of the sewage flow rate β reduction coefficient. In turn, an opposite tendency was observed with high levels of the location of the overflow edge. The lower the required usable volume of tank *ZR*, the smaller values of the inlet channel's bottom fall.

Another dependence, confirmed by the tests, demonstrates the fact that the decrease in the inlet channel's bottom fall results in an increase in both cubic ΔV and percentage % difference of the required usable volume of storage reservoir *ZR* in all analyzed design versions. It is worth mentioning that this dependence is greater when the sewage network is more compact. The most rational effects were obtained when the channels were laid with a minimum fall $i = 1$ ‰ and their layout was characterized by the most compact shape (catchment area I).

Dependences confirmed by tests result from the fact that when the inlet channel's bottom fall decreases, the backwater resulting from the impoundment of sewage covers an increasing area which makes it possible to rationally use the storage volume of the sewage network channels situated therein. This is particularly visible in compact catchment areas which include a large number of sewage network collectors on a small area. The highest value of the difference in cubic volume of storage reservoir ΔV was achieved each time in catchment area I, with the lowest examined sewage flow rate reduction coefficient $\beta = 0.3$ and the highest level of the location of the edge between chambers ($h = D + 1.0$ m) by, respectively: (1) $\Delta V = 553$ m^3 with channel bottom fall $i = 3.0$ ‰, (2) $\Delta V = 1{,}018$ m^3 with channel bottom fall $i = 2.0$ ‰ and (3) $\Delta V = 1{,}513.0$ m^3 with channel bottom fall $i = 1.0$ ‰.

It may be stated on the basis of results obtained from the model test that especially the adopted level of the location of the edge between chambers has decisive impact on the volume of the impact of channel holding and the observed reduction in usable volume of tank *ZR*. It turns out that the increase in the foundation height of the crown of overflow between chambers results in the fact that the multi-chamber storage reservoir *ZR* operating in a gravitational system requires the reservation of a significantly smaller required usable holding volume.

Based on the results of hydrodynamic simulations presented in Tables 3, 4 and 5, it was noticed that for the first two foundation heights of the overflow edge ($h = 1/4\ D$, $h = 1/2\ D$) a very small reduction in the required storage volume of tank *ZR* is observed as compared to the situation when the retention capacity of the inlet channel are not taken into account when determining the tank's required holding volume. The observed percentage differences in the required usable volume of the tank with these two lowest levels of the location of the overflow edge ranged between 0.14% and 7.69%, when analyzing all design options, and the differences in cubic volume range from 3 m^3 to 59 m^3.

When analyzing the results of simulation tests with the use of hydrodynamic modelling with subsequent levels of the location of the edge between chambers, a very dynamic growth in the reduction of the required usable volume of the multi-chamber tank *ZR* is observed. The possibility of a total abandonment of the need to build a storage reservoir, at least in one design case, is observed with the two highest levels of the location of the overflow edge ($h = D + 0.5$ m, $h = D + 1.0$ m).

Another important element of the tests is the confirmation that the growth in the β coefficient decreases the degree of reduction in the usable cubic volume ΔV of the storage reservoir caused by the use of channel holding of the inlet channel to the tank *ZR*. In turn, the degree of the reduction in usable volume expressed as a percentage value increases along with the growth in the β coefficient value.

The basic task performed each time after sewage storage in the storage reservoir is the removal of contamination deposited on the tank's bottom. The removal is connected with expenses as part of the tank's operation. It is also worth noting that the storage volume of the inlet channel to the storage reservoir is used in the first place during the holding process, which is presented in Figure 2. The excess of sewage flows over the overflow edge to the accumulation chamber of the storage reservoir only after the tank's storage capacity is exhausted. Therefore, the use of the inlet channel's storage capacity makes it possible to reduce the volume of sewage directed to the accumulation chambers of the storage reservoir and during heavy rain it also makes it possible to completely avoid using them. This enables a considerable reduction in costs related to using the tank, due to the reduction in the frequency of works related to cleaning its chambers.

However, it should be emphasized that the effective use of usable channel holding before the supply of storm water to multi-chamber tanks consists only in locating the edge between chambers at a given level. For this reason, the usable volume of used storage reservoirs with their overflow edge situated at a low level as compared to the inlet channel's bottom may be significantly increased in a very simple and cheap manner. In the case of the designed multi-chamber tanks, already at the stage of determining their necessary usable volume, it is possible to consider a reduction in their volume by the amount of a properly determined volume of channel holding which will take place in the inlet collector to the storage reservoir.

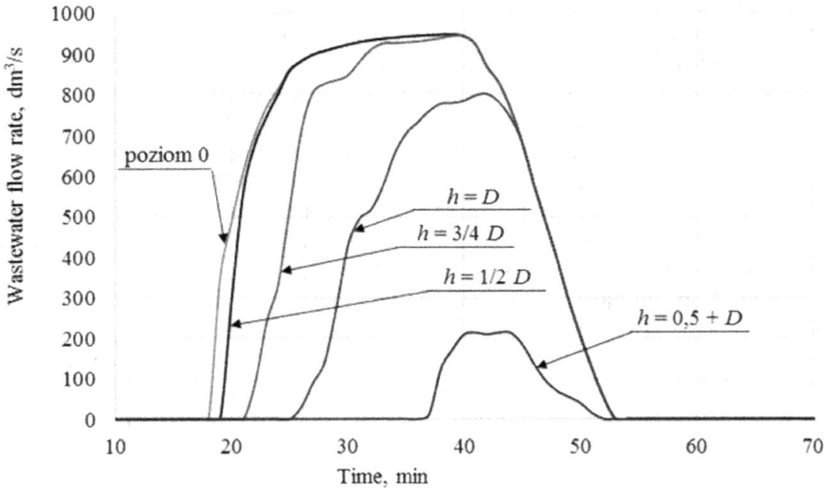

Figure 2. Intensity of sewage supply to the accumulation chamber of a multi-chamber tank *ZR*.

4 SUMMARY

The results of the conducted simulation tests confirm the relevance of the implementation of the scientific task. The basic conclusion may be reduced to the statement that the use of the storage volume of the inlet channel to the multi-chamber tank makes it possible to significantly reduce its required usable volume, and taking this storage capacity into account is reasonable, regardless of the hydrological parameters of the analyzed catchment area and the hydraulic parameters of the designed and/or used storm water drainage system or a combined sewage system.

Another important regularity was confirmed by the tests, and it includes a tendency which indicates that the decrease in the fall of the placement of the channel's bottom of the sewage network leads to the reduction in the required volume of the storage reservoir, taking into account the use of the usable storage volume of the inlet channel. With small falls of the inlet channel's bottom, the sewage backwater reaches further and further into the sewage network which makes it possible to the storage capacity of a longer section of the sewage network.

Based on the conducted tests, it was stated that the obtained degree of reduction in the required storage volume of the tank depends on the following three parameters: (1) the height on which the level of the overflow edge between chambers is located, (2) the adopted sewage flow rate β reduction coefficient and (3) the layout of sewage networks channels.

The most measurable benefits are observed, however, when the layout of the sewage network near the storage reservoir is compact, the overflow edge between chambers is at the maximum permitted height, and the sewage flow rate β reduction coefficient is high. It is also worth noting that the use of the storage capacity of the inlet channel in many analyzed cases made it possible to totally abandon the construction of the storage reservoir.

The use of channel holding in the case of multi-chamber storage reservoir consists only in locating the edge between chambers at a given level. For this reason, the investment costs may be significantly decreased in a very simple and cheap manner. In addition, taking the storage capacity of the inlet channel into account makes it possible to reduce costs related to using the holding facility due to the occasional use of the holding chambers.

Therefore, based on the conducted tests, it may be stated that one of the priorities when designing cubature facilities is taking the practical use of the storage capacity of the inlet channel into account for multi-chamber storage reservoirs.

REFERENCES

Bartoszek, L., Koszelnik, P., Gruca-Rokosz, R. & Kida, M. 2015. Assessment of Agricultural Use of the Bottom Sediments from Eutrophic Rzeszaw Reservoir, Rocznik Ochrona Środowiska 17, 396–409.

Benzerra, A., Cherrared, M., Chocat, B., Cherqui, F. & Zekiouk, T. 2012. Decision support for sustainable urban drainage system management: A case study of Jijel, Algeria. Journal of Environmental Management, vol. 101, 46–53.

Błaszczyk, W., Stamatello, H. & Błaszczyk, P. 1983. Kanalizacja T.1 Sieci i pompownie. Arkady, Warszawa.

Dziopak, J. & Słyś D. 2007. Modelowanie zbiorników klasycznych i grawitacyjno-pompowych w kanalizacji. Oficyna Wydawnicza Politechniki Rzeszowskiej, Rzeszów.

Dziopak J. & Starzec M. 2016. Wpływ układu hydraulicznego zbiornika na wymaganą pojemność użytkową układu zbiorników retencyjnych w kanalizacji. Czasopismo Inżynierii Lądowej, Środowiska i Architektury JCEEA, z. 63 (2/II/16), 105–120.

Gogate, N.G., Kalbar P.P. & Raval P.M. 2017. Assessment of stormwater management options in urban contexts using Multiple Attribute Decision-Making. Journal of Cleaner Production, vol. 142, 2046–2059.

Kaźmierczak B. & Kotowski A. 2012. Weryfikacja przepustowości kanalizacji deszczowej w modelowaniu hydrodynamicznym. Oficyna Wydawnicza Politechniki Wrocławskiej, Wrocław.

Kordana, S., Słyś D. & Dziopak J. 2014. Rationalization of water and energy consumption in shower systems of single-family dwelling houses, Journal of Cleaner Production, vol. 82, 58–69.

Kordana S. 2017. SWOT analysis of wastewater heat recovery systems application, E3S Web of Conferences, vol. 17, 00042.

Mazurkiewicz K., Skotnicki M. & Sowiński M. 2016. Opracowanie hietogramów wzorcowych na potrzeby symulacji odpływu ze zlewni miejskich. Monografie Komitetu Gospodarki Wodnej PAN, z. 39, 33–47.

Pochwat K., Słyś D. & Kordana S. 2017. The temporal variability of a rainfall synthetic hyetograph for the dimensioning of stormwater retention tanks in small urban catchments. Journal of Hydrology, vol. 549, 501–511.

Pochwat K. 2017. Hydraulic analysis of functioning of the drainage channel with increased retention capacity, E3S Web of Conferences, vol. 17, 00075.

Przybyła C., Bykowski J., Mrozik K. & Napierała M. 2011. The Role of Water and Drainage System Infrastructure in the Process of Suburbanization. Rocznik Ochrona Środowiska, vol. 13, 769–786.

Słyś, D. & Dziopak, J. 2011. Development of mathematical model for sewage pumping-station in the modernized combined sewage system for the town of Przemyśl, Polish Journal of Environmental Studies, vol. 20, No. 3, s. 743–753.

Słyś, D. & Kordana, S. 2014. Financial analysis of the implementation of a Drain Water Heat Recovery unit in residential housing, Energy and Buildings, vol. 71, 1–11.

Stec, A. & Słyś, D. 2014. Optimization of the hydraulic system of the storage reservoir hydraulically unloading the sewage network, Ecological Chemistry and Engineering S—Chemia i Inżynieria Ekologiczna, 21, s. 215–228.

Stec, A., Kordana S. & Słyś, D. 2017. Analysing the financial efficiency of use of water and energy saving systems in single-family homes, Journal of Cleaner Production, vol. 151, 193–205.

Suligowski, Z. & Orłowska-Szostak, M. 2013. Retencja w warunkach aglomeracji miejskich—zbiornik rurowy. Instal, nr 10, 72–75.

Szeląg B. & Mrowiec M. 2016. The methods of evaluating storage volume for single-chamber reservoir in urban catchments. Archives of Environmental Protection, vol. 42, 20–25.

Wiśniowska-Kielian B., Niemiec M. & Arasimowicz M. 2013. Przydrożne zbiorniki ścieków opadowych jako element ochrony jakości wód. Inżynieria Ekologiczna, nr 34, 62–75.

Zeleňáková M., Markovič G., Kaposztásová D. & Vranayová Z. 2014. Rainwater management in compliance with sustainable design of buildings. Procedia Engineering, vol. 89, 1515–1521.

Underground Infrastructure of Urban Areas 4 – Madryas et al. (Eds)
 ISBN 978-1-138-55953-0

Designing a retention sewage canal with consideration of the dynamic movement of precipitation over the selected urban catchment

M. Starzec, J. Dziopak & D. Słyś
Department of Infrastructure and Water Management, Rzeszow University of Technology, Poland

ABSTRACT: This paper discusses the influence of the direction and velocity of precipitation wave on the required retention capacity of the innovative *Retention Sewage Canal* solution. Simulations were performed in a real municipal catchment equipped with a storm water drainage network, with the use of SWMM 5.1 hydrodynamic software. The formulated research task was realized after determining boundary conditions at efflux of storm water effluents from the drained catchment, assuming that the maximum intensity of effluent efflux from the *Retention Sewage Canal* is 600 dm^3/s. Results of simulation tests allow to determine that both the direction, as well as velocity of movement of precipitation wave have a small impact on the determined, required retention capacity of the *Retention Sewage Canal* in the analyzed design variant. It was also determined that the highest required retention capacity of the object is always caused by precipitation moving from the west to the east at 3,0 m/s velocity. However, from the performed tests it may be concluded that the taking of precipitation movement over the assumed municipal catchment into account does not cause any changes in the value of calculational precipitation duration for dimensioning the *Retention Sewage Canal*, which in the created design concept assumes a value of 100 minutes. Comparing the results of model tests, determined for stationary precipitation and precipitation moving over the catchment, it was demonstrated that the differences in the required capacity of *Retention Sewage Canal* are inversely proportional to the duration of precipitation. For this reason, when adopting higher values of sewage flow reduction coefficient β in the *Retention Sewage Canal*, the determined differences in its required retention capacity would be higher than if dynamic rainfall for adopted for analysis.

Keywords: sewerage systems; analytical flow; precipitation wave; storage reservoirs

1 INTRODUCTION

The increased concentration of building objects and the surface of drained catchments (Kordana 2017) through their sealing continues to have adverse effect on the undesirable change in the sewage balance, which results in increased storm water volumes flowing in to existing sewerage systems (Przybyła et al. 2011). The consequence of this phenomenon are more frequently occurring hydraulic overloads of sewerage systems, which effect in the occurrence of local flooding (Zeleňáková et al. 2014). This, in turn, necessitates increasing outlays on removing their effects, and paying claimed damages (Stec & Słyś 2014).

The lack of open space and high infrastructural congestion in urban agglomerations limit the availability of free spaces for constructing cubic sewage infrastructure (Suligowski & Orłowska-Szostak 2013). This causes the necessity of searching for advanced techniques of controlling the flow of effluents, at minimum utilization of catchment space and modification of the existing sewerage system (Dziopak & Słyś 2007). Serious challenges have been posed in the last decades on the designers of drainage systems, which necessitate system approach and taking the local conditions into account. Therefore, it is deemed justified to consider a number of design variants, in which the technological progress in terms of innovative solutions and methodologies for dimensioning drainage networks and objects must be taken into account (Słyś & Dziopak 2011).

The choice of the right design variant should be preceded by analyzing a number of factors, including technical, financial (Kordana et al. 2014, Słyś & Kordana 2014) and environmental analysis (Bartoszek et al. 2015), taking into account the technical life of the sewerage system.

In order to fulfill all the above mentioned requirements, it is required to use innovative, and at the same time efficient technical solutions (Pochwat 2017, Pochwat et al. 2017). The technical object that ensures both effective storage, as well as controlling the volume of storm effluent stream is the unique, patented *Retention Sewage Canal* solution. This solution is a highly competitive alternative to the presently used cubic objects, aiming at relieving hydraulic ducting systems.

The *Retention Sewage Canal* is an integral part of sewerage network, which features specific length and adequately designed cross section, and is located at the sewerage system route, or operating in by-pass. Its geometry is defined relative to the cross section of the designed sewers, and is connected with the ground surface through (inspection) chambers. In practice, the diameters of *Retention Sewage Canal* may be larger, equal, or sometimes even smaller than the sewers dimensioned for calculational rain, depending on the local conditions and design principles. The characteristic feature of the innovative solution is dividing the space inside the sewer into a specific number of storage chambers through the use of damming baffles, which are most preferably positioned in water traps or chambers, perpendicular to the sewage flow direction. A single damming baffle comprises an outlet hole of specific geometry, located at the sewer bottom, while an auxiliary overflow hole is located in its upper part (Słyś & Dziopak 2010).

Determining the required retention capacity of the *Retention Sewage Canal* requires determining the parameters of critical rain for its dimensioning. In most cases, the calculational methodology assumes that the rain has static nature, without any speed and direction of dislocation. This assumption significantly simplifies hydrodynamic calculations, since it allows to assign only a single precipitation to all the analyzed catchments. This simplification is too extensive, however, since it fails to reflect the actual conditions; as is commonly known, each precipitation, being a random event, is characterized by direction and velocity of movement. In practice, this means that a drained catchment is subjected to precipitation load in a number of stages. In the period shortly before the beginning of rainfall, only small part of the drained catchment is covered by its reach. As time from the beginning of rainfall moves on, its reach is smaller and smaller, covering a larger a larger area.

As it has been demonstrated in papers (Berne et al. 2004, Dziopak & Starzec 2014, Zawilski & Brzezińska 2014), the adoption of dynamic rainfall leads to determining reliable hydrograms of effluent inflow into storage objects, the trends of which are more critical, and the designed networks and sewerage objects require larger geometries. The magnitude of these differences, which the differences result from adopting precipitation as static or dynamic, is dependent on a number of factors, the main of which are: directions of effluent flow in the analyzed catchment, its shape, storm water velocity and direction. In catchments that feature elongated shape, and especially that feature a single, principal rain effluent flow direction, largest differences in the determined flows are observed when the direction and velocity of rainfall movement is considered.

2 METHODOLOGY AND ADOPTED CALCULATIONAL VARIANTS

A scientific task was formulated aiming at determining the level of influence of precipitation wave direction and velocity on the required retention capacity of the innovative *Retention Sewage Canal* solution. Simulations were performed with the use of SWMM 5.1 hydrodynamic software, based on a created hydrodynamic model of a real municipal catchment, located in central Poland, the diagram of which is presented in Figure 1. The drained catchment adopted for analysis comprises a storm water drainage network that includes 120 sections of length from 13.35 to 130 meters. The sewerage system interceptor slopes range from 1,0 to 65,8 ‰. Total area of the drained catchment is 53,91 ha. Runoff coefficient ranges from 0,24 to 0,80. Total tight surface of reduced catchment, to which the storm water efflux is directed is 27.21 ha.

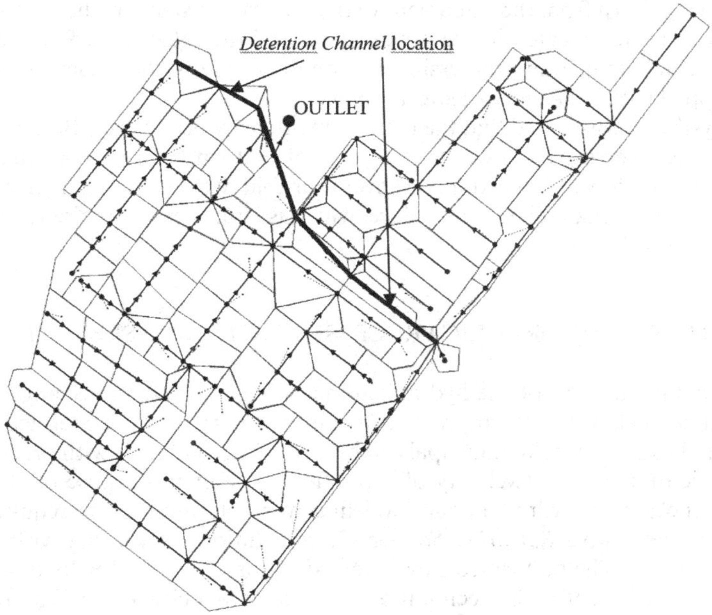

Figure 1. Diagram of arrangement of real sewerage network in the analyzed drained catchment.

The simulations were performed with the use of synthetic fall. The Błaszczyk formula was used for determining unit precipitation intensity (Błaszczyk et al. 1983), which describes the dependence between precipitation intensity and duration; and so:

$$q = \frac{6{,}631\sqrt[3]{H^2 \cdot c}}{t_d^{2/3}} \tag{1}$$

where q – unit rain intensity; H – amount of mean annual precipitation; c – precipitation frequency; and t_d – duration of precipitation.

The analysis of hydraulic functioning of the *Retention Sewage Canal* assumed average annual precipitation level at H = 600 mm/year, and precipitation frequency at $c = 2$ years. The tests took into account precipitation that featured fixed intensity throughout its entire duration.

The effect of precipitation movement was achieved by assigning each catchment an individual rain file in SWMM 5.1 program. Adoption of variable precipitation starting times at the selected sub-catchment allowed to simulate its movement over the catchment area. Value of the precipitation wave velocity was related to the difference in precipitation start between test points. Simulations were performed at precipitation wave velocity of fixed value throughout its entire duration. Three precipitation movement velocities were adopted for the tests, i.e. 1,5; 3 and 6 m/s. The rainfall that starts at the entire catchment simultaneously was attributed infinitely large velocity.

In the analysis it was also assumed that the direction of rainfall movement is fixed in the entire cycle of the simulation. It was also assumed that the rainfall face may be represented as a straight line that is perpendicular to the direction of its movement. In practice, this means that at points lying on lines perpendicular to the assumed rainfall movement direction, rain starts and ends at the same time. Four main directions of the precipitation wave dislocation were assumed in the simulations, i.e. E-W, W-E, N-S and S-N.

Geometry of individual chamber of the *Retention Sewage Canal* was reflected in SWMM software with the use of *Couduit* function. Drainage hole was described with the function *Orifice link*, while the emergency outflow was described as function *Outlet*.

Sewage flowing down from the catchment to the sewerage system are directed to the receiving body that is a ditch located in the northern part of the catchment. For the purpose of protecting the catchment from progressing erosion of its banks, the presented sewerage system was equipped with a *Retention Sewage Canal.*

In the analysis it was assumed that the maximum intensity of sewage efflux from the drainage system to the receiving body will have a value of 600 dm^3/s. In practice this means that from the calculation flow of 2.609 dm^3/s an over four-fold decrease in the intensity of rainfall waste at the retention channel will take place, which is equivalent to the sewage flow reduction coefficient β of 0,23.

3 SIMULATION OF HYDRAULIC PROCESSES AND ANALYSIS OF TEST RESULTS

The article includes analysis of the hydraulic functioning of *Retention Sewage Canal* based on the determined balance of storm water effluents transported by the operated storm water network located in a real-world municipal catchment, when loading it with precipitation that features specific direction and velocity of movement. The main purpose of the performed analysis with the use of hydrodynamic modeling was determining the required retention capacity of the innovative *Retention Sewage Canal* solution co-working with a real-world municipal catchment. The performed simulations demonstrated that the loading of the analyzed municipal catchment with precipitation of variable velocity and different wave movement direction causes marginal differences in the volume of the determined useful capacity of *Retention Sewage Canal.* The determined values of the required retention capacity of the designed *Retention Sewage Canal* are presented in Table 1.

The peak value of 2.729 m^3 of the required useful retention capacity of the *Retention Sewage Canal* was achieved at precipitation moving from the east to the west (E-W) at a speed of 3,0 m/s, at a duration of t_d = 100 minutes. It turns out that the peak value of the required

Table 1. Required useful retention capacity of *Retention Sewage Canal* co-working with analyzed storm water drainage network at the tested municipal catchment.

		Direction of movement of rainfall wave over the catchment											
		S-N			E-W			W-E			N-S		
		Velocity of the movement of rainfall wave over the catchment v, m/s											
Duration of rainfall, t_d	Static rainfall	**1.5**	**3**	**6**	**1.5**	**3**	**6**	**1.5**	**3**	**6**	**1.5**	**3**	**6**
10	**991**	1003	1011	998	1057	**1089**	1055	**894**	936	978	937	966	982
15	1301	1350	1335	1315	1340	**1352**	1338	**1267**	1281	1292	1273	1285	1293
20	1545	1561	1551	1547	1564	**1566**	1562	**1520**	1529	1539	1522	1527	1542
25	1750	1761	1761	1756	1763	**1765**	1760	**1710**	1724	1729	1719	1726	1732
30	1923	1933	1932	1929	1937	**1939**	1931	**1878**	1896	1911	1891	1904	1914
40	2170	2175	2174	2172	2179	**2181**	2178	**2119**	2128	2143	2122	2133	2147
50	2354	2360	2360	2358	2368	**2373**	2362	**2309**	2327	2341	2328	2339	2348
60	2485	2488	2489	2487	2496	**2501**	2492	**2456**	2463	2469	2464	2466	2473
70	2595	2594	2599	2598	2607	**2610**	2603	**2556**	2571	2584	2577	2582	2589
80	2660	2662	2664	2662	2671	**2673**	2668	**2613**	2629	2642	2638	2641	2646
90	2698	2699	2701	2700	2711	**2713**	2708	**2651**	2676	2688	2682	2686	2691
100	2716	2717	2719	2720	2728	**2729**	2723	**2684**	2697	2708	2698	2707	2713
110	2714	2715	2717	2718	2724	**2726**	2721	**2683**	2696	2706	2697	2704	2711
120	2713	2712	2714	2715	2722	**2724**	2719	**2682**	2695	2706	2695	2703	2710
130	2690	2691	2693	2696	2701	**2703**	2699	**2659**	2671	2683	2669	2681	2686
140	2651	2650	2652	2653	2660	**2661**	2657	**2622**	2634	2644	2635	2642	2648
150	2600	2601	2601	2602	2610	**2611**	2609	**2576**	2591	2595	2589	2594	2598
160	2543	2544	2545	2546	2552	**2553**	2549	**2518**	2528	2537	2529	2536	2541

capacity of the retention object determined by the performed hydrodynamic simulations is higher only by 13 m^3 compared to the peak retention capacity defined for static rainfall.

When analyzing the results of the performed tests presented in Table 1 it was noted that at two directions of precipitation movement, i.e. E-W and S-N, higher required retention capacities of the *Retention Sewage Canal* were observed compared to static conditions.

With the two other directions, i.e. W-E and N-S, conversely, smaller capacities of the retention object were required. It was determined that the highest required retention capacities of the *Retention Sewage Canal* are necessitated by precipitation wave approaching from E-W movement, at 3.0 m/s velocity. What is particularly worth mentioning, this tendency occurs irrespective of the analyzed precipitation duration value. On the other hand, in the opposite direction, i.e. E-W, and at precipitation wave velocity of 1.5 m/s, the lowest required retention object capacities were observed, irrespective of the analyzed precipitation duration.

Analyzing the adopted design variants, it is evident that the value of the obtained differences in useful cubic capacity of the Retention Sewage is also dependent on the assumed precipitation wave dislocation velocity. At E-W precipitation movement, a tendency was noted of the initial increase in the required retention capacity of the Retention Sewage, with a decrease in velocity of the precipitation wave movement. This tendency is not constant, however, and at some point a reverse situation is observed. The peak values of the required retention capacity of *Retention Sewage Canal* at this precipitation wave direction were observed whenever the precipitation movement velocity was equal 3.0 m/s. This is an important observation, since in the designing process the critical precipitation movement velocity at which the required retention object capacity is the highest may be determined. On the other hand, with S-N direction, the precipitation wave movement velocity effecting in the highest required *Retention Sewage Canal* capacity assumes different values with changing precipitation duration.

For N-S and W-E direction, the change in the precipitation movement direction has impact on reducing the required *Retention Sewage Canal* capacity—regardless of the precipitation duration assumed for simulation.

In the process of dimensioning retention objects, the parameter that determines their required capacity is the analytical rainfall duration time. When analyzing the data presented in Table 1 it was observed that the time at which the highest required volume of the *Retention Sewage Canal* is observed is $t_d = 100$ minutes, regardless of the precipitation wave velocity and direction. From the detailed simulation tests performed it was concluded that in the analyzed drained catchment the speed and direction of precipitation movement does not impact the determining of duration of calculational rain for dimensioning this retention object. Thus, the performed tests using hydrodynamic modeling finally confirmed an important conclusion of cognitive and applicational significance.

Continuing the analysis of the test results, minor differences in the required useful capacities of *Retention Sewage Canal* were obtained. However, at all times, the highest value of the required retention capacity of that object was determined in conditions that took into account the precipitation wave migration over the assumed urbanized catchment. It turns out that the minor differences result from assuming a significantly higher precipitation duration that is determined methodically to determine the critical useful cubic capacity of the retention object compared to the duration of calculational rain calculated for dimensioning the analyzed sewerage system. The trends of sewage inflow hydrogram to the *Retention Sewage Canal* when loading the catchment with calculational rail for its dimensioning of duration $t_d = 100$ min are presented on Figure 2.

When analyzing the curves from Figure 2 it can be demonstrated that taking into account the precipitation wave movement over the analyzed catchment at rainfall duration $t_d = 100$ minutes does not cause an increase in the peak value of effluent efflux intensity in the sewerage system. It turns out that the precipitation migration over the catchment effects in only marginal changes to the hydrogram curve reflecting the inflow of storm water effluent from the catchment to the *Retention Sewage Canal* from the analyzed storm drainage system.

The results of simulation tests from Table 1 confirm the tendency that the largest differences in the required *Retention Sewage Canal* capacity are observed at the occurrence of precipitation of the minimum assumed duration of $t_d = 10$ minutes. In this case, the highest achievable retention capacity of the *Retention Sewage Canal* is 1.089 m^3, and is 98 m^3 higher

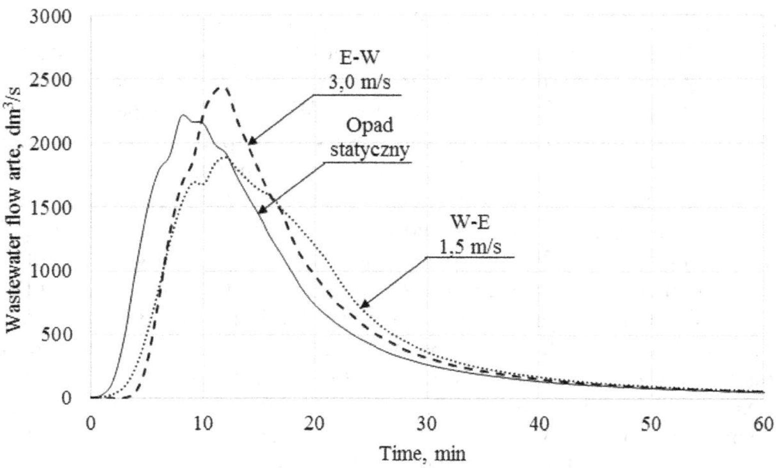

Figure 2. Hydrograms of intensity of effluent inflow to *Retention Sewage Canal* at precipitation duration $t_d = 100$ minutes and variable precipitation wave movement conditions.

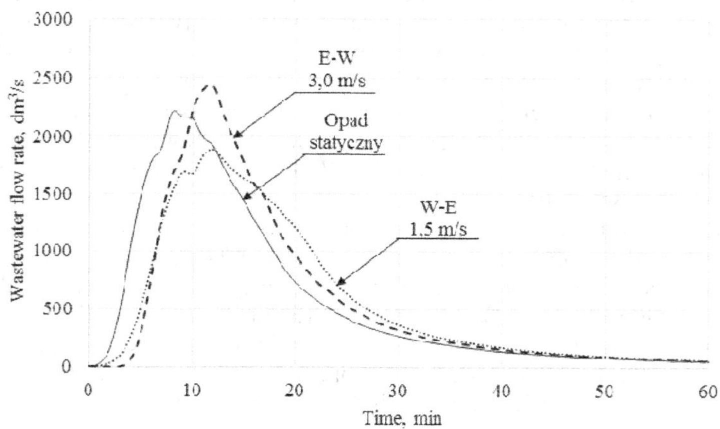

Figure 3. Hydrograms of intensity of effluent inflow to *Retention Sewage Canal* at precipitation duration $t_d = 10$ minutes and variable precipitation wave movement conditions.

compared to the required capacity of this object, determined as a result of loading the catchment with static rainfall.

Analyzing the trend of the curves in Figure 2 it is evident that the largest differences in the shape of the sewage inflow to the *Retention Sewage Canal* were achieved at precipitation duration of $t_d = 10$ minutes And so, it can be assumed that in case when the calculational time for the dimensioning of *Retention Sewage Canal* would assume a value nearing 10 minutes, the achieved differences in its required retention capacity would be much higher.

The trends of effluent outflow hydrograms from the operated sewerage system at precipitation $t_d = 10$ minutes are presented in Figure 3, in consideration of different directions and velocity of precipitation over the drained catchment.

It is worth mentioning that the analyzed catchment features condense arrangement of drains, and the outflow of effluents into the receiving body takes place virtually from all directions, while the value of β effluent flow reduction coefficient adopts a low value. As has been demonstrated in papers (Dziopak & Starzec 2014, Zawilski & Brzezińska 2014), at such hydrological parameters of the catchment and the hydraulic parameters of the sewerage system, small differences in the required capacity of retention objects are observed between

Table 2. Maximum (calculational) effluent outflow intensity from the sewerage system before equipping it with the *Retention Sewage Canal.*

		Direction of movement of rainfall wave over the catchment											
		S-N			**E-W**			**W-E**			**N-S**		
		Velocity of the movement of rainfall wave over the catchment v, m/s											
Duration of rainfall, t_d	Static rainfall	**1.5**	**3**	**6**	**1.5**	**3**	**6**	**1.5**	**3**	**6**	**1.5**	**3**	**6**
10	2207	2398	2412	2411	2389	**2421**	2325	**1890**	1934	2054	1912	1962	2067
15	2520	2511	2541	**2586**	2496	2540	2544	**2105**	2214	2384	2166	2234	2402
20	2582	2542	2598	**2630**	2500	2564	2603	**2317**	2416	2506	2435	2522	2538
25	2417	2449	**2477**	2457	2384	2408	2414	**2288**	2342	2378	2323	2377	2389
30	2329	2311	**2338**	2324	2320	2334	2317	**2186**	2242	2291	2242	2263	2301
35	2250	2231	2242	2248	2244	**2255**	2233	**2116**	2161	2203	2167	2193	2212
40	2169	2150	2160	2162	2171	**2178**	2152	**2059**	2097	2127	2090	2114	2139

static and dynamic precipitation, which allows to simplify the calculational procedure when dimensioning the *Retention Sewage Canal* being the subject of the innovative solution.

When analyzing the results of the performed hydrodynamic simulations it was noted that the *Retention Sewage Canal* not always required reserving its highest useful retention capacity in a situation when, at the same time, the highest value of effluent outflow intensity from the operated sewerage system was observed. Such cases were observed at precipitation of duration $t_d = 15 \div 30$ minutes. When dimensioning retention objects, the factor determining the value of the required retention capacity is the trend of variability of effluent inflow intensity versus time. And so, a hydrogram characterized by highest momentary flow intensity is not always indicative when dimensioning retention objects. The values of maximum effluent outflow intensity from the sewerage system before equipping it with the *Retention Sewage Canal* are presented in Table 2.

When analyzing the results of the performed simulations it was noted that the highest effluent outflow intensity from the sewerage system of 2.630 dm^3/s was observed at precipitation duration of $t_d = 20$ minutes, when it was moving at 6.0 m/s southwards from the north. At the same time, regardless of the precipitation wave movement direction and velocity, the highest effluent outflow intensity from the analyzed sewerage system was observed at precipitation duration of $t_d = 20$ minutes.

4 SUMMARY

The performed simulation tests employing hydrodynamic modeling in consideration of boundary conditions and assumed design parameters allowed to determine the required useful capacity of the *Retention Sewage Canal* of volume 2.729 m^3. The critical retention capacity was determined at loading the assumed real-world catchment with rain fall of duration $t_d = 100$ minutes, which was moving westwards from the east at 3.0 m/s velocity. At the same time, it was noted that when loading the catchment with static rainfall, the required useful retention capacity of the object is only marginally smaller than the volume determined during the occurrence of precipitation in which the direction and velocity of precipitation wave were taken into account.

Hence, from the performed tests it may be concluded that, in the analyzed design variant, assuming only static precipitation for dimensioning the required retention capacity of *Retention Sewage Canal* would cause no material under-dimensioning of the object. It must be noted, however, that in the analyzed case the sewage flow reduction coefficient β assumed small value, and sewage outflow from the catchment took place from all directions. Normally in such conditions, only minor impact of the precipitation wave on the value of peak intensity of sewage flow and the volume of retention objects was observed.

The tests also demonstrated another pattern that shows that even at favourable design parameters, major differences in the trend of effluent inflow hydrograms to the *Retention*

Sewage Canal are observed even when moving, at short fall durations, over the drained precipitation catchment. And so, at high values of sewage flow reduction coefficient β, the differences in the required retention capacity of the object would be much higher.

On the other hand, taking the precipitation wave movement into account in the analyzed design variant did not have any impact on the change in the calculational time determined for dimensioning the *Retention Sewage Canal*, as well as calculational time for dimensioning the sewerage system sewers.

The performed tests confirm important tendencies and dependencies, being among others that the established differences in determining the required retention capacity of objects, resulting from taking the direction and velocity of precipitation wave movement into account, are largely dependent on the catchment shape, sewage outflow direction in the sewerage system, and adopted value of sewer flow reduction coefficient β.

The tests confirmed an important conclusion of practical significance, namely that the directions of movement of precipitation wave must always be taken into account, since neglecting this phenomenon often leads to designing insufficient retention object geometries. Designing retention objects with the use of dynamic precipitation is especially recommended when they co-operate with sewerage networks from which the efflux of effluents takes place from a single, principal direction, and the effluent flow reduction coefficient β has high values.

REFERENCES

Bartoszek, L., Koszelnik, P., Gruca-Rokosz, R. & Kida, M. 2015. Assessment of Agricultural Use of the Bottom Sediments from Eutrophic Rzeszaw Reservoir, Rocznik Ochrona Środowiska 17, 396–409.

Berne, A., Delrieu, G., Creutin, JD. & Obled, C. 2004. Temporal and spatial resolution of rainfall measurements required for urban hydrology, Journal of Hydrology, vol. 299, 166–179.

Błaszczyk, W., Stamatello, H. & Błaszczyk, P. 1983. Kanalizacja T.1 Sieci i pompownie. Arkady, Warszawa.

Dziopak, J. & Słyś D. 2007. Modelowanie zbiorników klasycznych i grawitacyjno-pompowych w kanalizacji. Oficyna Wydawnicza Politechniki Rzeszowskiej, Rzeszów.

Dziopak J. & Starzec M. 2014. Wpływ kierunki i prędkości fali deszczu na kubaturę użytkową wielokomorowych zbiorników retencyjnych, JCEEA Czasopismo Inżynierii Lądowej, Środowiska i Architektury, Oficyna Wydawnicza Politechniki Rzeszowskiej, nr 61, s. 83–94.

Kordana, S., Słyś D. & Dziopak J. 2014. Rationalization of water and energy consumption in shower systems of single-family dwelling houses, Journal of Cleaner Production, vol. 82, 58–69.

Kordana S. 2017. SWOT analysis of wastewater heat recovery systems application, E3S Web of Conferences, vol. 17, 00042.

Pochwat K., Słyś D. & Kordana S. 2017. The temporal variability of a rainfall synthetic hyetograph for the dimensioning of stormwater retention tanks in small urban catchments. Journal of Hydrology, vol. 549, 501–511.

Pochwat K. 2017. Hydraulic analysis of functioning of the drainage channel with increased retention capacity, E3S Web of Conferences, vol. 17, 00075.

Przybyła C., Bykowski J., Mrozik K. & Napierała M. 2011. The Role of Water and Drainage System Infrastructure in the Process of Suburbanization. Rocznik Ochrona Środowiska, vol. 13, 769–786.

Słyś, D. & Dziopak, J. 2011. Development of mathematical model for sewage pumping-station in the modernized combined sewage system for the town of Przemyśl, Polish Journal of Environmental Studies, vol. 20, No. 3, s. 743–753.

Słyś, D. & Dziopak, J. 2010. Retencyjny kanał ściekowy. Patent Application P–391198. The Patent Office of The Republic of Poland. Warsaw. Poland.

Słyś, D. & Kordana, S. 2014. Financial analysis of the implementation of a Drain Water Heat Recovery unit in residential housing, Energy and Buildings, vol. 71, 1–11.

Stec, A. & Słyś, D. 2014. Optimization of the hydraulic system of the storage reservoir hydraulically unloading the sewage network, Ecological Chemistry and Engineering S – Chemia i Inżynieria Ekologiczna, 21, s. 215–228.

Suligowski, Z. & Orłowska-Szostak, M. 2013. Retencja w warunkach aglomeracji miejskich—zbiornik rurowy. Instal, nr 10, 72–75.

Zawilski, M. & Brzezińska, A. 2014. Areal rainfall intensity distribution over an urban area and its effect on a combined sewerage system, Urban Water Journal, vol. 11, 532–542.

Zeleňáková M., Markovič G., Kaposztásová D. & Vranayová Z. 2014. Rainwater management in compliance with sustainable design of buildings. Procedia Engineering, vol. 89, 1515–1521.

Underground Infrastructure of Urban Areas 4 – Madryas et al. (Eds)
 ISBN 978-1-138-55953-0

The impact of land use and urbanization on drainage system

A. Stec & D. Słyś
Department of Infrastructure and Water Management, Rzeszów University of Technology, Poland

ABSTRACT: A growing number of people around the world and their migration to cities is changing the zoning of the areas and enforces the need for their expansion. Intensive development of urban areas is a challenge for the designed, in particular for the existing rainwater drainage and the combined sewerage systems. Urbanization, which causes changes in the way of the catchment usage, affecting the growth of the sealing of the area surface, also contributes to intensify the outflow of rainwater. The result of this, commonly found in urban areas phenomenon, is the dissipation of the cubic reserves of the existing channels, flooding the sewage facilities, the occurrence of so-called "urban flooding" and flooding events in urban streams. The article presents the results of hydrodynamic simulations performed for the actual urban catchment. In the research, conducted on the SWMM software, the impact of expansion of the city and the way of the rainwaters usage on its area on the existing drainage system was analysed.

1 INTRODUCTION

Currently, almost 54% of the world's population lives in urbanized areas, and this number is going to increase up to 66% by 2050 (UN, 2014). The United States of America (82%), South America (80%) and Europe (73%) are the most urbanized regions in the world. Progressive urbanization brings new challenges in ensuring broadly understood residents' safety and an appropriate standard of living. Increasing migration of people to cities forces them to expand by increasing their territorial coverage or intensifying existing ones. This results in an increase in the degree of sealing the area, which was so poorly developed and covered with vegetation. The change in the way of catchment management causes some disturbances in the natural hydrological cycle where there is a quantitative balance between precipitation and the processes of runoff, sinking and evaporation of precipitation. The increase in surface sealing due to urbanization contributes to the reduction of the roughness coefficient of the substrate and the shortening of the runoff concentration to the change in precipitation conditions from the catchments. Human ingenuity in the natural water balance results in many adverse effects, both for the natural environment and the functioning of the city technical infrastructure. Among them the most important are: excessive drying of the ground, lowering of groundwater level, flooding of rainwater sewers, occurrence of so-called urban floods and hydraulic overload of sewage networks and sewage treatment plants (Pochwat et al., 2017, Kaźmierczak and Kotowski, 2014; Fletcher et al., 2013; Todeschini, 2016; Lu et al., 2014; Du et al., 2012). Cities are also the major source of surface water and groundwater. The solution to these problems will not be possible without changing the paradigm of water management that underpins the functioning and social and economic development of urban areas. This was reflected, inter alia, in the provisions of the UNESCO International Hydrological Program 2014–2021, "Water and human settlements of the future" refers directly to cities (UNESCO, 2012). Rainwater in urbanized areas should not be seen solely as a source of potential hazards, but especially as a valuable substitute for water and an attractive element of the landscape. A contemporary approach to urban planning and management is based on the idea of sustainable development, whose the main goal is to maintain proper relations between economic development, quality of life and environmental care. Therefore, water management in line with the concept of sustainable development thanks to, inter alia, limiting the emergence of flood hazards and improving the quality and quantity of available surface and groundwater resources will bring tangible benefits to urban residents.

A full use of water ecosystem services enables synergies to be created in the multifunctional management of urban space (Januchta-Szostak, 2012).

In Poland, the most commonly used solution for the management of precipitation water is its collection and discharge of rainwater or combined sewage system to the receiver. However, such a model of rainwater management is not compatible with the standards of modern and environmentally friendly water and sewage management. According to the guidelines of the Framework Directive of the European Union, rainwater should be managed in accordance with the concept of sustainable development (Directive, 2000). The main purpose of the action in this regard is the retention and management of as much rainwater as possible in areas where precipitation has occurred and where there is a limitation of surface runoff into rivers and lakes. There are applied various devices for groundwater retention and infiltration facilities that are part of Sustainable Urban Drainage Systems (SUDS) (Elliott and Trowsdale, 2007), which can be the systems of distributed rainwater management or collective systems. The distributed systems are usually the individual devices used on individual investor plots. These include the following (Adamiak-Burszta and Stec, 2017; Zeleňáková et al., 2014; Suligowski, 2008; Słyś, 2013; Słyś et al., 2012): unsealed and perforated surfaces, drainage wells and drainage basins, rigolas and drainage drains, drainage boxes and chambers, rain gardens, green roofs and dams and installations for the commercial use of precipitation waters.

Collective systems, on the other hand, are solutions that are located on sewer systems or are run by rainwater management. They cover larger areas such as residential estates. In this case, most often (Starzec et al., 2015; Stec and Słyś, 2014; Dziopak and Słyś, 2007): retention tanks, retention and storage tanks and absorbent tanks and basins are applied.

2 RESEARCH PROBLEM

Taking into account the above the research has been conducted whose aim was to determine the impact of urban drainage development and an increase and sealing of its surface on the functioning of the existing sewage system. The research was done on a hydrodynamic model built using the Storm Water Management Model (SWMM). This model includes the left bank of the city of Przemyśl located in south-eastern Poland on the San River. It is an area of the Slaughter District with an area of 632 hectares and a population of about 35,000 (Figure 1).

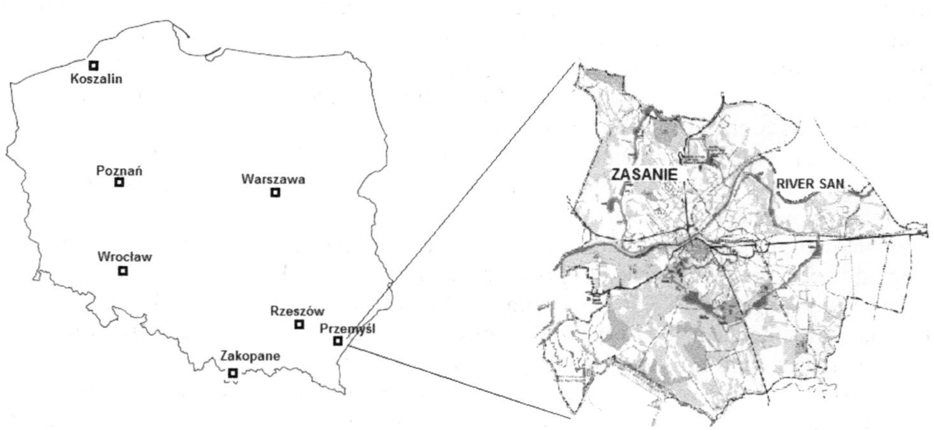

Figure 1. Location of a case study city in Poland.

In the analyzed area the sewage and drainage sewage is discharged to the gravitational sewage system. The sewage is discharged through the main sewers located along the San River and it is transported to a pumping station from which they are pumped to the other side of the river, where it is connected to the urban sewage treatment plant. The wastewater is discharged to the San River. The city of Przemyśl is characterized by very diverse topographical conditions which have a decisive influence on the functioning of the city sewage system. In addition, due to an expansion of the city and an increase in the number of impervious areas, there has been recently an intense increase in the amount of rainwater discharged into the sewerage system, which adversely affects the hydraulic conditions in the main sewers, especially during heavy rainfall.

The hydrodynamic model of the Zasanie district developed in the SWMM was used to carry out simulations that determined the hydraulic parameters of the sewerage network in the current state and variants that extended the area by adding new basins. The hydrodynamic model of the sewage network includes 81 main sections with a total length of 177,15 km. These are circular and oval concrete sewers. Due to the age and condition of the conductors, the Manning's roughness coefficient was used in the range from 0,013 to 0,018. Figure 2 shows the system of the analyzed catchment with the existing sewage system and the new 5 catchments: S-I, S-II, S-III, S-IV and S-V. Each new catchment has an area of 10 hectares. At present, these areas are covered with vegetation and unpaved surfaces, from which the rainwater is absorbed by a large amount of land, and only a small part of it forms surface runoff. In the case of the extension of the city borders over these areas, it was assumed that they would be developed in a similar way to the neighboring existing basins. Therefore, it has been assumed that new housing estates will be built there, along with streets, car parks, pavements and bike paths and large commercial buildings. The development of the Zasanie district is in line with the current Study of Conditions and Directions of Spatial Development of Przemyśl.

The study looked at different ways in which the new drainage areas were integrated, which affected the degree of surface sealing of each of them. For existing basins it is in the range

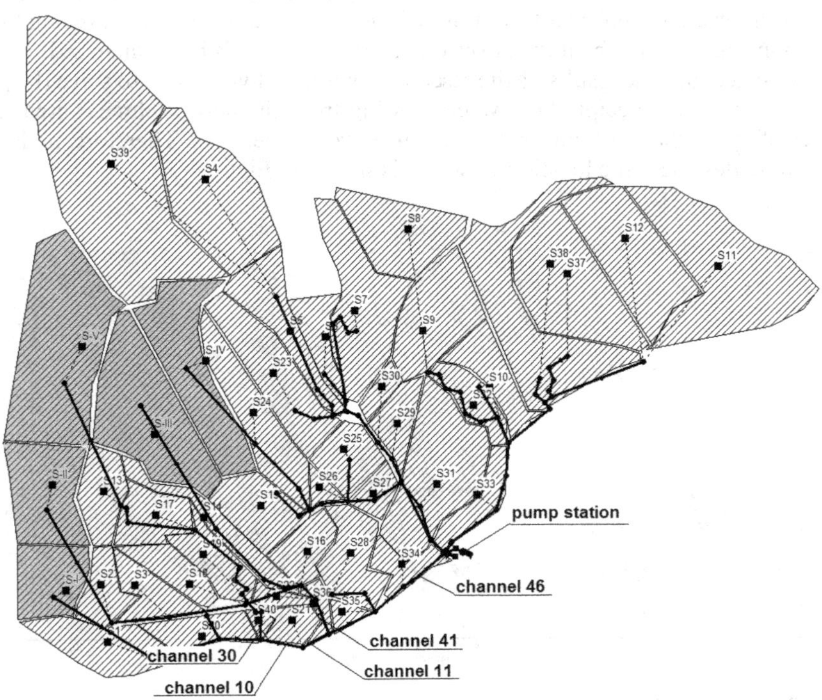

Figure 2. Hydrodynamic model of the Zasanie district.

of about 30% to 60%. Consequently, the calculated runoff coefficients for the new catchment bases were 0,3, 0,4, 0,5 and 0,6. The value of these coefficients is not only due to the participation in the area of individual green and hardened basins, but also from the use of facilities and equipment used for retention and decomposition of precipitation waters. This allowed to make a detailed analysis of the hydraulic conditions prevailing in the existing network after it brought rainwater from the attached catchment. The simulations were carried out with the use of several dozen rainfall phenomena from 2007–2008, which were registered by three rain gauges located in the Zasanie district.

3 RESULTS OF THE ANALYSIS AND DISCUSSION

The analysis of the results of hydrodynamic simulations indicates that the addition of new catchments to the existing sewage system will cause, inter alia, the hydraulic overload of the main collectors, manifested by periodic pressure sewage flows. The selected research results in this area are shown in Figure 3. It is the main sewer no. 30 that collects waste water from several catchment areas and feeds them to a collector located along the San River. This fragment of the existing network functions gravitationally. Bringing sewage from the new S-V catchment to it will result in pressurized sewage flows regardless of the degree of sealing of the surface of this catchment area. This means that this sewer does not have sufficient retention capacity to accommodate additional waste water.

A similar situation was observed in section 41, in addition to sewage from the existing basin, sewage from the surface of the S-II and S-III catchment flew (Fig. 4) and sewer no. 46 located before the sewage pumping station (Fig. 5). The research conducted for several dozen precipitation phenomena has shown that in this case, however, already in the existing state there are hydraulic overloads resulting in short-term pressurized sewage flows. The results for selected rainfalls are shown in Figure 6.

The study also analyzes the amount of wastewater flow in the existing sewers after the catchment area is expanded by additional areas. Figure 7 shows the flow rate of wastewater through the sewer to the pumping station depending on the duration of the rain. At the culminating point where the effluent flow is about 400 dm^3/s greater than the existing flow rate, as is shown in Figure 5, the hydraulic overload of the collector is being analyzed.

Taking into account the results of the research, an attempt was made to solve the problem of the development of precipitation waters flowing from the new catchment areas and the occurrence of hydraulic overloads on the existing sewerage network. Therefore, underground reservoirs were designed, the location of which is shown in Figure 8.

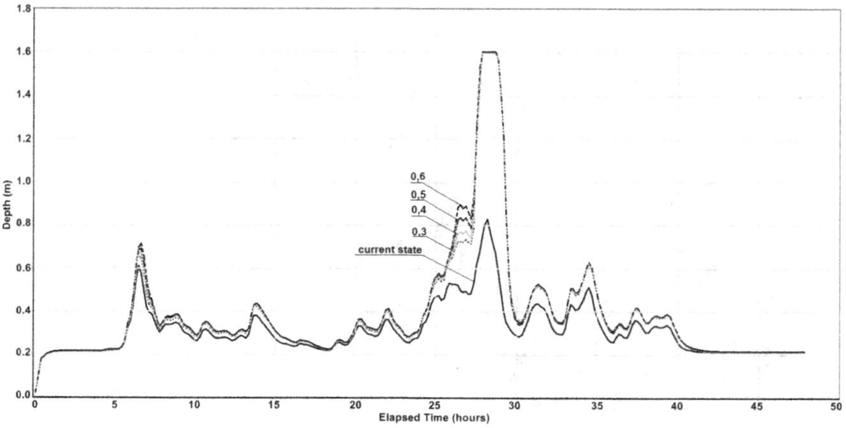

Figure 3. The height of filling up the main sewer no. 30 for precipitation from 4th and 5th September 2007.

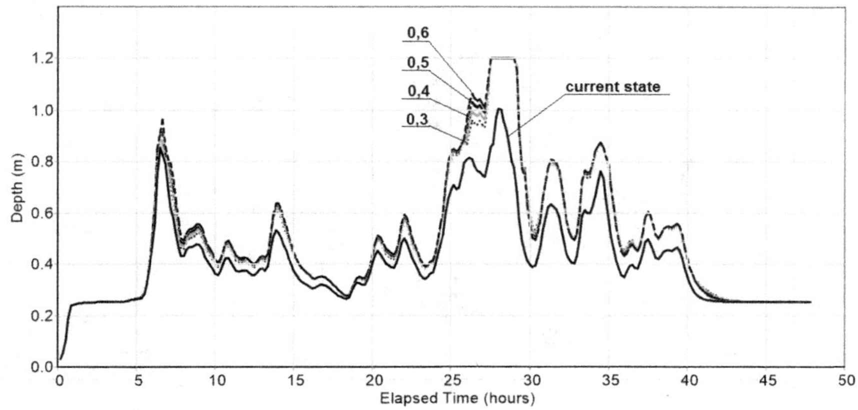

Figure 4. The height of filling up the main sewer no. 41 for precipitation from 4th and 5th September 2007.

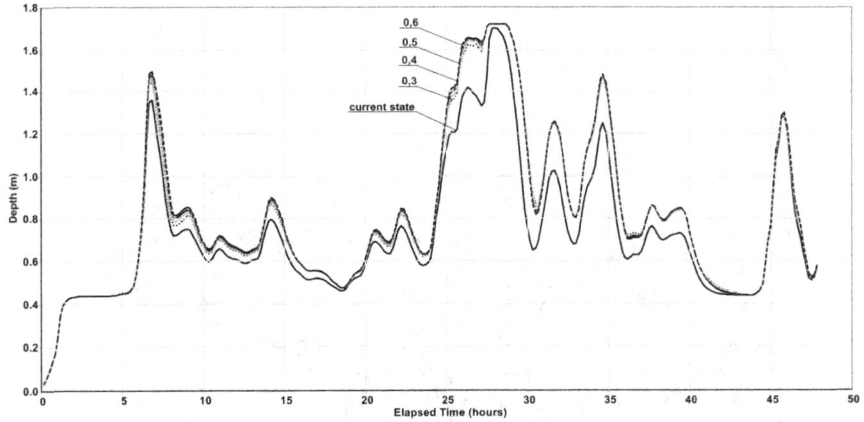

Figure 5. The height of filling up the main sewer no. 46 for precipitation from 4th and 5th September 2007.

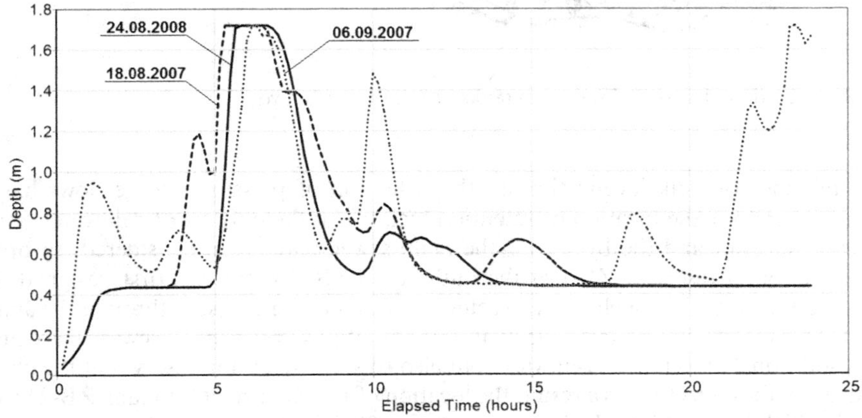

Figure 6. Fill height of sewer for pumping stations for selected precipitation from 2007–2008 - existing condition.

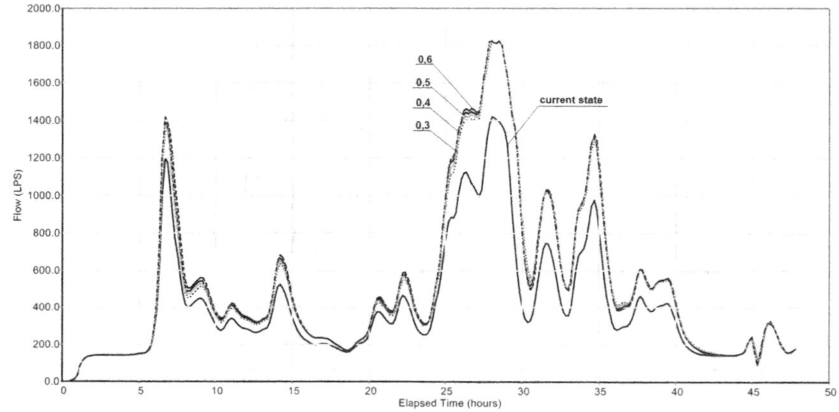

Figure 7. Flow rate of the main sewer no. 46 leading to waste water pumping stations for 4th and 5th September 2007.

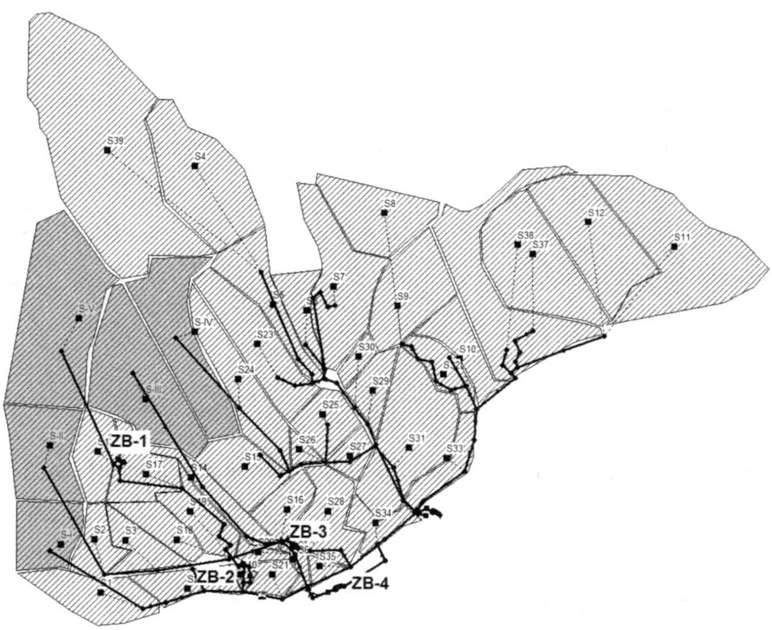

Figure 8. Location of designed retention tanks on the sewage network.

The influence of tank localization on the reduction of pressure sewage flows has been analyzed. Therefore, the location of retention facilities at the outflow of rainwater from the attached catchment and the bypass of the main sewers have been considered. In order to relieve sewer 30, a reservoir ZB-1 at the outlet of the S-V basin was first designed. However, such a location did not eliminate wastewater flows under pressure through the analyzed channel. This means that unfavorable hydraulic conditions occurring in sewer 30 are due not only to additional amounts of rainwater but also to small retention reserves in the channels located above the collector. As a result, the location of the ZB-1 retention tank ZB-2 has been changed, which is located on the bypass of section 29. The capacity of this reservoir is 3000 m^3 on the basis of the simulation. Figure 9 shows the sewerage in the main sewer no. 30 for variants z and without reservoir.

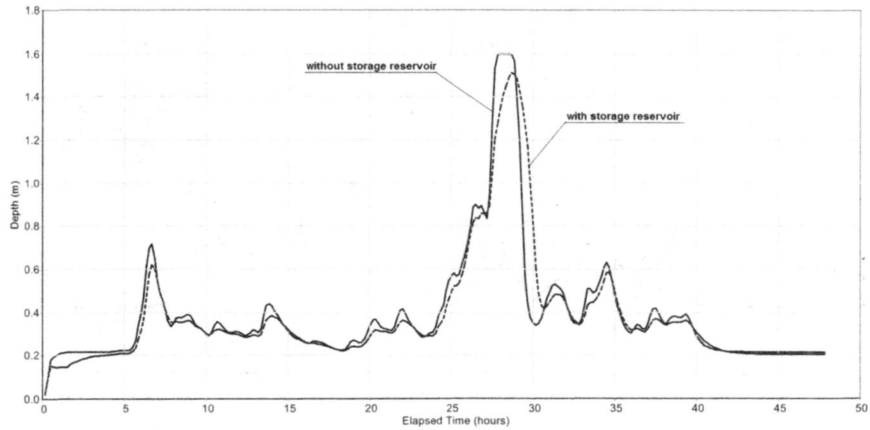

Figure 9. The height of sewage in the main sewer no. 30 in the conditions of the expansion of the existing basin for precipitation from 4th and 5th September 2007.

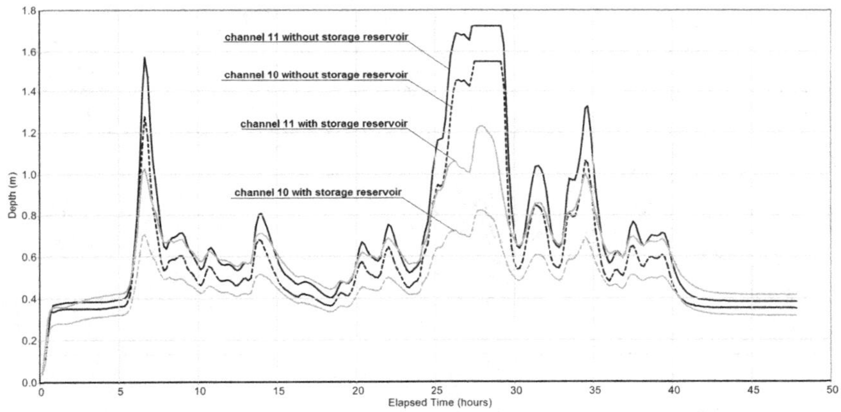

Figure 10. The height of sewage in the main sewers no. 10 and 11 in the conditions of the expansion of the existing basin for precipitation from 4th and 5th September 2007.

The use of the ZB-2 tank has also benefited the functioning of the main sewers located downstream along the banks of the San River that transport waste water to the pumping station. Figure 10 shows the filling height of sewage sections 10 and 11. There is a significant reduction in the level of wastewater for the expansion of the existing catchment area for reservoir retention. Waste water treatment, for the analyzed rainfall, solved the problem of the occurrence of pressure flows in these sections of the network.

Similar modernization measures on the existing network are planned at its section 41, to which additionally rainwater flows from the two associated catchments: S-II and S-III. In this case, a 5000 m^3 ZB-3 tank was located on the by-pass of the sewer 41 (Fig. 8). Retention of sewage in the course of intense rainfall has affected the discharge of the sewer analyzed. The level of sewage in this sewer for the variant without reservoir and with reservoir ZB-3 is shown in Figure 11.

Taking into account the ZB-1, ZB-2 and ZB-3 retention tanks in the analyzed sewerage system reduced the time of the pressure flows with the sewer no. 46, but they did not evaporate. Considering this, it was necessary to design a retention tank to relieve this sewer. Due to the high level of the existing sewerage system in this part of the catchment area, which did not permit the use of a gravity reservoir, a ZB-4 gravity pumping tank with a capacity of

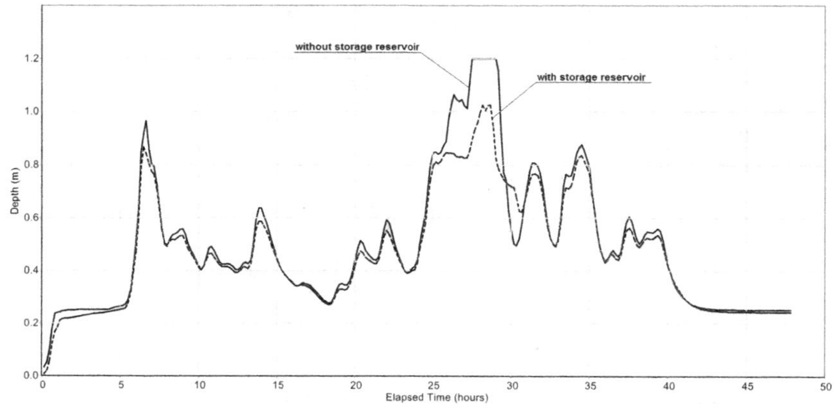

Figure 11. The height of sewage in the main sewer no. 41 in the conditions of the expansion of the existing basin for precipitation from 4th and 5th September 2007.

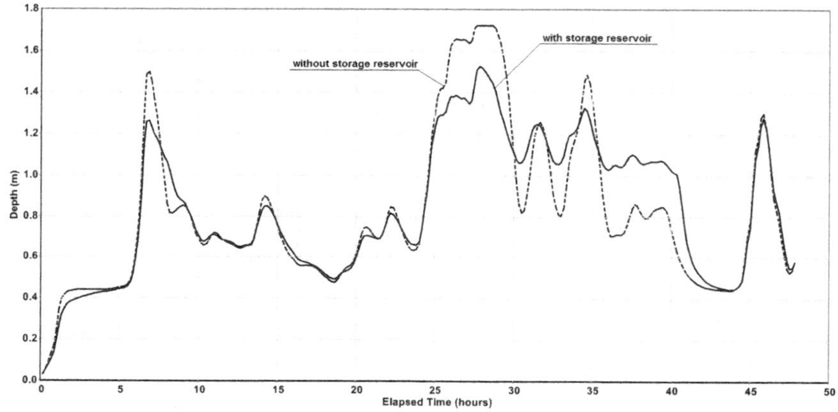

Figure 12. Sewage filling height in the main sewer no. 46 supplying wastewater to the pumping station under conditions of expansion of the existing catchment basin on September 4th and 5th, 2007.

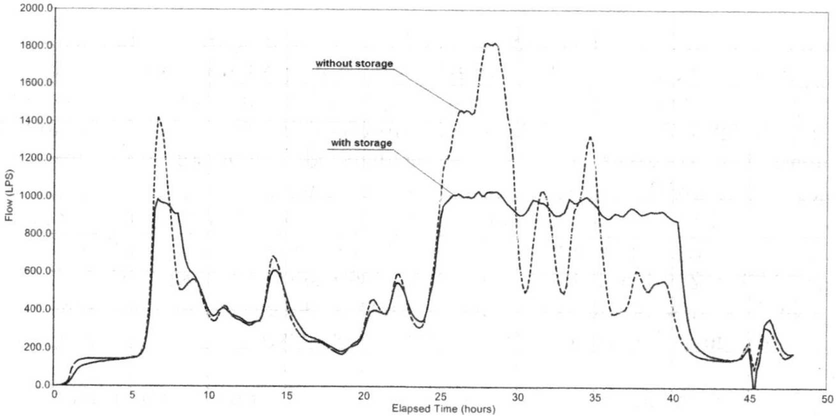

Figure 13. Wastewater flow rate in the main sewer no. 46 leading to sewerage pumping stations under conditions of expansion of existing catchment basin on September 4th and 5th, 2007.

5000 m^3 was designed. The results of the simulations are shown in Figures 12 and 13. Operation of the ZB-4 tank has reduced the flow rate of sewer 46 by about 800 dm^3/s, which not only contributed to the hydraulic conditions present in this sewer, but also to the pumping station.

4 SUMMARY

The hydrodynamic studies carried out on the model of the actual urban catchment showed that the way of catchment management and associated rainwater drainage had a significant influence on the functioning of the sewage system. The expansion of the analyzed catchment area, in addition to new areas, resulted in the presence of pressure sewage flows in the main sewage collectors. To solve this problem, simulations were carried out for the case where the reservoir retention was included in the existing sewer system. The designed retention tanks in an extensive catchment allow the hydraulic channels to be drained hydraulically through which sewage flows under pressure.

The results of the study confirmed that urban development planning should take into account that analysis, especially if the rainwater from the attached catchment is discharged to an already existing sewerage network. Often, especially in old parts of cities, sewerage systems do not have adequate cubature reserves to accommodate additional waste water. In this case, the implementation of the hydrodynamic model and the simulation would make it possible to select the most advantageous option of the modernization of the sewage network.

REFERENCES

Burszta-Adamiak E. & Stec A. 2017. Impact of the rainfall height on retention and runoff delay from green roofs. Journal of Civil Engineering, Environment and Architecture 64 (1/17): 81–95.

Du J., Qian L., Rui H., Zuo T., Zheng D., Xu Y. & Xu C.Y. 2012. Assessing the effects of urbanization on annual runoff and flood events using an integrated hydrological modeling system for Qinhuai River basin, China. Journal of Hydrology 464–465:127–139.

Dyrektywa 2000/60/WE Parlamentu Europejskiego i Rady z dnia 23 października 2000 roku ustanawiająca ramy wspólnotowego działania w dziedzinie polityki wodnej.

Dziopak J. & Słyś D. 2007. Modelowanie zbiorników klasycznych i grawitacyjno-pompowych w kanalizacji, Politechnika Rzeszowska, Rzeszów 2007.

Elliott A.H. & Trowsdale S.A. 2007. A review of models for low impact urban stormwater drainage. Environmental Modelling & Software 22:394–405.

Fletcher T.D., Andrieu H. & Hamel P. 2013. Understanding, management and modelling of urban hydrology and its consequences for receiving waters: A state of the art. Advances in Water Resources 51:261–279.

Januchta-Szostak A. 2012. Usługi ekosystemów wodnych w miastach [w:] Przyroda w mieście. Usługi ekosystemów—niewykorzystany potencjał miast. Fundacja Sendzimira.

Kaźmierczak B. and Kotowski A. 2014. The influence of precipitation intensity growth on the urban drainage systems designing. Theor Appl Climatol 118:285–296.

Lu H.W., He L., Du P. & Zhang Y.M. 2014. An Inexact Sequential Response Planning Approach for Optimizing Combinations of Multiple Floodplain Management Policies. Pol J Environ Stud 23:1245–1253.

Pochwat K., Słyś D. & Kordana S. 2017. The temporal variability of a rainfall synthetic hyetograph for the dimensioning of stormwater retention tanks in small urban catchments. Journal of Hydrology, dostęp 18 kwiecień 2017, w druku.

Słyś D. 2013. Zrównoważone systemy odwodnienia miast. Dolnośląskie Wydawnictwo Edukacyjne, Wrocław.

Słyś D., Stec A. & Zeleňáková M. 2012. A LCC analysis of rainwater management variants. Ecological Chemistry and Eng. 19:359–372.

Starzec M., Dziopak J. & Alexeev M.I. 2015. Effect of the sewer basin increasing to necessary useful capacity of multichamber impounding reservoir. Water and Ecology 1: 41–50.

Stec A. and Słyś D. 2014. Optimization of the hydraulic system of the storage reservoir hydraulically unloading the sewage network. Ecol Chem Eng S 21(2):215–228.
Suligowski Z. 2008. Alternatywa dla wód opadowych. Wodociągi i Kanalizacja 4.
Todeschini S.: Hydrologic and Environmental Impacts of Imperviousness in an Industrial Catchment of Northern Italy. Journal of Hydrological Engineering 21.
UNESCO. 2012. Water security: responses to local, regional, and global challenges. Strategy plan, Paris: UNESCO IHP.
United Nations, Department of Economic and Social Affairs, Population Division. World Urbanization Prospects: The 2014 Revision, Highlights.
Zeleňáková M., Markovič G., Kaposztásová D. and Vranayová Z. 2014. Rainwater Management in Compliance with Sustainable Design of Buildings. Procedia Eng. 89:1515–1521.

Underground Infrastructure of Urban Areas 4 – Madryas et al. (Eds)
 ISBN 978-1-138-55953-0

Mechanized tunneling technologies for weak rocks of Middle East, revisited.

J.B. Stypulkowski & F.G. Bernardeau
CDM Smith, Doha, Qatar

T.D. Sandell
The Robbins Company, Seattle, USA

ABSTRACT: This paper discusses the tunneling technology applicable to weak/soft rocks of Middle East with particular emphasis on Qatar. Qatar's rapid expansion of infrastructure translated into significant growth in the tunneling industry. The authors will discuss common rock types of the region and rapid tunneling methodology available globally which is used to safely and economically excavate the tunnels.

1 INTRODUCTION

1.1 *Tunneling project in middle east*

Increasing infrastructure needs of Middle Eastern countries have led to recent construction of tunnels for water as well as for transportation. Tunneling in UAE involved 13 km of subway construction in Dubai as well as 41 km of sewer tunnels in Abu Dhabi. The State of Qatar is currently undertaking an ambitious plan involving 112 km of tunnels for transportation and 123 km of tunnels for water, both sewer and storm. This paper will concentrate on those two areas and a USA-based project in similar rock conditions.

1.2 *UAE briefing*

Dubai metro involved 13 km of tunnel 8.5 m I.D. facilitating twin track tunnel beneath central Dubai and the Dubai Creek. Earth Pressure Balance (EPB) Tunnel Boring Machines (TBMs) were used with spoil conditioning using foam injected at the face. Mitsubishi supplied three 9.56 m diameter machines that completed ten separate drives. Average progress was 10 Rings (15 m/day) and the maximum production was 21 Rings (31.5 m in 24 hour). Construction costs were AED 28 billion/US$ 7.6 billion. Red Line opened in 2009 and Green Line in 2011.

The Strategic Tunnel Enhancement Program (STEP) is a 40 km long wastewater tunnel, which will upgrade the Emirate's previously strained sewage network in Abu Dhabi, UAE. The whole project includes three contracts: T-01, T-02 and T-03 (16 + 14.5 + 10.5 = 41 km). The contracts T-02 and T-03 are excavated with EPB TBMs. The first 15 km of the T-02 Contract has been already completed. STEP T-02 Contract included three Herrenknecht EPB TBMs. On April 2011 the first TBM was launched in a 50 m depth shaft, with breakthrough on April 1, 2012. The second TBM was launched by the end of June 2011 concluding the drive on May 1, 2012. The third TBM from the T-02 Contract started to excavate at the end of July 2011 and emerged at the end of May 2012.

Contract T-01 is 16.1 km long, 5.32 m diameter and built by 3 identical full face Kawasaki EPB TBMs. A high volume of water inflow was encountered during tunneling from WS2 to WS1, over a 615 m length with a maximum flow of an approximately 43 m^3/hr.

1.3 *Qatar briefing*

At the time of submission of the paper all tunnels for Qatar Rail and Abu Hamour Phase I water tunnels have been completed. Detailed tunneling data correlated to geology generated by Qatar Rail has not been published. Only generalized logistical data was recently made available out of 112 km of tunnels and 21 EPB TBMs. Maximum performance was reported on Green Line to be 283.2 m/week for 7 m diameter Herrenknecht EPB TBM. Abu Hamour (Musaimeer) Surface & Ground Water Drainage Tunnel—Phase I (AHSO) is 9.5 km long with 3.7 m ID storm water tunnel about 30 m below ground surface which was completed in early 2017. Contract included two Herrenknecht EPB TBMs. Average progress was 12 Rings (15.6 m/day) and the best week was 19 Rings (24.7 m/day) in the first one and 13 Rings (16.9 m/day)/21 Rings (27.3 m/day) in the second one.

1.4 *California, USA's San Francisco central subway briefing*

California, USA's San Francisco Central Subway project is a challenging modern example of urban tunneling in mixed ground conditions with parallels to projects in the Middle East. Two 6.3 m diameter Earth Pressure Balance Machines (EPBs) excavated parallel 2.5 km long tunnels under low cover. The small launch site situated between an interstate and an off-ramp, curved tunnel alignment, and complex, mixed geology are particular challenges. These elements required customized tunnel and machine design, from TBM shipment and assembly to launch and excavation.

2 TYPICAL ROCKMASS

2.1 *Regional geology*

The Qatar peninsula and UAE are geologically a part of the Arabian Gulf Basin. It forms a part of the Arabian shelf between the Arabian shield and Iranian mobile belt. Basement rocks in the area under consideration are overlain by different age sandstones, salts, limestones and shales.

2.2 *Geology of Dubai metro tunnels*

Ground comprised loose to medium dense, high permeability marine sand, cemented sand Sabkha deposits with some hard concretions and voids, underlain by calcareous sandstone interbedded with cemented or uncemented sands, underlain by interbedded gypsiferous sandstone and then siltstone.

2.3 *Geology of Abu Dhabi tunnels*

The project area geology consists of marine sands and silts, which in addition of wind erosion, capillary action and evaporation has led to hypersaline Sabkha deposits. These deposits overlay alternating beds of carbonate rocks which are predominantly dolomite and they consist of mudstone, claystone, siltstone, calcarenite, carbonate sandstone, gypsum and cemented sands. The subsurface are considered to be in a karst terrain with carbonate based rocks susceptible of dissolution. The T-02 Contract shafts penetrated the superficial deposits into saturated silty sands followed by a sequence of mudstone and gypsum. The T-01 Contract rocks encountered interbedded layers of mudstone and gypsum as well as sandstone, calcarenite, siltstone and claystone. The rocks ranged in strength from being extremely weak to weak.

2.4 *Geology of Abu Hamour Doha, Qatar*

The Qatar peninsula forms a part of the Arabian shelf between the Arabian shield and Iranian mobile belt. Thickness of sediments in the Qatar region is estimated to be about 10 km.

The post Cretaceous sedimentation is basically a sequence of shallow marine limestone with occasional shale and evaporates in a shallow basin. Structurally, Qatar appears to be an anticline or dome with N-S axis from Al Majidah to Shahaniyah. The formations encountered in Doha region comprise of Quaternary marine, aeolian and Sabkha deposits. They are: Rus of Lower Eocene, Lower Dammam and Upper Dammam of Middle Eocene and Lower Dam of Lower Miocene.

During tunneling and shaft excavation for Abu Hamour three rock domains were encountered: Simsima Limestone—which contains Dolomite of Upper Dammam formation and Midra Shale and Rus of Lower Dammam formation. The three formations were distributed roughly as follows:

- Midra Shale: ~5.9 km
- Simsima Limestone: ~2.4 km
- Rus: ~1.2 km

2.5 *Geology of San Francisco central subway*

The geology of the tunnel alignment is mixed and it ranges from mud deposits, sand and clay to sandstone, mudstone, and shale. The TBMs encountered three disparate geological formations: the Colma, Old Bay and Franciscan. However, stating that there are only three types of ground is a vast oversimplification. There was a wide range of diverse and fluctuating ground conditions within these three generalized formations.

The three formations were distributed roughly as follows:

- Colma Formation (Surficial soils): ~1,750 m
- Undifferentiated Old Bay Deposits: > 500 m Interspersed
- Franciscan Bedrock (sandstone): ~750 m
- Maximum rock strength (UCS): 27 MPa

Along the southern third of the alignment approximately 4.6 m of loose to medium dense sand fill overlies 7.6 m (25 ft) of medium dense sand with clay and clayey sand. Underlying the sand and clayey sand units is the 14 m thick Colma Formation, a dense to very dense sand with silt to silty sand and beds of clay. The Colma sand is underlain by very stiff clay and dense to very dense silty sand or clayey sand referred to as Undifferentiated Old Bay deposits. Below that is the highly variable Franciscan Formation, bedrock comprising weathered to fresh beds of fractured sandstones, shale and a mélange unit made up of the same rock types in a weak, crushed matrix. The middle third of the alignment of each tunnel traverses the Franciscan Formation. Groundwater levels vary from 2 m to 9.8 m below ground surface in the soft-ground portions of the alignment.

3 REPORTED ROCK PROPERTIES

3.1 *General*

The ground investigations were conducted for all projects under consideration. The following information the authors were able to compile from available documents: Uniaxial Compressive Strength (UCS), Brazilian Tensile Strength (BTS), UCS/BTS ratio, Shear Wave Velocity, Elastic Modulus, point load strength index $I_s(50)$, UCS/$I_s(50)$ ratio, Slake Durability, moisture content, Cerchar Abrasiveness Index (CAI), Abrasion Value and Abrasion Value Cutter Steel (AVS). In addition to testing typically prescribed for rock mass selected soil tests like for example undrained shear strength tests (UU) was reported on SF Central Subway. No spoil conditioning test results were reported.

3.2 *Rock strength and elastic properties*

The tunnel stability depends amongst other factors on the shear parameters the deformation characteristics. Mechanical resistance of the excavated material can be measured by standard laboratory tests like compressive strength and tensile strength, or measured with specific load test like the "Point Load Test". Sample preparation for weak rock materials is more difficult compared to that of hard rock materials. Often methods of sample preparation and laboratory testing for weak rocks are similar to soils. The International Society for Rock Mechanics (ISRM) suggested a length/diameter (L/D) ratio of 2.5 for samples in compression tests. However it has been reported in literature that it is hard to obtain a sufficient number of samples with an L/D ratio bigger than 2, as the drilling program into weak rock is usually difficult, and it might only obtain 10–20 percent of the drilling length. Thus, the ratio could be less than 2.5. This also leads to collection of best samples for testing not the representative ones therefore results as shown in Table 1 will likely be higher than they are in-situ.

`The results in Table 1 when compared to criteria set out by Santi (2006) in Table 8 when considering the sampling bias discussed above classifies the rocks as weak in term of UCS strength.

Average Splitting Tensile Strength of the softer samples was reported for STEPS T-02 to be approximately 0.5 MPa. Mudstone brittleness (UCS/BTS) of approximately 6 was reported on the same project. Splitting tensile strengths on other projects in the region fall into the same brackets however brittleness appears to vary significantly as shown in Table 2.

The deformation characteristic depends on the Elastic Modulus and Poisson coefficient. Shear wave velocity and associated elastic modulus measurements were made in all investigated regions and the results are presented in Table 3. However Poisson coefficients were not reported.

The results in Table 3 when compared to criteria set out by Santi (2006) in Table 8 when considering the shear wave velocity classifies the rocks as weak.

Due to some degree of technical difficulties, the uniaxial compression test for soft rock is often replaced by a point load strength index test. This test is simpler in procedures than that of the uniaxial compression test, and the test does not need necessarily cylindrical samples. A conversion factor can be applied to the point load strength index, $I_S(50)$, for estimating the Uniaxial Compressive Strength (UCS). Point load strength index test results are presented in Table 4.

When comparing Tables 2 and 4 ratios the lack of consistency between results have been noted. San Francisco Central Subway reported series of undrained shear strength tests (UU) on weathered sandstone (0.06–0.27 MPa) and on weathered shale (0.13–0.71 MPa).

Table 1. Uniaxial compressive strength test results for selected projects.

Project	Rock type	UCS		
		Min MPa	Max MPa	Avg MPa
STEP T-02	Mudstone	0.04	33.00	3.30
	Calcarenite	1.00	2.00	1.07
	Gypsum	0.70	27.00	12.05
Abu Hamour	Simsima Limestone Zone C	2.49	46.42	11.87
	Simsima Limestone Zone B	5.04	65.30	21.67
	Simsima Limestone Zone A	4.43	34.22	17.62
	Midra Shale Series (Shale)	4.77	19.50	10.80
	Midra Shale Series (Limestone)	10.16	39.88	23.84
	Rus Formation	4.22	47.03	15.88
SF Central Subway	Colma Formation (Surficial soils)			
	Undifferentiated Old Bay Deposits			
	Franciscan Bedrock (intact sandstone) *estimated	7	27	

*interbedded dolomitic limestone strata below tunnel horizon.

Table 2. Brazilian tensile strength test results for selected projects.

Project	Rock type	BTS Min Mpa	BTS Max Mpa	BTS Avg MPa	UCS/BTS
STEP T-02	Mudstone			0.5	6
	Calcarenite				
	Gypsum				
Abu Hamour	Simsima Limestone Zone C	0.30	1.90	0.84	13.8
	Simsima Limestone Zone B	0.46	4.57	1.70	14.2
	Simsima Limestone Zone A	0.55	4.18	2.36	5.4
	Midra Shale Series (Shale)	0.90	2.03	1.31	5.3
	Midra Shale Series (Limestone)	2.14	9.11	4.45	5.4
	Rus Formation	0.10	5.60	1.35	20.1
SF Central Subway	Franciscan Bedrock (intact sandstone)	11	14		
	Franciscan Bedrock (intact shale)*	3.4	5		

*fresh.

Table 3. Shear wave velocity and elastic modulus test results for selected projects.

Project	Rock type	Vs (m/s)	E MPa
STEP T-02	Mudstone	862	292
	Calcarenite		344
	Gypsum		6560
Abu Hamour	Simsima Limestone Zone C	800	3660
	Simsima Limestone Zone B	1000	5720
	Simsima Limestone Zone A	1300	9667
	Midra Shale Series (Shale)	800	3660
	Midra Shale Series (Limestone)		
	Rus Formation	800	3660
SF Central Subway	Franciscan Bedrock (intact sandstone)	625	57–297*
	Franciscan Bedrock (intact shale)		104*

*reload moduli from pressuremeter test.

Table 4. $I_s(50)$ and $UCS/I_s(50)$ test results for selected projects.

Project	Rock type	$I_s(50)$ Min MPa	$I_s(50)$ Max MPa	$I_s(50)$ Avg MPa	$UCS/I_s(50)$
STEP T-02	Mudstone	0.02	1.57	0.09	
	Calcarenite	0.02	0.96	0.15	
	Gypsum	0.02	2.19	0.92	
Abu Hamour	Simsima Limestone Zone C	0.15	7.54	2.30	4.8
	Simsima Limestone Zone B	0.16	6.28	2.34	8.6
	Simsima Limestone Zone A	0.21	3.15	1.38	8.1
	Midra Shale Series (Shale)	0.61	4.91	2.78	50.0
	Midra Shale Series (Limestone)	0.03	3.33	0.47	
	Rus Formation	0.06	2.32	0.47	42.1
SF Central Subway	Franciscan Bedrock (intact sandstone)	0.62	3.22		
	Franciscan Bedrock (intact shale)*	0.23	2.78		

*slightly to moderately weathered

Table 5. Slake durability test results for selected projects.

Project	Rock type	Slake durability		
		Min	Max	Avg
STEP T-02	Mudstone	5	79	37
	Calcarenite	2	98	45
	Gypsum	7	87	70
Abu Hamour	Simsima Limestone Zone C			
	Simsima Limestone Zone B			93
	Simsima Limestone Zone A			
	Midra Shale Series (Shale)			98
	Midra Shale Series (Limestone)			
	Rus Formation			90
SF Central Subway	Franciscan Bedrock (intact sandstone)			97.7*
	Franciscan Bedrock (intact shale)			21.3 to 98.2

*one test only.

Table 6. Moisture content test results for selected projects.

Project	Rock type	Moisture content		
		Min	Max	Avg
STEP T-02	Mudstone	0.1	53.5	26.2
	Calcarenite	13.9	29.9	21.9
	Gypsum	0.6	37.0	15.4
Abu Hamour	Simsima Limestone Zone C			
	Simsima Limestone Zone B	0.9	18.2	4.9
	Simsima Limestone Zone A			
	Midra Shale Series (Shale)	0.7	17.6	5.2
	Midra Shale Series (Limestone)			
	Rus Formation	1.5	13.6	4.3
SF Central Subway	Franciscan Bedrock (intact sandstone)	0.2	14	
	Franciscan Bedrock (intact shale)	1	18	

3.3 *Rock porosity and friability*

The slake-durability test is as a simple test for assessing the influence of weathering on rock. Average test results for STEPS show low to medium durability while for Abu Hamour results indicate generally high results (ISRM, 1977).

Porosity is a measure of the void spaces in a material and from the Table 6 it appears evident that rock tests on Abu Hamour show much lower porosity than on STEP.

The results in Table 6 when compared to criteria set out by Santi (2006) in Table 8 when considering the moisture content classifies the rocks as weak.

3.4 *Rock abrasivity*

Only Cerchar Abrasiveness Index (CAI) values were reported for both projects and the test results presented in Table 7 indicate that rock at STEP were not very abrasive while at Abu Hamour are slightly abrasive. Quartz content from petrographic analysis has been reported only on SF Central Subway for sandstone ranging between 14 and 55 percent.

Table 7. Cerchar Abrasiveness Index (CAI), abrasion value and Abrasion Value cutter Steel (Avs) test results for selected projects.

Project	Rock type	CAI	AV	AVS
STEP T-02	Mudstone	0.2–0.4	0	
	Calcarenite	0.1–0.3	1	
	Gypsum	0.1		0.5
Abu Hamour	Simsima Limestone Zone C	0.6–0.7		
	Simsima Limestone Zone B			
	Simsima Limestone Zone A			
	Midra Shale Series (Shale)			
	Midra Shale Series (Limestone)			
	Rus Formation			
SF Central Subway	Franciscan Bedrock (intact sandstone)*	2.5–3.2	n/a	2.5–4.5
	Franciscan Bedrock (intact fresh shale)**	1.35–2.63	n/a	1.5–2.0***
	Franciscan Bedrock (intact weathered shale)	0.38–0.47	n/a	

*Sievers' J-value 22.2–72.3, ***Sievers' J-value 26–47.8, ***fresh to weathered.

Table 8. Summary of engineering properties of weak rock (from Santi 2006).

Test or property	Value or range for weak rock
Compressive strength	1–20 MPa
Standard penetration test	50–300 blows per ft. (15–90 blows per m)
Rock quality designation RQD	< 25–75 percent
Hammer rebound	> = category 4
Seismic wave velocity	< 7,000 ft. per second* (2100 m/sec)
Ratio of weathered matrix to unweathered blocks	> 75 percent matrix
Jar slake	< = 4
Slake durability, Id(2), ASTM D4644–87	< 90 percent
Free swell	> 3–4 percent
Natural moisture content	> 1 percent for igneous and metamorphic rocks > 5–15 percent for clayey rocks
Dearman weathering classification	> = category 4
CSIR rock mass rating (Bieniawski, 1976)	< 35–60
Norwegian Geotechnical Institute "Q" rating	< 2

*Lower velocities may be appropriate, depending on equipment used and degree of rock fracturing.

3.5 *Weak rock*

The common definitions for hard rock tunneling include the following: UCS exceeding 50 MPa (ISRM 1980), can't be excavated economically by roadheader, hard, consolidated load bearing which has to be removed by blasting, rock sample that requires more than one blow by a geological hammer to split, metamorphic and igneous. The upper limits for soils have been established by Therzaghi and Peck (1967) to be STP of 50 and UCS of 0.4 MPa. So by deduction everything in between would fall into soft/weak rock category. However as reported in literature sharp limits are hard to specify for all rocks without additional criteria.

Technical Commission on Soft Rock of ISRM has been established but it hasn't published anything yet. First to delineate soft rock was International Society of Soil Mechanics and Geotechnical Engineering (ISSMGE). Most early publications on this subject were by International Association of Engineering Geology (IAEG) sponsored events. Kanji (2014) in a recently published article concludes after a state-of-the-art review that the "concept of soft rocks is not yet well defined".

Therefore the authors decided to use the Santi (2006) approach. The published summary of research and analysis evaluating classification and testing methods for weak rocks is aiding in soft/weak recognition by assessing their engineering properties. According to Santi and Doyle (1997), weak materials are "intact, unweathered to slightly weathered materials that have low compressive strength or are highly fractured". A summary of engineering properties potentially identifying weak rock and the selection of ranges and cutoff values is given in Table 8. In general, these ranges and cutoff values recognize the influence of fractures, decreased strength, creation of weathered matrix material, and reaction to water demonstrated by weak and weathered rock materials.

While not all of the parameters were available for evaluation from Table 8, the one which are reported herein appears to classify rocks as "weak" as per Santi 2006.

4 MECHANIZED TUNNELING

4.1 *General*

There are different ways of classification of tunneling machines. ITA working Group 14 on mechanized excavation is working on establishing terminology and guidelines to aide in optimum choice of the machine. In general division into soft ground and hard rock technology still applies. However, it is more often recognized that the same tunneling machine may have to negotiate through soft as well as hard conditions. TBM are a somewhat inflexible but very fast method of excavation if successful.

4.2 *Soft ground review*

Mechanized tunneling in soft ground usually involves use of a shielded TBM. In ground conditions that require a higher level of support than the basic Brunel or digger shield, closed face tunneling machines need to be used. There are two options in this category: the EPB and the slurry face machine (SFM).

There are some distinct differences between EPB and SFM. In the EPB the pressure is transmitted to the face mechanically. Technical Manual for Design and Construction of Road Tunnels by FHWA states the following: "Control is obtained by matching the volume of soil displaced by forward motion of the shield with the volume of soil removed from the pressurized face by that screw conveyor and deposited (at ambient pressure) on the conveyor or muck car. Clearly the range of natural geologic conditions that will result in suitably plastic material to transfer the earth pressure to the face and, at the same time, suitably frictional to form the "sand plug" in the screw conveyor is rather limited—generally only combinations of fine sands and silts".

"The SFM transmits pressure to the face hydraulically through a viscous fluid-formed by the material cut and trapped at the face and mixed with slurry (basically bentonite and water). In this case the pressure transmitted can be controlled by means of pressure gages and control valves in a piping system."

4.3 *Hard rock review*

Mechanized tunneling in rock usually involves use of the roadheader or TBM. The roadheader is a versatile method of rock excavation allowing non circular excavation and is adaptable to various underground conditions. This includes large or inclined openings, highly abrasive or high-strength rock conditions, as well as highly variable rock conditions. The benefit of this technology is reduced vibrations at the surface when comparing to TBM mining commonly causing ground borne vibrations. The Roadheader method can be used for a short tunnel option, connection to access ramps and in construction of all necessary enlargements. If the circular shape is a good choice and the tunnel is long, a TBM is usually used. There are many various types of TBMs in use today for the mechanized tunneling in hard rock. If a higher level of face support is required a TBM with full face excavation is likely to be used. There are three main types of full face TBMs: open-type gripper, double shield with gripper, and single shield. The type of TBM to be used largely depends on stand-up time and rock fracturing.

The open-type gripper TBM uses its grippers to react thrust from the surrounding rocks to move forward. Tunnel lining may consist of a combination of ring beams, wire mesh, rock bolts, steel straps, shotcrete, and/or cast-in-situ concrete depending on the level of ground support needed. Shielded rock TBMs can utilize either a Double Shield configuration that allows the TBM to push off tunnel walls in Double Shield Mode, or a Single Shield configuration. The Single Shield tunneling method allows the assembly of the segmental lining in the rear portion of the machine, as excavation progresses. For forward movement the shield machine uses shield jacks taking reaction from the previously constructed segments.

4.4 *Weak rock review*

As described in a guideline for best practice of closed-face tunneling machines, "weak rock may be regarded effectively as a soft ground environment because systems used to excavate soft-ground types may also be applied to weak rock materials" (BTS with ICE, 2005). Therefore, in self-supporting weak rock closed face systems may not be required; however, if there is a danger of high ground water inflows closed face TBM is effective method of tunneling.

Since EPB can advance much faster in good ground conditions than slurry it is frequently used if risks are acceptable. Therefore, it is reasonable to ask what the risks are and how to evaluate them to know if they are acceptable. The authors therefore turn to guidelines for assistance. The most recent are "Recommendations for selecting TBMs", revised in October 2010 (in German) and for EPB Table 3.7 includes "weak rock" under hard rock. The following classification parameters are listed for EPB (SM-V5): compressive strength, rock quality designation (RQD), rock mass ratio (RMR), water inflow, abrasiveness (CAI), swelling behavior and supporting pressure. For the range of parameters shown in the tables herein it is clear that if rock strength is below 50 MPa, RQD below 75 percent, RMR below 60, CAI less than 4, fair swelling and supporting pressure less than 2 bars the EPB application is possible. According to the same guideline EPB is applicable specifically in rock with RQD less than 25 percent, RMR less than 20, CAI less than 1 and for supporting pressure under 1 bar. The same reference in Table 3.6 finds slurry type of machine SFM (SM-V4) applicable to all conditions except high swelling. The main field of applications of slurry in rock is for supporting pressure up to 4 bars, with CAI less than 1 and for no/poor swelling of rock.

Institute of civil engineers (ICE) issued guidelines for closed face tunneling machines in 2005 where discusses ground types and their applicability to EPB or slurry technology. However, they only reference work by Whittaker and Frith from 1990 where the chief selection criteria is the amount of fines (passing through 200 sieve) and gradation curves. While weak

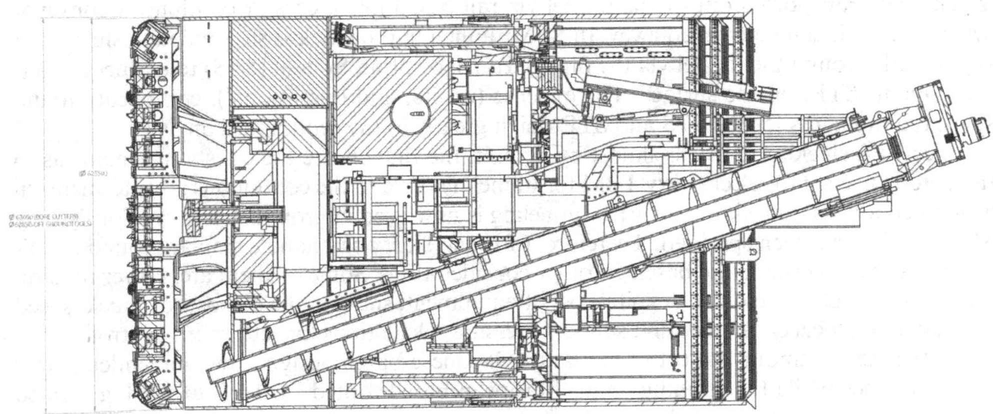

Figure 1. SF central subway EPB TBM.

rock is mentioned and described as "effectively soft-ground" and close face is recommended there is no discussion on EPB vs. slurry.

In North American practice FHWA on line Technical Manual for Design and Construction of Road Tunnels covers selection between EPB and slurry under soft ground tunneling and uses the same approach as ICE.

AFTES—French Tunnelling Society in 1999 issued recommendations on choosing mechanized tunneling techniques where parameters required for consideration were listed. The first step in a two-step approach is called elementary selection, where influence of basic function (support, hydrostatic pressure, excavation, mucking) on tunneling techniques is evaluated using the following scale: 0 – no effect, 1 – some effect, 2 decisive and SO—n/a. In step two using the same scale, the degree to which elementary selection parameters affect tunneling techniques is evaluated. If only mechanical properties of the ground are used in evaluation there is no difference between EPB and SFM.

The ITA withdrew its "Recommendations and Guidelines for Tunnel Boring Machines (TBMs)", (2000) and now recommends use of various national documents dealing with mechanized tunneling.

Japan Society of Civil Engineering translated its 2006 version of Standard Specification for Tunneling into English. In the volume about shielded tunnels a comprehensive flow chart on how to select tunneling method is presented. While geotechnical conditions are listed there are no qualifying parameters that could limit EPB or slurry technologies. In the volume about mountain tunnels there is a list of parameters influencing TBM type choice: UCS, RQD, elastic wave velocity as well as information about fractures, faults and water issues. There is no quick guide; a full engineering study is expected before the decision can be made.

So far only "Recommendations for selecting Tunnel Boring Machines" (2010), published in German lists the parameters and provides clear limits for applicability, and concludes that for weak rocks as presented here both technologies (EPB + SFM) are applicable.

5 EPB VS SFM

Review of the criteria available leads to the conclusion that evaluation of non-geotechnical measures is very important. the lower capital costs and the simpler operation of EPB shields creates a bias in favor of the latter, although slurry shields can allow for a finer control of, and the ability to maintain higher ground pressures particularly under mixed face conditions or when tunneling through weak rock. Since EPB can advance much faster in good ground conditions than slurry it is frequently used if risks are acceptable. Underground, both types of machines are very similar in: structure, propulsion, and segment handling. The main difference is in how settlement is controlled and how the excavated material (muck) is removed from the heading, and the equipment required on the surface to handle the muck. EPB muck is generally transported out of the tunnel via rail based muck cars or continuous conveyor system where it can be trucked away. In SFM tunneling the muck is suspended in slurry and is pumped through the tunnel via the Slurry Transportation System (STS) to a Slurry Treatment Plant (STP) on the surface. The expense (capital and operational), complication and space requirements for the STS and STP weigh greatly in the economical analysis.

Since EPB shields are often, almost by default, the first choice the question remains as to the criteria needed to select slurry TBM for tunneling. Due to the continuous advancements in ground conditioning technology, EPB tunneling is now used in ground where previously only SFM would have been specified. Therefore, EPB and slurry can handle similar ground conditions so selection criteria are based on other considerations. Some of them are: power requirements, availability of additives, capital cost of equipment, site size, the disposal of muck, speed of excavation, local experience and support pressure calculations for settlement control.

On the Abu Hamour project in Doha, the Japanese based designer recommended slurry type machines while the winning team, European based, opted for EPB technology which was successfully used to complete the project.

Figure 2. SF central subway EPB TBM.

It seems apparent from the type of geotechnical documents which are being prepared on those projects that designers responsible for preparing recommendations are not sure which guideline to follow. When defining weak rock they treat it as rock, running all rock specific testing. Meanwhile, they employ soft ground technology without proper definition of the properties typically needed to decide on the most suitable form of mechanized technology. While Santi 06 helps in classifying weak rocks, a connection to mechanized excavation hasn't been developed yet.

SF Central Subway GBR appears to be slightly different and attempts to merge soil and rock specific testing requirements.

6 CONCLUSIONS

The review of the test data reported for tunneling in weak rock in Middle East and North America shows differences in the approach. Middle Eastern projects tend to be design build without baseline parameters therefore there is less incentive on the designer's side to define them. This leads to design build teams assuming more risks than would normally be acceptable in North American practice.

ACKNOWLEDGEMENTS

The authors would like to acknowledge Salini-Impregilo for providing geotechnical reports from STEP.

REFERENCES

AFTES – (1999), Recommendations on Choosing Mechanized Tunneling Techniques, French Tunnelling Society.

Anagnostoul G, Rizos D, (2009) Geotechnical And Contractual Aspects Of Urban Tunnelling With Closed Shields, ITA-AITES World Tunnel Congress 2009, "Safe Tunnelling For The City and Environment", Budapest.

Atkins, Dubai Metro Presentation—on line.
Abu Zeid, M.M. (1991) – Lithostratigraphy and framework of sedimentation of the subsurface Paleogene succession in northern Qatar, Arabian Gulf. N.Jb. Geol. Paliaiont. Mh.:191–204.
Bappler, K., Schrader, D., "Accomplishing Extraordinary Task as a Machine Supplier for Metro Doha", in RETC—Rapid Excavation & Tunneling Conference and Exhibit, June 4–7, 2017, San Diego, CA, pp 2–13.
Bernardeau, F.G., Stypulkowski, J.B., "Engineered and Safe Approach to Segment Lining Installation with Dowelled-In Connectors on the First TBM Tunnel in Qatar", in RETC—Rapid Excavation & Tunneling Conference and Exhibit, June 4–7, 2017, San Diego, CA, pp 694–411.
Closed-Face Tunneling Machines and Ground Stability, BTS with ICE, 2005.
Del Negro, E., et al, A. Boscaro, C. Campinoti, D. Michelis (2013) "Ground conditioning: STEP Abu Dhabi sewer project", The First Arabian Tunneling Conference & Exhibition, Dubai, United Arab Emirates, 10–11 December 2013.
ISRM, Suggested Method for Determining Water Content, Porosity, Density, Absorption and Related Properties and Swelling and Slake-Durability Index Properties – 1977.
Jafari, M.R., Bernardeau, F.G., Stypulkowski, J.B., Siyam, A.F.M., Abu Hamour Outfall Tunnel Project In Doha, Qatar", in RETC—Rapid Excavation & Tunneling Conference and Exhibit, June 7–10, 2015, New Orleans, LA, pp 401–411.
Kanji, M.A., (2014), Critical issues in soft rocks", 2014 ISRM Conference on Soft Rocks, Beijing, China, June 6–7, 2014.
Lovat, R.P., TBM Design Considerations: Selection of Earth Pressure Balance or Slurry Pressure Balance Tunnel Boring Machines.
Pathak, A.K., Stypulkowski, J.B, Bernardeau, F.G., "Supervision of Engineering Geological Activities during Construction of Abu Hamour Surface and Ground Water Drainage Tunnel Phase-1 Doha, Qatar", in International Conference on Engineering Geology in New Millennium at IIT, New Delhi 27–29 Oct 2015. Special Publication, J of EG October 2015, pp 467–486.
Peila D., Borio L., Pelizza S., Dal Negro E., Boscaro A., Schulkins R. (2011). Laboratory test for EPB ground conditioning. In: T&T INTERNATIONAL n. September 2011, pg. 48–50.
Recommendations for selecting Tunnel Boring Machines (German only), revised version as of October 2010 (Taschenbuch Tunnelbau 2011).
Santi, P.M., February 2006, Field Methods for Characterizing Weak Rock for Engineering, Environmental & Engineering Geoscience, Vol. XII, No. 1, pp. 1–11.
Santi, P.M., Doyle, B.C., 1997, The locations and engineering characteristics of weak rock in the U.S. In Santi, P.M. and Shakoor, A. (Editors), Characterization of Weak and Weathered Rock Masses, Association of Engineering Geologists Special Publication #9: Association of Engineering Geologists, Denver, CO, pp. 1–22.
SFMTA Central Subway, Geotechnical Baseline Report, Contract 1252 – Tunnels, February 11, 2011.
Standard Specification for Tunneling – 2006: Shield Tunnels (2007), Japan Society of Civil Engineers.
Stypulkowski, J.B., Pathak, A.K., Bernardeau, F.G., " Abu Hamour, TBM Launch Shaft—A Rock Mass Classification Attempt for a Deep Shaft in Doha, Qatar." in EUROCK14, ISRM International Symposium, May 27–29, 2014, Vigo, Spain, CRC Press, Taylor & Francis Group, pp 117.
Stypulkowski, JB, Pathak, AK, Bernardeau, FG, "Engineering geology for weak rocks of Abu Hamour surface and ground water drainage tunnel Phase-1 Doha, Qatar", International Conference on Recent Advances in Rock Engineering, Specialized Conference of ISRM in Bengaluru, India, 16–18 November 2016, RARE2016, pp 85–90.
Stypulkowski, J.B., Siyam, A.A.F.M., Bernardeau, F.G. and Al Kuwari, N.G., "Abu Hamour Drainage Tunnel, First TBM Mined Tunnel in Doha, Qatar". The First Arabian Tunnelling Conference & Exhibition, Dubai, United Arab Emirates, 10–11 December 2013, pp. 300–314.
Technical Manual for Design and Construction of Road Tunnels, FHWA, https://www.fhwa.dot.gov/bridge/tunnel/pubs/nhi09010/index.cfm.
The Dubai Metro tunnels, 30 January 2009, Tunnels and Tunneling.
Whittaker, BN & Frith, RC (1990). "Tunnelling: Design, Stability and Construction". Institution of. Mining and Metallurgy: London. 460 pp.
Youngjin Shin, Grahame Bunce, Willie Cannon, Chris Phillips, Hyoungjin Lee, Yonghee Kim (2013) "Abu Dhabi STEP project Design and Construction for TO1", The First Arabian Tunneling Conference & Exhibition, Dubai, United Arab Emirates, 10–11 December 2013.
Ziad R. Beydoun, "Arabian plate oil and gas: Why so rich and so prolific", Episodes, Vol. 21, No 2, June 1998, pp 74–81.

Underground Infrastructure of Urban Areas 4 – Madryas et al. (Eds)
© 2018 Taylor & Francis Group, London, ISBN 978-1-138-55953-0

Author index

9781138559530